ENNIS AND NANCY HAM LIBRARY
ROCHESTER COLLEGE
800 WEST AVON ROAD
ROCHESTER HILLS, MI 48307

FUNDAMENTALS OF TOOL DESIGN

A PUBLICATION IN THE A.S.T.M.E.
MANUFACTURING ENGINEERING SERIES

Coordinated by:

Anthony R. Konecny

Associate Professor in Mechanical Engineering
Washington University
St. Louis, Missouri

Willis J. Potthoff

Tooling Supervisor
Emerson Electric Manufacturing Company
St. Louis, Missouri

FUNDAMENTALS OF TOOL DESIGN

CONCERNING THE THEORY, PRINCIPLES, AND TECHNIQUES FOR THE MODERN DESIGN OF CUTTING TOOLS, CUTTING AND FORMING DIES, FIXTURES, AND OTHER RELATED TOOLING

Prepared Under Auspices of:

TEXTBOOKS SUBCOMMITTEE
TECHNICAL PUBLICATIONS COMMITTEE
AMERICAN SOCIETY OF TOOL AND MANUFACTURING ENGINEERS

FRANK W. WILSON, Editor-in-Chief

PRENTICE-HALL, INC.
Englewood Cliffs, New Jersey
1962

© 1962 by American Society of Tool and Manufacturing
Engineers, Detroit, Michigan

Library of Congress Catalog Card Number: 62-16660

Printed in the United States of America
C—34486

PREFACE

American industries utilize millions of men, production tools, machines, processes, material handling devices, buildings, other related facilities, and billions of dollars in order to shape and produce materials to meet the needs of mankind. The competitive system forces a methodical selection and utilization of the factors of production in the manufacture of high quality products at a low cost. The many alternative processes available to change the size and shape of materials require complementary tooling. Ingenuity is required in the design of this tooling to facilitate scheduled and economic machining, casting, joining and pressworking of the many engineering materials.

The field of tool and manufacturing engineering encompasses a wide variety of industries. It is concerned with the manufacture of airplanes, food handling equipment, glassware, refrigerators, communications equipment, sewing machines, machine tools, textiles, electronic equipment, sporting goods, automobiles, stoves, furniture, packaging equipment, missiles, farm equipment, space capsules, and so on. It is a necessary function in unit or high volume production and in large or small enterprises. The tool and manufacturing engineer articulates in an environment which requires a thorough understanding of scientific and engineering principles. He must understand the broad manufacturing aspects of the industry in which he is employed and he must also be able to design specific production tooling.

This book has been written to meet a need for a fundamental textbook on the subject of tool design. It is a textbook which describes the basic principles of the design of tools for the material removal, pressworking, casting, joining and inspection processes. Engineering and scientific principles have been considered in all sections of the book in an attempt to explain why tools work the way they do. The various sections of the textbook have been written by authors who are experts in their respective fields of tool and manufacturing engineering.

The allocation of time for a specific course of study depends largely on the instructor, the prerequisite courses, and the time available for the course. An approximate time allocation based on the importance of the subject and teaching ease can be stated as follows:

Tool design for the material removal processes	40 per cent
Tool design for the pressworking of materials	30 per cent
Tool design for the inspection and gaging processes	15 per cent
Tool design for the material joining processes	10 per cent
Tool design for the material casting processes	5 per cent

This time allocation is appropriate for one or more courses on the general subject of tool design.

The Textbooks Subcommittee is indebted to the many authors and reviewers who generously contributed time from their already busy schedules.

The continued interest of Mr. Dale Long, as former President of the American Society of Tool and Manufacturing Engineers, inspired the Textbooks Subcommittee to forge ahead during the difficult periods.

Our thanks also to all the Directors and Officers of the Society for their support and encouragement and to the members of the Technical Publications Committee for their guidance and active participation. The tireless efforts of Mr. W. J. Potthoff, Emerson Electric Manufacturing Company, who helped to coordinate the various phases of the book are indeed appreciated. The editing assistance of Mr. F. W. Wilson, Technical Director of A.S.T.M.E., and his staff is highly valued by all. Thanks also to Mr. R. L. Perlewitz of The George J. Meyers Company, Mr. C. E. Lane of The David Ranken School of Mechanical Trades, Mr. R. E. Nauth of the Detroit Engineering Institute, Professor E. Laitala of Clemson College, Mr. D. J. McKeon of the Pioneer Central Division, Bendix Aviation Corporation, Professor A. B. Draper of the Pennsylvania State University, Mr. Stanley Snorek, Western Electric Company, and Mr. Gilbert Stafford of the Dixon Corporation, who assumed responsibilities for major sections of the book. We are grateful for the patience of our spouses and families who were frequently neglected while the textbook was being written.

A. R. Konecny, *Chairman*
Textbooks Subcommittee,
A.S.T.M.E.

Members of A.S.T.M.E. TECHNICAL PUBLICATIONS COMMITTEE

RAYMOND E. GARISS, *Chairman*
Douglas Aircraft Company
Long Beach, Calif.

FRANCIS L. EDMONDSON, *Vice Chairman*
General Dynamics, Fort Worth Division
Fort Worth, Texas

ROBERT E. NAUTH
Detroit Engineering Institute
Detroit, Mich.

ANTHONY R. KONECNY, *Vice Chairman*
Washington University
St. Louis, Mo.

SYDNEY H. PARSONS, JR.
Wasatch Div., Thiokol Chemical Corp.
Tremonton, Utah

RAYMOND H. MECKLEY, *Vice Chairman*
Flinchbaugh Products, Inc.
Red Lion, Pa.

RICHARD N. WILEY
General Electric Company
Somersworth, N. H.

PHILIP R. MARSILIUS, *Ex Officio Member*
Producto Machine Company
Bridgeport, Conn.

Members of TEXTBOOKS SUBCOMMITTEE

ANTHONY R. KONECNY, *Chairman*
Washington University
St. Louis, Mo.

WALTER F. COLES
Pittsburgh Engineering & Machine Co.
Glassport, Pa.

RAYMOND H. MECKLEY
Flinchbaugh Products, Inc.
Red Lion, Pa.

RAYMOND E. GARISS
Douglas Aircraft Company
Long Beach, Calif.

ROBERT E. NAUTH
Detroit Engineering Institute
Detroit, Mich.

EVERETT LAITALA
Clemson College
Clemson, S. C.

WILLIS J. POTTHOFF
Emerson Electric Manufacturing Co.
St. Louis, Mo.

LIST OF CONTRIBUTORS

WILFORD H. ABRAHAM, *Douglas Aircraft Co., Santa Monica, Calif.*
ROSS L. BEAULIEU, *Oneida, Ltd., Oneida, N. Y.*
SETH J. BECK, *Western Electric Co., Inc., Burlington, N. C.*
LOUIS H. BENSON, *Bendix Corporation, Davenport, Iowa*
BERNARD R. BETTER, *Scully-Jones Company, Chicago, Ill.*
JOSEPH S. BLACHUT, *Apex Corporation, Detroit, Mich.*
SHANNON BOYD, *St. Ambrose College, Davenport, Iowa*
ERNEST G. BRIND, *Long Beach State College, Long Beach, Calif.*
R. H. CADDELL, *University of Michigan, Ann Arbor, Mich.*
D. BERNARD CARDINAL, *Bendix Corporation, Davenport, Iowa*
PETER CARBONE, *Brown & Sharp Manufacturing Co., Providence, R. I.*
CARL H. CEDARBLAD, *Swan Engineering Co., Bettendorf, Iowa*
PAUL E. CHARRETTE, *Ford Motor Car Co., Dearborn, Mich.*
CHARLES E. CLARK, *Perfecting Service Co., Charlotte, N. C.*
WALTER F. COLES, *Pittsburgh Engineering & Machine Co., Glassport, Pa.*
HARRY CONN, *Scully-Jones Co., Chicago, Ill.*
EDWARD T. DRABIK, *United Township High School, East Moline, Ill.*
FRED R. DRAKE, *Ex-Cell-O Corporation, Springfield, Ohio*
ALAN B. DRAPER, *Pennsylvania State University, University Park, Pa.*
FRANCIS L. EDMONDSON, *General Dynamics, Fort Worth Div., Fort Worth, Texas*
WILLIAM A. EHLERT, *Line-Material, Inc., Milwaukee, Wis.*
KEITH D. ENTREKIN, *A-C Spark Plug Co., Milwaukee, Wis.*
JOSEPH C. FOGARTY, *Forester Manufacturing, Strong, Maine*
NORMAN G. FOSTER, *Western Electric Co., Inc., Greensboro, N. C.*
HOWARD A. FRANK, *Centralab, Milwaukee, Wis.*
RAYMOND E. GARISS, *Douglas Aircraft Co., Long Beach, Calif.*
ROGER L. GEER, *Cornell University, Ithaca, N. Y.*
RALPH D. GLICK, *Douglas Aircraft Co., Santa Monica, Calif.*
FLOYD D. GOAR, *Moline Sr. High School, Moline, Ill.*
WILLIAM H. GOURLIE, *W. H. Gourlie Company, West Hartford, Conn.*
FRED D. HITTER, *General Dynamics, Fort Worth Div., Fort Worth, Texas*
HOWARD HOLDER, *Boice Gage Co., Inc., Hyde Park, N. Y.*
DONALD G. JOHNSON, *John Deere Co., Moline, Ill.*
KENNETH M. KELL, *Uchoff Co., Davenport, Iowa*
DONALD R. KING, *Consulting Engineer, Milwaukee, Wis.*
JOSEPH A. KLANCINK, *Bendix Corporation, Davenport, Iowa*
ANTHONY R. KONECNY, *Washington University, St. Louis, Mo.*
JOSEPH C. KOPECK, *Harley Davidson, Inc., Milwaukee, Wis.*

List of Contributors

ROBERT C. KRISTOFEK, *Borg and Beck, Chicago, Ill.*
EVERETT LAITALA, *Clemson College, Clemson, S. C.*
CLARENCE E. LANE, *Ranken School of Mechanical Trades, St. Louis, Mo.*
ROBERT M. LARSON, *International Harvester Co., East Moline, Ill.*
CHARLES O. LOFGREN, *Sundstrand Corp., Rockford, Ill.*
HARRY J. LUND, *American Paper Bottle Co., Walled Lake, Mich.*
LINCOLN MAGER, *Newton Insert Co., Los Angeles, Calif.*
LOUIS J. MAHLMEISTER, *Sheffield Corp., Dayton, Ohio*
JAMES C. MANGUS, *Columbia Research and Development Corp., Columbus, Ohio*
DAN J. MCKEON, *Bendix Aviation, Davenport, Iowa*
RAYMOND H. MECKLEY, *Flinchbaugh Products Co., York, Pa.*
KARL H. MOLTRECHT, *General Electric Company, Detroit, Mich.*
JOSEPH MUNDBROT, *Centralab, Milwaukee, Wis.*
ROBERT E. NAUTH, *Detroit Engineering Institute, Detroit, Mich.*
HARRY B. OSBORN, JR., *Ohio Crankshaft Co., Cleveland, Ohio*
CARL OXFORD, JR., *National Twist Drill & Tool Co., Rochester, Mich.*
CLAYTON F. PAQUETTE, *Dana Corporation, Ecorse, Mich.*
ROY B. PERKINS, *University of Illinois, Chicago, Ill.*
RALPH L. PERLEWITZ, *George J. Meyer Co., Milwaukee, Wis.*
WILLIS J. POTTHOFF, *Emerson Electric Manufacturing Co., St. Louis, Mo.*
ROBERT J. QUILLICI, *Scully-Jones Co., Chicago, Ill.*
FRED T. RICHTER, *Western Electric Co., Chicago, Ill.*
EDWARD S. ROTH, *Sandia Corporation, Albuquerque, N. M.*
DONALD M. SATAVA, *Pipe Machinery Corp., Wickliffe, Ohio*
GEORGE H. SHEPPARD, *DoALL Company, DesPlaines, Ill.*
RICHARD A. SMITH, *Pratt & Whitney Co., Hartford, Conn.*
STANLEY J. SNOREK, *Western Electric Co., Inc., Chicago, Ill.*
FRED L. SPAULDING, *University of Illinois, Urbana, Ill.*
JOHN D. SPRINKEL, *Long Beach City College, Long Beach, Calif.*
HOWARD R. SWANSON, *Sargent and Co., New Haven, Conn.*
GILBERT S. STAFFORD, *Dixon Corporation, Monroe, N. C.*
RONALD F. STEWARD, *Lindberg Steel Treating Co., Melrose Park, Ill.*
FRANK SWANEY, *Fairchild Stratos Corp., Hagerstown, Md.*
EDWARD A. TOBLER, *Douglas Aircraft Co., Long Beach, Calif.*
EDWIN M. VAUGHN, *St. Ambrose College, Davenport, Iowa*
CLIFFORD C. VOGT, *J. I. Case Co., Rock Island, Ill.*
JOHN T. VUKELICH, *Square D Company, Milwaukee, Wis.*
EMMETT J. WELKY, *Interstate Drop Forge Co., Milwaukee, Wis.*
ERNIE W. WHEELER, *Carrier Corp., Syracuse, N. Y.*
CHARLES W. WILLIAMS, *Fletcher Aviation, San Gabriel, Calif.*
BEVERLY D. WILSON, *International Harvester Co., East Moline, Ill.*
GERALD C. WOYTHAL, *Cardinal Designers, Milwaukee, Wis.*
LESTER C. YOUNGBERG, *Swan Engineering, Bettendorf, Iowa*
RAYMOND J. ZALE, *H. K. Porter Co., Alliquippa, Pa.*

CONTENTS

1 DESIGN OF MATERIAL-CUTTING TOOLS — 1

Single-Point Tools	2
Basic Principles of Multiple-Point Tools	56
Linear-Travel Tools	66
Axial-Feed Rotary Tools	69
Control of the Causes of Tool Wear and Failure	81
Problems	86

2 WORKHOLDING DEVICES — 89

Elements and Types of Fixture Design	115
Evolution of Workholders	144
Fixture Design Summary	160
Problems	162

3 DESIGN OF PRESSWORKING TOOLS — 166

Power Presses	166
Cutting (Shearing) Operations	169
Types of Die-Cutting Operations	199
Piercing-Die Design	200
Blanking-Die Design	202
Compound-Die Design	204
Scrap-Strip Layout for Blanking	206
Commercial Die Sets	208
Evolution of a Blanking Die	211
Evolution of a Progressive Blanking Die	213
Problems	219
References	220

4 BENDING, FORMING AND DRAWING DIES — 221

Bending Dies	221
Forming Dies	225
Drawing Dies	246
Evolution of a Draw Die	256
Progressive Dies	259
Selection of Progressive Dies	260

4 BENDING, FORMING AND DRAWING DIES (cont.)

Strip Development for Progressive Dies	260
Evolution of a Progressive Die	269
Examples of Progressive Dies	273
Extrusion Dies	277
Tool Design for Forging	284
The Forging Process	293
Forging Design	296
Drop Forging Dies and Auxiliary Tools	303
Upset or Forging Machine Dies	316
Problems	322
References	324

5 DESIGN OF TOOLS FOR INSPECTION AND GAGING — 325

Workpiece Quality Criteria	327
Basic Principles of Gaging	332
Gage Types and Applications	338
Amplification and Magnification of Error	345
Gaging Positionally Toleranced Parts	357
Problems	370
References	371

6 TOOL DESIGN FOR THE JOINING PROCESSES — 372

Tooling for Physical Joining Processes	373
Tooling for Soldering and Brazing	382
Tooling for Mechanical Joining Processes	390
Problems	404

7 TOOLING FOR CASTING — 406

Sand Casting	407
Shell Mold Casting	426
Metal Mold Casting	427
Die Casting	431
Problems	436
References	437

8 GENERAL CONSIDERATIONS IN TOOL DESIGN — 439

Safety as Related to Tool Design	443
Tool Materials	445
Heat Treating	456
Surface Roughness	460
Fits and Tolerances	463
Tooling Economics	469
Material Handling at the Workplace	476
Rules for Good Design	478
Problems	484
References	484

INDEX — 485

FUNDAMENTALS
OF TOOL DESIGN

1

DESIGN OF

MATERIAL-CUTTING TOOLS

The primary method of imparting form and dimension to a workpiece is the removal of material by the use of edged cutting tools. An oversize mass is literally carved to its intended shape. The removal of material from a workpiece is termed *generation of form by machining,* or simply *machining.*

Form and dimension may also be achieved by a number of alternate processes such as hot or cold extrusion, sand casting, die casting, and precision casting. Sheet metal can be formed or drawn by the application of pressure. Metal removal can be accomplished by chemical or electrical methods. A great variety of workpieces may be produced without resorting to a machining operation. Economic considerations, however, usually dictate form generation by machining, either as the complete process or in conjunction with another process.

Elements of the Machining Process. Material removal by machining involves interaction of four elements: the cutting tool, the toolholding and/or guiding device, the workholder, and the workpiece. The cutting tool may have a single cutting edge or may have many cutting edges. It may be designed for linear or rotary motion. The geometry of the cutting tool will depend on its intended function. The toolholding device may or may not be used for guiding or locating. Toolholder selection will be governed by tool design and intended function.

The physical composition of the workpiece will greatly influence the selection of the machining method, the tool composition and geometry, and the rate of material removal. The intended shape of the workpiece will influence the selection of the machining method and the choice of linear or rotary tool travel. The composition and geometry of the workpiece will to a great extent determine the workholder requirements. Workholder selection will also depend on forces produced by the tool and toolholder on the workpiece. The workholder must hold, locate, and support the workpiece. Tool guidance may be incorporated into the workholding function.

Successful design of tools for the material-removal processes requires above all a complete understanding of cutting-tool function and geometry. This knowledge will enable the designer to specify the correct tool for a given task. The tool in turn will govern the selection of toolholding and guidance methods. The tool forces will govern selection of the workholding device. Although the process involves interaction of the four elements, everything begins with and is based on what happens at the point of contact between the workpiece and the cutting tool.

SINGLE-POINT TOOLS

The Basic Tool Angles

Cutting tools are designed with sharp edges to minimize rubbing contact between the tool and workpiece. Variations in the shape of the cutting tool influence tool life, surface finish of the workpiece, and the amount of force required to shear a chip from the parent metal. The various angles ground on a tool bit are called the *basic tool angles,* and compose what is often termed the *tool geometry.* The *signature* is a sequence of numbers listing the various angles, in degrees, and the size of the nose radius. This numerical method of identification has been standardized by the American Standards Association, and is illustrated in Fig. 1-1, together with the elements that make up the tool signature.

Back Rake Angle. This is the angle between the face of the tool and a line that is parallel to the base of the toolholder. It is measured in a plane that is parallel to the side cutting edge and perpendicular to the base. Variations in the back rake angle affect the direction of chip flow. As this angle is increased while other conditions remain constant, tool life will increase slightly and the cutting force required will decrease. Because continual regrinding of this angle reduces the thickness of the tool with its resultant weakening, steep rake angles are usually obtained by alterations in the side rake rather than the back rake angle.

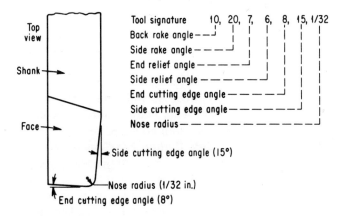

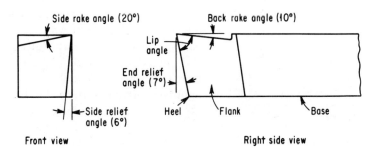

Fig. 1-1. A straight-shank, right-cut, single-point tool, illustrating the elements of the tool signature as designated by the ASA. Positive rake angles are shown.

Side Rake Angle. This angle is defined as the angle between the tool face and a plane parallel to the tool base. It is measured in a plane perpendicular to both the base of the holder and the side cutting edge. Variations in this angle affect the direction of chip flow. As the angle is increased, reductions in cutting force, increased tool life, and improvement in surface finish usually result.

End Relief Angle. This is the angle between the end flank and a line perpendicular to the base of the tool. The purpose of this angle is to prevent rubbing between the workpiece and the end flank of the tool. An excessive relief angle reduces the strength of the tool, so the angle should not be larger than necessary.

Side Relief Angle. This is the angle between the side flank of the tool and a line drawn perpendicular to the base. Comments regarding end relief angles are applicable also to side relief angles. For turning operations, the side relief angle must be large enough to allow for the feed-helix angle on the shoulder of the workpiece.

End Cutting Edge Angle. This is the angle between the edge on the end of the tool and a plane perpendicular to the side of the tool shank. The purpose of the angle is to avoid rubbing between the edge of the tool and the workpiece. As with end relief angles excessive end cutting angles reduce tool strength with no added benefits.

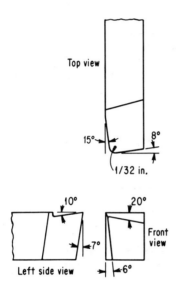

Fig. 1-2. A left-cut tool. All other aspects are identical with Fig. 1-1.

Side Cutting Edge Angle. This is the angle between the straight cutting edge on the side of the tool and the side of the tool shank. This side edge provides the major cutting action and should be kept as sharp as possible. Increasing this angle tends to widen the thin chip and influences the direction of chip flow. An excessive side cutting edge angle may cause chatter and should be avoided. As the angle is increased, increased tool life and minor improvement in surface finish can be expected. However, these benefits will usually be lost if chatter occurs, so an optimum maximum angle should be sought.

Nose Radius. The nose radius connects the side and end cutting edges and should blend smoothly into each to facilitate grinding. Although straight chamfers are sometimes ground to form the nose, most satisfactory results are obtained when the nose is in the form of an arc. Sharp-pointed tools have a nose radius of zero. Increasing the nose radius from zero avoids high heat concentration at a sharp point. Improvement in tool life and surface finish and a slight reduction in cutting force usually result as nose radius is increased. There is, however, a limit to radius size that must be considered. Chatter will result if the nose radius is too large; an optimum maximum value should be sought.

Tool Signature. The seven elements that comprise the signature of a single-point cutting tool are always stated in the following order: back rake angle, side rake angle, end relief angle, side relief angle, end cutting edge angle, side cutting edge angle, and nose radius. Figure 1-1 illustrates and lists the signature of a single-point tool as 10, 20, 7, 6, 8, 15, $\frac{1}{32}$. It is usual practice to omit the symbols for degrees and inches, simply listing the numerical value of each component. Unless specified, the rake angles are understood to be positive as shown. Negative rake angles are shown in Fig. 1-3.

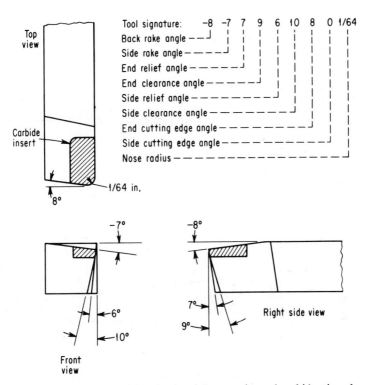

Fig. 1-3. A straight shank, right-cut, sintered-carbide tipped, single-point tool. Rake angles are negative and secondary clearance angles on end and side are illustrated.

A comparison of Figs. 1-1 and 1-2 illustrates the difference between a right- and left-cut tool. Tools are usually ground as the right-cut type. In some cases, secondary relief or clearance angles are employed as illustrated and specified in Fig. 1-3. The additional angles are added to the basic tool signature as shown.

Figure 1-4 illustrates the effect of using a holder that positions the base of the tool in a plane nonparallel with the plane of feeding motion. A 15° toolholder is used, and the tool signature indicates the angles that result when the tool is positioned in the holder. The specified back rake angle of 0° requires grinding a negative 15° back rake angle on the tool prior to placing the tool in the holder.

To avoid confusion, no two angles shown in the illustrations are the same. Actual tool signatures often have several angles that are equal. Relief angles are often ground to the same magnitude, whereas they are shown with different values in the illustrations.

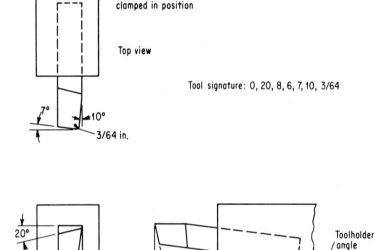

Fig. 1-4. A right-cut, single-point tool mounted in a 15-deg toolholder. Note that the tool signature lists the angles that result when the tool is clamped in the holder.

Tables 1-1, 1-2, and 1-3 give the recommended angles for single-point tools of high speed steel, carbide, and cast alloys respectively.

Chip Breaker. The chip breaker is a small step or groove ground into the face of a tool or a separate piece fastened to the tool or toolholder to cause the chip to curl and break into short sections.

When machining operations are being performed, the high tensile strength and ductility of steel and some other workpiece materials result in a continuous chip which is hazardous and difficult to remove from the vicinity of the cutting edge. Ground chip breakers provide an effective

TABLE 1-1. RECOMMENDED ANGLE FOR HIGH-SPEED-STEEL SINGLE-POINT TOOLS

Material	Side-relief angle, deg	Front-relief angle, deg	Back-rake angle, deg	Side-rake angle, deg
High-speed, alloy, and high-carbon tool steels and stainless steel	7 to 9	6 to 8	5 to 7	8 to 10
SAE steels:				
1020, 1035, 1040	8 to 10	8 to 10	10 to 12	10 to 12
1045, 1095	7 to 9	8 to 10	10 to 12	10 to 12
1112, 1120	7 to 9	7 to 9	12 to 14	12 to 14
1314, 1315	7 to 9	7 to 9	12 to 14	14 to 16
1385	7 to 9	7 to 9	12 to 14	14 to 16
2315, 2320	7 to 9	7 to 9	8 to 10	10 to 12
2330, 2335, 2340	7 to 9	7 to 9	8 to 10	10 to 12
2345, 2350	7 to 9	7 to 9	6 to 8	8 to 10
3115, 3120, 3130	7 to 9	7 to 9	8 to 10	10 to 12
3135, 3140	7 to 9	7 to 9	8 to 10	8 to 10
3250, 4140, 4340	7 to 9	7 to 9	6 to 8	8 to 10
6140, 6145	7 to 9	7 to 9	6 to 8	8 to 10
Aluminum	12 to 14	8 to 10	30 to 35	14 to 16
Bakelite	10 to 12	8 to 10	0	0
Brass, free-cutting	10 to 12	8 to 10	0	1 to 3
Red, yellow, bronze—cast, bronze—commercial	8 to 10	8 to 10	0	−2 to −4
Bronze, free-cutting	8 to 10	8 to 10	0	2 to 4
Hard phosphor bronze	8 to 10	6 to 8	0	0
Cast iron, gray	8 to 10	6 to 8	3 to 5	10 to 12
Copper	12 to 14	12 to 14	14 to 16	18 to 20
Copper alloys:				
Hard	8 to 10	6 to 8	0	0
Soft	10 to 12	8 to 10	0 to 2	0
Fiber	14 to 16	12 to 14	0 to 2	0
Formica	14 to 16	10 to 12	14 to 16	10 to 12
Nickel iron	14 to 16	10 to 12	6 to 8	12 to 14
Micarta	14 to 16	10 to 12	14 to 16	10 to 12
Monel and nickel	14 to 16	12 to 14	8 to 10	12 to 14
Nickel silvers	10 to 12	10 to 12	8 to 10	0 to −2
Rubber, hard	18 to 20	14 to 16	0 to −2	0 to −2

TABLE 1-2. RECOMMENDED ANGLES FOR CARBIDE SINGLE-POINT TOOLS

Material	Normal end-relief, deg	Normal side-relief, deg	Normal back-rake, deg	Normal side-rake, deg
Aluminum and magnesium alloys	6 to 10	6 to 10	0 to 10	10 to 20
Copper	6 to 8	6 to 8	0 to 4	15 to 20
Brass and bronze	6 to 8	6 to 8	0 to −5	+8 to −5
Cast iron	5 to 8	5 to 8	0 to −7	+6 to −7
Low-carbon steels up to SAE 1020	5 to 10	5 to 10	0 to −7	+6 to −7
Carbon steels SAE 1025 and above	5 to 8	5 to 8	0 to −7	+6 to −7
Alloy steels	5 to 8	5 to 8	0 to −7	+6 to −7
Free-machining steels SAE 1100 and 1800 series	5 to 10	5 to 10	0 to −7	+6 to −7
Stainless steels, austenitic	5 to 10	5 to 10	0 to −7	+6 to −7
Stainless steels, hardenable	5 to 8	5 to 8	0 to −7	+6 to −7
High-nickel alloys (Monel, Inconel, etc.)	5 to 10	5 to 10	0 to −3	+6 to +10
Titanium alloys	5 to 8	5 to 8	0 to −5	+6 to −5

TABLE 1-3. CUTTING ANGLES FOR CAST ALLOY TOOLS *

Material	Back-rake angle, deg	Side-rake angle, deg	Side-relief angle, deg	Front-relief angle, deg	Side-cutting-edge angle, deg	End-cutting-edge angle, deg
Steel	8–20 †	8–20 †	7	7	10	15
Cast steel	8	8	5	5	10	10
Cast iron	0	4	5	5	10	10
Bronze	4	4	5	5	10	10
Stainless steel	8–20 †	8–20 †	7	7	10	15

* Stellite 98M2-turning tools.
† Angle depends on grade and type of steel. Boring tools use the same rake but greater relief to clear the work.

Chap. 1 Design of Material-cutting Tools   9

means of curling and breaking the chips into small segments and disposing of them. Figure 1-4A and Table 1-4 give the form and pertinent dimensions for general use.

TABLE 1-4. DIMENSIONS FOR PARALLEL- AND ANGULAR-TYPE CHIP BREAKERS

Depth of cut	Feed	0.006–0.012	0.013–0.017	0.018–0.025	0.028–0.040	Over 0.040
	R *	0.010–0.025	0.035–0.065	0.035–0.065	0.035–0.065	0.035–0.065
	T *	0.010	0.015	0.020	0.030	0.030
1/64–3/64	W	1/16	5/64	7/64	1/8	
1/16–1/4	W	3/32	1/8	5/32	3/16	3/16
5/16–1/2	W	1/8	5/32	3/16	3/16	3/16
9/16–3/4	W	5/32	3/16	3/16	3/16	3/16
Over 3/4	W	3/16	3/16	3/16	3/16	1/4

* All dimensions are in inches.

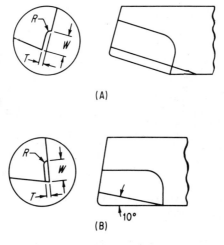

(A)

(B)

Fig. 1-4A.

Chip Formation

The majority of metal-cutting operations involve the separation of small segments or chips from the workpiece material to achieve the required shape and size of manufactured parts. Chip formation involves three basic requirements: (1) there must be a cutting tool that is harder and more wear resistant than the workpiece material; (2) there must be interference between the tool and the workpiece as designated by the feed and depth of cut, and (3) there must be a relative motion or cutting velocity between the tool and the workpiece with sufficient force and power to overcome the resistance of the workpiece material. As long as these three conditions exist, the portion of the material being machined that interferes with free passage of the tool will be displaced to create a chip.

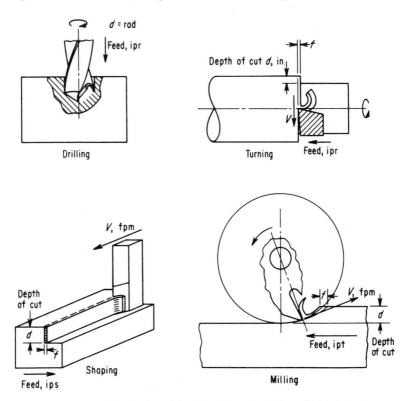

Fig. 1-5. Examples of feed depth and velocity relationships for several chip-formation processes.

Many possibilities and combinations exist that may fulfill the requirements. Variations in tool material and tool geometry, feed and depth of cut, cutting velocity, and workpiece material have an effect not only upon the formation of the chip, but also upon cutting force, cutting horsepower, cutting temperatures, tool wear and tool life, dimensional stability, and the quality of the newly created surface. The interrelationship and the interdependence among these "manipulating factors" constitute the basis for the study of machinability—a study which has been popularly defined as the response of a material to machining.[1] *

* Superior numbers indicate specific references at the end of this chapter.

Chap. 1 Design of Material-cutting Tools   11

Types of Chips. Figure 1-5 illustrates the necessary relationship between cutting tool and the workpiece for chip formation in several common machining processes. Although it is apparent that different general shapes and sizes of chips may be produced by each of the basic processes, all chips regardless of process are usually classified according to their general behavior during formation.

The three most common types of chips are illustrated by the photomicrographs of Fig. 1-6; (a) discontinuous or segmental, (b) continuous without built-up edge, and (c) continuous with built-up edge. The type of chip is generally a function of the work material and the cutting condi-

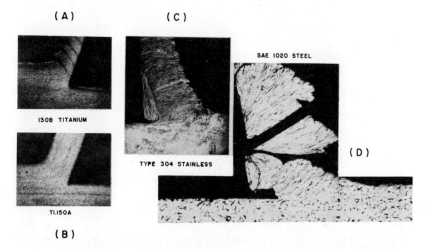

Fig. 1-6. Examples of types of chips: (A) segmental chip, (B) continuous chip without built-up edge, (C) continuous chip with built-up edge. Part (D) shows the "piling up" or heavy distortion common to soft, ductile materials which have large capacity for plastic flow.

tions. The discontinuous or segmented chip is typical of the more brittle materials, while the continuous chips are typical of ductile materials.

Mechanism of Chip Formation.[2, 3, 4] Observations during metal cutting reveal several important characteristics of chip formation: (1) the cutting process generates heat; (2) the thickness of the chip is greater than the thickness of the layer from which it came; (3) the hardness of the chip is usually much greater than the hardness of the parent material; and (4) the above relative values are affected by changes in cutting conditions and in properties of the material to be machined, to give chips that range from small lumps to long continuous ribbons.

These observations indicate that the process of chip formation is one of deformation or plastic flow of the material, with the degree of deformation dictating the type of chip that will be produced.

Plastic flow takes place by means of a phenomenon called *slip,* along what are referred to as *slip planes.* The capacity for plastic flow depends upon the number of these slip planes that are available. The number of planes, in turn, depends upon the crystal lattice structure of the material and upon prior treatment. When the resisting stresses in a material exceed its elastic limit, a permanent relative motion occurs between those adjacent slip planes which are most favorably oriented in the direction of the applied force. Once this motion or slip takes place, these particular planes are strengthened and resist further deformation in preference to other, now weaker, planes that are available. This strengthening is called *work* or *strain hardening,* and is characteristic of all but a few of the common materials.

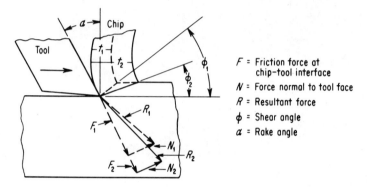

Fig. 1-7. Effect of friction force upon shear angle and upon amount of chip distortion. Force polygons show effect of friction force upon magnitude and direction of resultant force.

As the tool advances into the workpiece, frictional resistance to flow along the tool face and resultant work hardening cause deformation by shear to take place ahead of the tool along a shear plane. This plane extends from the vicinity of the cutting edge toward the free surface of the workpiece at some shear angle ϕ, as illustrated in Fig. 1-7. If the workpiece material is brittle and has little capacity for deformation before fracture, when the fracture shear stress is reached separation will take place along the shear plane to form a discontinuous or segmental chip. Ductile materials, however, contain sufficient plastic flow capacity to deform along the shear plane without rupture. Strain hardening permits a transfer of slip to successive shear planes, and the chip tends to flow in a continuous ribbon along the face of the tool and away from the work surface. The chip is highly worked, and much harder than the material from which it is taken.

Figure 1-8 shows the relative distortion in grid specimens of brittle and ductile materials. The ductile specimen (B) shows evidence of great chip distortion in that the grid lines are completely obliterated. In contrast, the grid lines on the brittle segment (A) are plainly visible and show little distortion. The rate of deformation in continuous-chip formation is extremely high since the shear plane is the boundary between relatively undisturbed and highly disturbed material.

The effect of a change in shear angle is shown in Fig. 1-7. For a given depth of cut, a smaller shear angle ϕ_2 causes greater chip cross section and, therefore, greater distortion than in the case of the larger shear angle ϕ_1. For a given material, the shear angle is a function of the tool rake angle and the coefficient of friction along the tool face. Materials that have fewer slip planes and low work-hardening capacity generally show higher shear angles than the more ductile materials, and have a lower ratio of chip thickness to the size of cut.

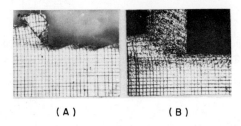

(A)   (B)

Fig. 1-8. Grid specimens showing amount of distortion in chip and adjacent area in (A) brittle material, and (B) ductile material. Grid lines 0.003 in. apart. Line of demarcation between chip and parent material is the shear plane. Wavy surface in (A) due to chatter.

Built-up Edge. Consideration of chip flow along the face of the tool in the formation of continuous chips is of prime importance. If the friction force that resists the passage of the chip along the tool face is less than the force necessary to shear the chip material, the entire chip will pass off cleanly as shown in Fig. 1-6B. This ideal case of chip formation may be approached but is seldom realized. It is generally associated with materials of high strength and of low work-hardening capacity and with low coefficients of friction—factors which lead to large shear angles. High cutting speeds are also favorable.

In most cases, however, it is virtually impossible to prevent some amount of seizure between the chip and the tool face. Unless surfaces are perfectly flat, contact is made along the high spots over only a fraction of the total area. As the chip passes over the tool face, cutting forces give rise to extremely high unit pressures, sufficient to form pressure welds. If these welds are stronger than the ultimate shear strength of the material, that portion of the chip which is welded to the tool shears off as the chip is displaced and

Fig. 1-9. Underside of a chip that had seized to the face of the cutting tool in the light area along the cutting edge. Part of the built-up edge is shown in the process of passing off with the chip. Part of it was also being forced over the cutting edge of the tool and would eventually be deposited on the work surface.

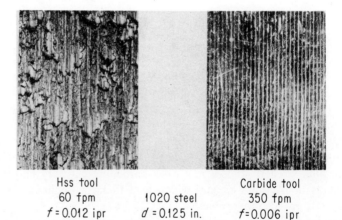

Hss tool		Carbide tool
60 fpm	1020 steel	350 fpm
$f = 0.012$ ipr	$d = 0.125$ in.	$f = 0.006$ ipr

Fig. 1-10. Surfaces produced on hot rolled AISI 1020 steel under conditions that resulted in continuous chip formation with built-up edge (left), and without built-up edge (right).

becomes what is called a *built-up edge*. Continuous chips with built-up edges are illustrated in Figs. 1-6 C, D.

The built-up edge is common to most metal-cutting operations. The edge builds up to a point where it eventually breaks down, part of it going off with the chip, and part of it being deposited on the work surface. This characteristic occurs at very rapid intervals and is exhibited in Fig. 1-9. Any change in cutting conditions that reduces or eliminates the built-up edge will usually improve surface quality (Fig. 1-10). Built-up edge affords some protection to the cutting edge to reduce wear, and a small amount may be desirable. The problem then becomes one of size control through the effects of the various manipulating factors.

Effect of Manipulating Factors

Certain manipulating factors provide some control of the metal-cutting characteristics. Some effects of these factors are illustrated in Figs. 1-11 through 1-15. The results shown were derived from turning cuts on an AISI 1020 hot rolled steel with sharp tools. The results cannot be listed as all-inclusive, since effects of tool wear have not been considered. However, they do represent the general trends of most metal-cutting operations even though built-up edge, surface roughness, and chip shapes may not be the same for each.

In studying the various examples, one should keep in mind the relationship between tool, chip, and surface appearance. The size and the degree of brittleness of the chip may be a good indication of the severity of the cutting operation. The back of the chip, which is in contact with the tool face, gives a fairly good indication of the built-up edge condition. In the absence of built-up edge, the back of the chip should be clean, smooth, and highly burnished. The workpiece surface should be correspondingly good. As the size of the built-up edge increases, more and more markings are evident on the chip. Generally the workpiece is affected in the same manner.

In observing the built-up edge on the tool, two factors should be considered: (1) the size and position relative to the cutting edge, and (2) the uniformity along the cutting edge. Large built-up edges are very unstable, do not slough off uniformly, and tend to leave large deposits on both chip and work. Uniformity along the cutting edge is very important, particularly in forming operations where the quality of the finished surface depends upon the condition of the cutting edge in contact with it. Irregularities in the surface can be matched with the irregularities in the built-up edge, particularly if that edge extends over the cutting edge of the tool.

Velocity. The effect of velocity upon the cutting process must be considered upon the basis of temperature. At low velocities, the temperature at the tool point is below the recrystallization temperature of the material. As a result, work hardening in the chip is retained and the built-up edge forms as previously described. If the velocity increases to the point where the cutting temperature is above the recrystallization temperature of the material, the chip material at the interface tends to recrystallize to form a weld not much stronger, if any, than the rest of the chip. Thus, very little shear will occur and the built-up edge will diminish or disappear as the entire chip passes off.

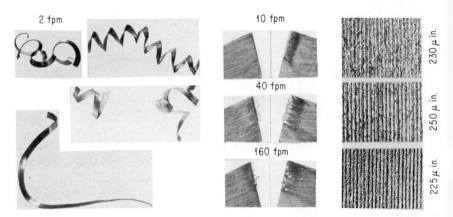

Fig. 1-11. Examples of chips, tools, and surfaces to show effect of velocity on chip formation, built-up edge, and surface quality. Approximate relative magnifications: chips, 1×; tools, 4×; surface, 10×. Note that velocity has practically no effect upon direction of chip flow as seen by markings on tool face. The relatively small effect of velocity on surface finish is due to the fact that practical permissible velocities with HSS tools are not high enough to completely eliminate built-up edge. Tool material—HSS. Tool shape—8, 21, 6, 6, 6, 15, 0. Work material—SAE 1020. Cutting fluid—dry.

Chip form or shape at high velocities can be very troublesome on ductile materials. The reduced resistance to chip flow and the resultant increase in shear angle gives a thinner, less distorted chip, but one which becomes longer and straighter as the velocity increases. Artificial chip breakers become necessary for safety and ease of chip disposal.

Size of Cut. Changes in the size of cut effectively change the cross-sectional area of chip contact (Fig. 1-15 *A, B*). How this area is changed determines the effect upon the cutting process. An increase in depth of cut for a constant feed merely lengthens the contact but does not change the thickness, and the force per unit length remains the same. Theoretically, the built-up edge should not change in either height or width. There should also be no change in surface quality. However, an increase in feed for a

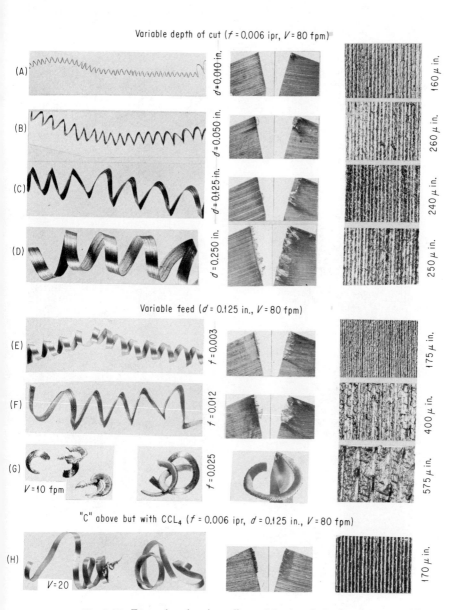

Fig. 1-12. Examples showing effect of feed and depth of cut on chip form, built-up edge, and surface quality. Note how direction of chip flow changes with size of cut. Tool material—HSS. Tool shape—8, 21, 6, 6, 6, 15, 0. Work material—SAE 1020. Cutting fluid—dry.

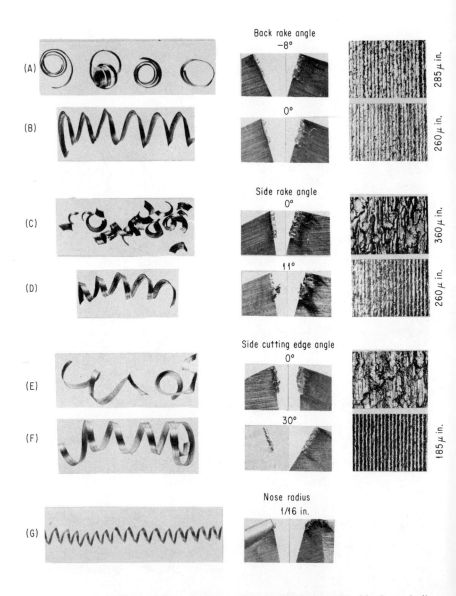

Fig. 1-13. Effect of changes in tool geometry upon chip form, built-up edge, and surface quality. Note also the effect upon direction of chip flow. Tool material—HSS. Work material—SAE 1020. Cutting fluid—dry. Basic tool shape—8, 21, 6, 6, 6, 15, 0.

Chap. 1 Design of Material-cutting Tools

given depth widens the area of contact and changes the force per unit length. This results in greater chip distortion and should alter the height and width of the built-up edge. These observations are substantiated by the results of Fig. 1-12.

There are several factors that affect surface quality to a greater degree than may be predicted. Lack of rigidity will permit greater deflections as a result of higher forces. Increases in feed and depth of cut may then cause chatter, poor surface quality, and loss of dimensional stability. Deep turning cuts on relatively small diameters have a greater percentage change in velocity along the length of the cutting edge. This might result in erratic built-up edge behavior with poorer surface quality.

The effect upon the chip is also much more pronounced with increases in feed than with increases in depth. Because of the greater distortion the chips tend to break up more readily. Actually, it is possible to form segmented chips with ductile materials with very heavy feeds and depth.

Effect of Tool Geometry. For given cutting conditions, changes in tool geometry have two direct effects on chip formation: (1) effect upon shear angle, and (2) effect upon chip thickness. The two are related in that a change in one usually affects the other.

The effects of side cutting edge angle and of nose radius can be explained in terms of the effect upon chip thickness. Figure 1-15C shows that an increase in the side cutting edge angle reduces the chip thickness for a given feed by a factor of the cosine of the angle. This, in effect, reduces the chip contact width to thin out the built-up edge. An increase in nose radius has the same general effect as seen in Fig. 1-15D. The shape of the contact area changes, but at the point of contact between the machined surface and the tool, the chip is very thin and the built-up edge is reduced very effectively. In comparison, the feed marks are much smoother than those left by a sharp-nosed tool.

In view of the above effects, reduction of chip thickness must be accompanied by an increase in the length of engagement of the cutting edge. This results in higher cutting forces, in spite of the thinner chip, and could lead to chatter in nonrigid setups.

The effects of changes in rake angles are shown in Fig. 1-16. The lower rake angles decrease the shear angle, cause greater chip distortion, and increase the resistance to chip flow. The size of the built-up edge increases to produce rougher and more work-hardened surfaces. At low or negative rake angles the chip is so highly distorted that it breaks up into short lengths. The side rake angle has much more effect than does the back rake angle. The effects on three-dimensional cutting must be described on the basis of the effective rake angle in the direction of chip flow.

Tool Material. One effect of tool material lies in its ability to sustain high cutting velocities, as for example between high-speed steel and car-

20   Design of Material-cutting Tools   Chap. 1

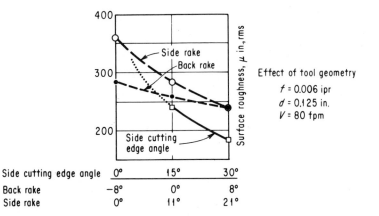

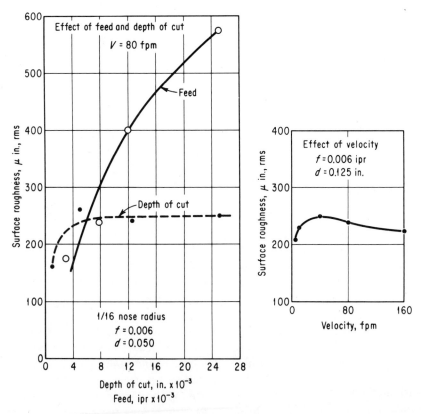

Fig. 1-14. Summary of surface roughness results given in Figs. 1-11 through 1-13. Standard tool geometry—8, 21, 6, 6, 6, 15, 0. HSS tools. Work material—AISI 1020 hot rolled steel.

bides. The effect of high velocity has already been described. Another factor is the coefficient of friction between chip and tool material. Usually, this is of little consequence with high-speed steel tools because the coefficient of friction does not change appreciably among the various grades. Sintered

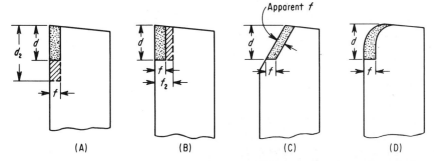

Fig. 1-15. Effect of size of cut and changes in tool geometry upon chip thickness: (A) change in depth, (B) change in feed, (C) effect of side cutting edge angle, and (D) effect of nose radius. Crosshatched portions represent increase in contact area.

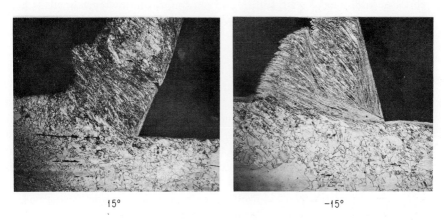

Fig. 1-16. Photomicrographs showing effect of rake angle upon shear angle, chip distortion, built-up edge, and work hardening of machined surface. Shaper tools set for some depth of cut. Apparent difference in depth due to higher separating force and greater tool deflection with negative rake tool. Work material—304 stainless steel.

carbide tools, however, are made with different compositions of carbide and binder materials, and these tools may behave quite differently in both wear and built-up edge characteristics for the same cutting conditions.

Cutting Fluid. Ideally, if a cutting fluid provides lubrication between the chip and the tool, the coefficient of friction will be reduced and the built-up edge will be minimized. However, effective lubrication may be difficult to achieve except possibly at very low cutting speeds. Effects vary with cutting conditions and with work materials. At high speeds fluids act principally as coolants, but may effectively stabilize the built-up edge to improve surface quality.

Workpiece Materials. Brittle materials give segmented or discontinuous chips with little or no built-up edge on the tool. Surface quality usually depends upon the spacing of the segments and is generally better in the direction of cutting than in the direction of feed. Cutting forces are usually lower than they would be for a ductile material of corresponding strength because of generally larger shear angles and because of lower resistance along the tool face.

Ductile materials produce continuous chips, normally with a built-up edge. With low friction and high cutting velocities, particularly with materials of low work-hardening capacity, a thinner, less distorted chip is produced and the built-up edge is reduced or eliminated. High frictional resistance to flow, low shear angles, and materials of high work-hardening capacity are associated with large distortions during cutting, and promote not only large built-up edges, but increased work-hardening of the work surface. Additions of lead, sulfur, and phosphorus to low-carbon steels help to break up chips, reduce built-up edge, and improve surface quality.

Tool Wear

For the sake of recognition and understanding of the fundamentals of metal cutting, the effects of changes in the manipulating factors have been described without regard to their influence upon such criteria as tool wear and tool life. Yet, there is no known tool material that can completely resist contact and rubbing at high temperatures and at high pressures, without some changes from its original contours over a period of time. It becomes necessary, therefore, to think of the effect of the manipulating factors not only upon the cutting process itself, but upon the performance of the cutting tool, which may, in turn, itself affect the cutting process.

Tool Failure. Failure of the cutting tool has occurred when it is no longer capable of producing parts within required specifications. The point of failure, together with the amount of wear that determines this failure, is a function of the machining objective. Surface quality, dimensional stability, cutting forces, cutting horsepower, and production rates may alone, or in combination, be used as criteria for tool failure. It may, for instance, take very little wear to affect the stability of the built-up edge, which in turn may affect surface quality, although the tool itself could continue to remove metal with little, if any, loss of efficiency. In contrast, only a few thousandths of an inch wear on a wide form tool might cause such a large increase in thrust or feeding forces that it would result in a loss of dimen-

Chap. 1 Design of Material-cutting Tools   23

sional stability, or require excessive power, in addition to a loss of surface quality.

Type of Tool Wear. Tool failure is associated with some form of breakdown of the cutting edge. Under proper operating conditions, this breakdown takes place gradually over a period of time. In the absence of rigidity, or because of improper tool geometry that gives inadequate support to the cutting edge, the tool may fail by mechanical fracture or chipping under the load of the cutting forces. This is not truly a wear phenomenon for it can be eliminated or at least minimized by proper design and application.

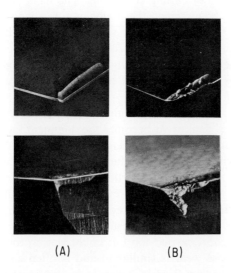

(A)   (B)

Fig. 1-17. Representative wear patterns on face, flank, and nose of cutting tool, typical of chip-removal processes on ductile materials. Crater on face of tool in (A) started well back of cutting edge. In (B) crater wear had progressed to point where weak cutting edge broke down under cutting forces.

As a result of the direct contact with the work material, there are three major regions on the tool where wear can take place: (1) face, (2) flank, and (3) nose (Fig. 1-17).

Face Wear. The face of the tool is the surface over which the chip passes during its formation. Wear takes the form of a cavity or crater which has its origin not along the cutting edge but at some distance away from it and within the chip contact area. As wear progresses with time, the crater gets wider, longer, and deeper and approaches the edges of the tool.

This form of wear is usually associated with ductile materials which give rise to continuous chips with built-up edge. If crater wear is allowed to proceed too far, the cutting edge becomes weak as it thins out, and breaks down suddenly. Usually, there is some preliminary breakthrough of the crater at the nose and at the periphery prior to total failure of the cutting edge. These preliminary breaks serve as focal points for the development of notches along the flank. These notches often lead to erratic built-up edges and have a serious effect upon surface quality. In general, crater wear develops faster than flank wear on ductile materials and is the limiting factor in determination of tool failure.

Flank Wear. Although crater wear is most prominent in the machining of ductile materials, flank wear is always present regardless of work and tool material, or even of cutting conditions. The flank is the clearance face of the cutting tool, along which the major cutting edge is located. It is the portion of the tool that is in contact with the work at the chip separation point and that resists the feeding forces. Because of the clearance, initial contact is made along the cutting edge. Flank wear begins at the cutting edge and develops into a wider and wider flat of increasing contact area called a *wear land*.

Materials that do not form prominent built-up edges promote little if any crater wear, and flank wear becomes the dominant factor in tool failure. In the case of most form tools and certain milling cutters, the wear land is in direct contact with the finished surface, and usually becomes the basis for failure even on ductile materials, particularly if surface finish specifications are the controlling factors in the process. Quite often, flank wear is accompanied by a rounding of the cutting edge, particularly in the machining of brittle or abrasive materials. This results in large increases of cutting and feeding forces which, if carried too far, could lead to tool fracture.

Nose Wear. Nose wear is similar to and is often considered a part of flank wear. There are times when it should be considered separately. Nose wear sometimes proceeds at a faster rate than flank wear, particularly when one is working on rather abrasive materials and using small nose radii. In finish turning operations, for example, the nose is in direct contact with the workpiece, and excessive wear might affect dimensional stability as well as surface roughness. Where sharp corners are to be maintained, the rounding or flattening of the nose can cause failure long before flank wear itself becomes a factor.

Mechanism of Tool Wear. Evidence indicates that wear is a very complex phenomenon and is influenced by many factors. The causes of wear do not always behave in the same manner nor do they always affect wear to the same degree in similar cutting conditions. The causes of wear are not fully understood. Much has yet to be learned about wear phenomena, and the problems appear to be complex. In recent years great strides have been made by various researchers. Even though there is some disagreement regarding the true mechanisms by which wear actually takes place, most investigators feel that there are at least five basic causes of wear:

1. by abrasive action of hard particles contained in the work material,
2. by plastic deformation of the cutting edge,
3. by chemical decomposition of the cutting tool contact surfaces,
4. by diffusion between work and tool material, and
5. by welding at asperities between work and tool.

The relative effects of these causes as a function of cutting velocity or cutting temperature are shown in Fig. 1-18. Investigations have also been made on other possible causes such as oxidation and electrochemical reactions in the tool-work contact zone.

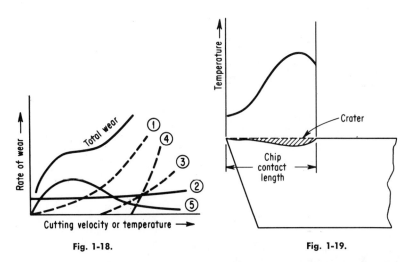

Fig. 1-18.   Fig. 1-19.

Fig. 1-18. Relative effects of various causes of tool wear: (1) abrasive wear, (2) plastic deformation of cutting edge, (3) chemical decomposition, (4) diffusion, (5) welding of asperities.

Fig. 1-19. Temperature distribution along tool-chip contact length.[6]

The most important factor influencing tool wear is cutting temperature. Of the five basic causes for wear, temperature has considerable effect in all but one. Cutting temperatures are important for two basic reasons: (1) most tool materials show rapid loss of strength, hardness, and resistance to abrasion above some critical temperature, and (2) the rate of diffusion between work and tool material rises very rapidly as temperature increases past the critical.

Analytical and experimental methods have been used [6,7] to show that the average peak temperatures at the tool-chip interface occur near the point where the chip leaves the tool surface (Fig. 1-19). Crater wear appears greatest at this point. The relationship between crater wear rate and average tool-chip interface temperature is shown in Fig. 1-20. The rate of wear increases very rapidly beyond a critical temperature. Flank temperatures were found to be maximum near the tool point as shown in Fig. 1-21.

The significance of temperature effects upon wear are associated with tool material properties. High-speed steel tools begin to lose their properties very rapidly at approximately 1100°F. Carbides show less drastic sensitivity to temperature up to about 1600°F. Chemical decomposition and diffusion will not occur at any appreciable extent until the critical temperatures are reached.

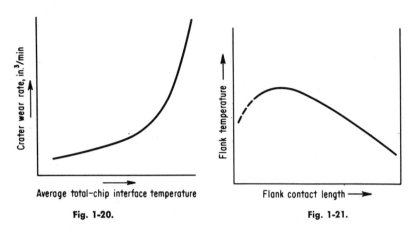

Fig. 1-20. **Fig. 1-21.**

Fig. 1-20. Relationship between rate of crater wear and average tool-chip interface temperature.[6]

Fig. 1-21. Distribution of flank temperature along flank-work contact length.[6]

Wear by Abrasive Action. This mechanism may be partly explained by the fact that hard particles (inclusions, carbides, etc.) in the workpiece material literally gouge or dislodge particles from the tool, causing continuous wear under any cutting condition. The rate of wear is thus dependent upon the number, size, distribution, and the hardness of the particles in the work material as well as the hardness of the cutting tool and the workpiece. At higher cutting speeds even some of the softer constituents may contribute the gouging action as a result of higher impact values and reduced tool resistance to abrasion.

Plastic Deformation of the Cutting Edge. This wear mechanism is believed to take place at all ranges of cutting temperatures; it arises from the high unit pressures imposed on the tool. This results in a slight rounding of the edge, similar to that shown in Fig. 1-22. The net effect is greater distortion during chip formation, with consequent rise in temperature, which accelerates wear even more.

Chemical Decomposition of the Cutting Tool. This wear mechanism occurs through localized chemical reactions at the tool-workpiece contact surfaces. These reactions are temperature-dependent and result in weakening the bond between minute tool segments and the segments surrounding them. This may occur either through formation of weaker compounds or, in the case of carbide tools, by a dissolving action of the bond between the binder and individual carbide particles. As a result of this weakening effect, the particles are pulled out from the main body of the tool by the chip or work as it moves past the contact surfaces. Once the critical temperature for this chemical action is reached, the rate of wear is relatively rapid.

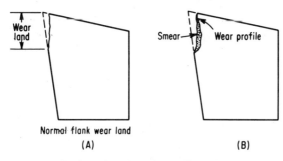

Fig. 1-22. Section of tool to show effect of smear and other causes on type of flank wear.

Diffusion. This very complex wear phenomenon between the work and the tool results in a very rapid breakdown of tool material once critical temperature is reached. Much has yet to be learned about this cause of wear, but basically it involves a change of composition at the tool-chip interface. There is an alloying effect which weakens the bond of the tool particles and permits them to be pulled out by the chip or built-up edge as it sloughs off. Within certain ranges of cutting temperatures, and to a certain degree, the built-up edge actually serves to protect the tool from certain forms of wear. It keeps the tool from actual contact with the workpiece and helps to reduce the temperatures at the tool point. Evidence of this may be seen in Fig. 1-9; the area on the tool face which was occupied by the built-up edge shows no heat discoloration although discoloration surrounds it. At higher temperatures, however, the built-up edge, which is welded to the tool face, serves as a source for diffusion. It remains in place until the bond at the interface is weakened by alloying to a point where the friction forces are sufficient to pull the built-up edge off the face of the tool.

Wear Through Welding of Asperities. This mechanism parallels that of the built-up edge. As shown in Fig. 1-18, the greatest rate of wear by this mechanism occurs at lower cutting velocities or temperatures. A built-up edge forms because of a high resistance to chip flow along the tool face, which causes a portion of the chip to shear off as it moves past the tool. This action is most prominent when cutting temperatures are below the recrystallization temperature for the material. Work hardening is retained and the built-up edge is harder and stronger than the rest of the chip. This same situation exists for wear through welding at the asperities.

The asperities on a tool are brittle and relatively weak in bending or tension. If welding takes place between the chip and the asperities because of the extremely high unit pressures, the work-hardened chip material is strong enough to pull these asperities off. However, if the temperature is near or beyond the recrystallization temperature, then the bond between the chip or built-up edge and the tool is no weaker than the material adjacent to it, because work hardening has not been retained. Therefore, the rate at which these asperities are pulled out diminishes.

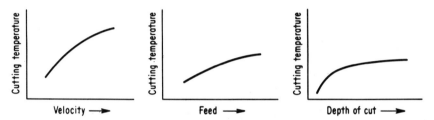

Fig. 1-23. Relative effects of velocity, feed, and depth of cut on cutting temperatures for a given work material and tool geometry.

If the cutting conditions are such that the resulting temperatures approach or go beyond the critical temperature for a given tool material, the reduced resistance of the tool, and the increasing tendency for alloying between work and tool material, cause a high rate of wear and very rapid failures. When the cutting temperatures are low, the processes of wear by abrasion and by welding at the asperities become most prominent.

Effects of Manipulating Factors Upon Tool Wear. The effects of the manipulating factors upon tool wear are concerned either with modifications that influence the cutting process directly for a given tool and workpiece material, or with inherent properties of materials that resist or promote wear. For a given tool and workpiece material combination, cutting temperatures are influenced most by cutting speed and feed (Fig. 1-23). Adjustments in speed or feed, or both, will affect tool wear. It may be possible to substitute another tool material that has inherently better temperature-resistant properties to maintain original or even higher production rates with less sensitivity to temperature failure. The cost of the second material may be higher than that of the first, but it may be more than justified by higher production rates for optimum tool life.

Changes in tool geometry that result in higher shear angles, less chip distortion, lower frictional resistance, and thinner chips will lower cutting forces and decrease cutting temperatures, and thus contribute to decrease the rate of tool wear for given cutting conditions. Within practical and design limitations, rake, relief, and side cutting edge angles, and nose radii, or their equivalents, should be as large as possible, yet small enough to prevent mechanical tool breakage and chatter. Heat transfer characteristics may also be adversely affected if the point of the tool is too thin as a result of high relief and rake angles. The heat at the point does not dissipate as rapidly and higher temperatures prevail.

Workpiece materials that have relatively high hardness, high shear strength, high coefficient of friction, and high work-hardening capacities, and contain hard constituents, promote more rapid wear for given cutting conditions. Materials such as titanium or stainless steels, which have poor thermal conductivity, do not dissipate heat from the cutting zone as rapidly as others, and temperature failures are more pronounced.

Effect of Wear on Machinability Criteria

The study of machinability involves certain criteria which play a prominent part in the evaluation of the cutting process. Machinability ratings for a given material are entirely relative in that some one material is used as a base. The ratings can vary not only among the machining processes, but also with the criterion used in the evaluation for a given process. Though many data are available about numerous materials, erroneous conclusions may result if the data are not interpreted or applied properly.

Many of the available data, particularly with respect to cutting forces, specific power requirements, and surface quality, are based upon the results of sharp-tool investigations. These investigations serve a valuable purpose in analysis of the cutting process and in determination of initial levels of performance, but they give no indication as to how long the initial level of performance will be maintained when tools begin to wear. Some materials that are given very high machinability ratings with sharp tools are extremely sensitive to tool wear. Performance may drop off very rapidly with time. On the other hand, very similar materials may have lower initial levels of performance but are less sensitive to increasing tool wear, maintain the original levels for longer periods of time, and actually receive higher production performance ratings.

Another sort of data that warrants some caution in direct application is the tool-life curve. These curves (see "Tool Life," below) are usually based on accelerated-wear tests. They are actually plots of cutting velocity versus cutting time, or of cubic inches of metal removal before failure for a given size of cut and otherwise constant cutting conditions. Failure can be identified in several ways. Absolute failure of the cutting tool is most common with high-speed steel tools. A preselected amount of flank wear may indicate failure in tests with carbide and ceramic tools. The quality of a machined surface may be used to denote failure of any tool material. Whenever wear is used in some form as the criterion of failure, the results can be plotted as a generally acceptable straight line on log-log coordinates.

For various reasons, mostly economic, the points used to establish the tool-life curve come from comparatively short tool lives, usually less than one hour of cutting time. If the velocity for some longer tool life is desired for a practical application, these curves are usually extended and the velocity for the desired tool life is extrapolated as indicated by the dotted line in Fig. 1-24. This practice can sometimes lead to very unsatisfactory results.

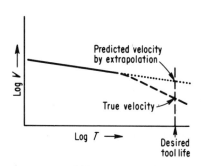

Fig. 1-24. Tool-life curve showing error that can be introduced in predicting velocity for longer tool lives by extrapolation of typical accelerated-wear results (solid line). Lower velocities affect the rate and type of tool wear to change slope of curve. Slopes approaching 0 indicate greater sensitivity to temperature while slopes approaching 1 indicate effect of abrasive wear.

High cutting velocities result in high cutting temperatures. When short tool lives are encountered, the cutting temperature is in or near the critical temperature ranges, which promote rapid tool wear by diffusion and chemical decomposition. In this region, small changes in velocity or temperature may have relatively large effects upon wear rate and, therefore, on tool life. The solid line in Fig. 1-24 represents a typical plot. The further the cutting temperature is removed from the critical, the less effective becomes the wear by diffusion and chemical decomposition, and the more prominent becomes the wear by abrasion and welding at the asperities. Since the total wear rate decreases at this point, there should be a lesser effect upon tool life for the same increment change in cutting velocity. The absolute slope of the curve should increase as represented by the dashed line in Fig. 1-24. The actual velocity for a specified tool life may be considerably lower than the predicted velocity by extrapolation from a curve founded on short-time tests. Whether all workpiece materials and tool materials exhibit this kind of behavior to any predictable degree is not fully known at this time. There is considerable evidence that accelerated-wear tests can give misleading information and that care should be exercised to make proper use of this information. Though some materials are much more sensitive to tool wear than others, the general trends are similar, although not to the same degree.

Numerous changes may take place during the life of a cutting tool. Whether these changes are important or not depends upon the machining objective. In the cutting of ductile materials, increases in crater and flank wear literally cause a change in the true tool geometry. This change in tool geometry usually affects chip formation (Fig. 1-25). Chip form changes can

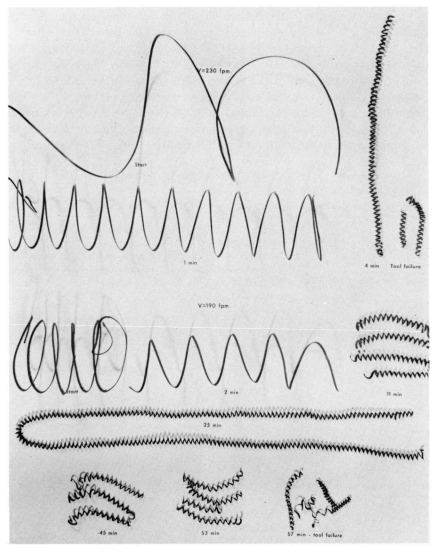

Fig. 1-25. Effect of tool wear upon chip form and shape. Times represent points at which changes occurred. Tool geometry—8, 14, 6, 6, 6, 0, 0.010. Tool material—HSS. Cutting fluid—oil with sulfur additive. Work material—C 1045.

be very sudden, particularly when the crater starts to break through the cutting edge.

One of the most notable effects of tool wear is the change in surface quality. During the test on AISI 1045 steel at 190 fpm, from which the chip samples in Fig. 1-25 were taken, there were five distinct changes in surface appearance before total tool failure occurred at 57 minutes. Table 1-5 lists these changes. The changes in surface quality reflect the unstable character of the built-up edge as wear progresses. Uneven wear or notching gives rise to large built-up edges which slough off erratically. With materials that are notorious for formation of built-up edge, surface roughness usually reaches unsatisfactory proportions long before actual cutting failure. In the previous example, if surface finish was the criterion of failure, tool life would have been about 11 minutes rather than the 57 minutes of actual cutting time. Figure 1-26 shows an actual application of failure on the basis of surface quality in the finish turning of large axles of AISI 1045 steel.

TABLE 1-5. EFFECTS OF TOOL WEAR ON SURFACE QUALITY

Cutting time, min.	Surface appearance
0–2	Clean, only minor traces of built-up edge
2–11	Dull and streaky
11–25	Numerous large deposits of built-up edge; unsatisfactory
25–45	Numerous small deposits of built-up edge; very streaky
45–57 (Failure)	Partially burnished at 45 minutes to very highly burnished at failure

Acceptable　　　　　Tool failed

Fig. 1-26. Example of tool failure based upon surface finish. Fax film reproductions show tearing and built-up-edge deposits which rendered surface unacceptable for functional requirements. Wear land at failure was only a few thousandths of an inch wide. Tool material—HSS. Work material—AISI 1045 steel.

Tool wear has a definite effect upon cutting forces; the ratio of increase can be beyond expectation. In production-type operations, feeding or thrust forces in straight forming can rise by a factor of from as low as 2 to as high as 40 times the sharp-tool values, depending upon the type of flank encountered. In one test,* the feeding force rose from a sharp-tool value of 11 lb to 505 lb for a flank wear land of 0.0084 in. In another test on a material of the same commercial grade but from another source, the feeding force rose from an initial value of 16 lb to only 180 lb for the same width of wear land. The cutting conditions were exactly the same in each case, but there was sufficient difference in material behavior (even though both materials were within commercial specifications for the grade) to cause flank wear patterns as illustrated in (A) and (B) of Fig. 1-22, the wear pattern at (B) producing the higher forces.

On the basis of the condition of the flanks, one might suspect that tangential cutting forces or cutting power would, or should, show the same general characteristics as the feeding forces. Actually, in the tests cited, there was not only comparatively little difference in power requirements for the operation in spite of the large difference in feeding force, but power requirements increased by a factor of only ½ in each case.

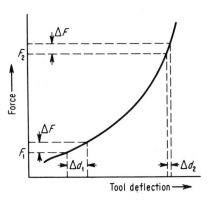

Fig. 1-27. Effect of forces upon machine-tool component deflections.

Although a large increase in feeding force may not appreciably affect power requirements, the effect upon machine tool deflections can be quite pronounced. In the example cited above, the formed diameters increased by 0.050 and 0.028 in., respectively, with the higher and lower forces. Because of clearances in machine tool assemblies, the typical force-deflection characteristics of a tool mounted on a slide controlled through a series of links can be represented by the curve shown in Fig. 1-27. It is seen that, initially, a small change in force can result in a rather large deflection. As the play between parts is taken up and elastic resistance increases, a comparable change in force results in a smaller deflection of the tool. Thus, it is apparent that a range of feeding force from a very low value to a high value would absorb the greatest deflection range. If the initial feeding force

---

* University of Michigan Research Institute Project No. 2575, Reynolds Metals Company.

is high, most of the tool deflection occurs prior to any effects created by tool wear, and the change in dimension is lower in magnitude.

The previous examples illustrate certain effects of tool wear, but more than that, they illustrate the ever-present variations and difficult-to-explain results that complicate metal-cutting practice.

Tool Life

The types and mechanisms of tool failure have been previously described. It was shown that excessive cutting speeds cause a rapid failure of the cutting edge; thus, the tool can be declared to have had a short life. Other criteria are sometimes used to evaluate tool life. These are:

1. change of the quality of the machined surface;
2. change in the magnitude of the cutting force resulting in changes in machine and workpiece deflections causing workpiece dimensions to change; and
3. change in the cutting temperature.

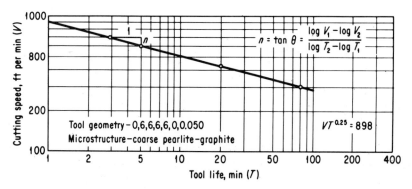

Fig. 1-28. Tool life vs. cutting speed. Tool material—Kennametal carbide. Tool geometry—0, 6, 6, 6, 6, 0, 0.050. Work material—gray cast iron, 195 Bhn.

The selection of the correct cutting speed has an important bearing on the economics of all metal-cutting operations. Fortunately, the correct cutting speed can be estimated with reasonable accuracy from tool-life graphs or from the Taylor Tool Life Relationship, provided that the necessary data are obtainable.

A tool-life graph is shown in Fig. 1-28. The logarithm of tool life in minutes is plotted against the logarithm of cutting speed in feet per minute.

Chap. 1 Design of Material-cutting Tools

The resulting curve is very nearly a straight line in most instances. For all practical purposes it can be considered a straight line. This curve is expressed by the following equation:

$$VT^n = C \qquad (1\text{-}1)$$

where V = cutting speed, feet per minute
T = tool life, minutes
C = a constant equal to the intercept of the curve and the ordinate or the cutting speed—actually it is the cutting speed for a one-minute tool life
n = slope of the curve $\left(n = \tan \phi = \dfrac{\log V_1 - \log V_2}{\log T_2 - \log T_1}\right)$

Example 1. A 2-in. diam bar of steel was turned at 284 rpm and tool failure occurred in 10 min. The speed was changed to 232 rpm and the tool failed in 60 min of cutting time.

Assuming a straight-line relationship exists, what cutting speed should be used to obtain a 30-min tool life (V_{30})?

Solution. Calculating the cutting speed,

$$V_1 = \frac{\pi DN}{12} = \frac{\pi(2)(284)}{12} = 149 \text{ fpm}$$

$$V_2 = \frac{\pi(2)(232)}{12} = 122 \text{ fpm}$$

The slope n of the tool-life curve can be determined from the equation shown above. If a log-log slide rule is available, a faster method of solution is shown as follows:

$$V_1 T_1^n = V_2 T_2^n = C$$

$$\frac{V_1}{V_2} = \left(\frac{T_2}{T_1}\right)^n$$

$$\frac{149}{122} = \left(\frac{60}{10}\right)^n$$

$$1.22 = 6^n$$

$$n = 0.11$$

From Eq. (1-1),

$$C = 149(10)^{0.11} = 149(1.288)$$
$$= 192$$

Thus

$$VT^{0.11} = 192$$

$$V_{30} = \frac{192}{(30)^{0.11}} = \frac{192}{1.455}$$

$$V_{30} = 122 \text{ fpm} \quad \textit{Answer}$$

TABLE 1-6. EQUATIONS SHOWING THE RELATION BETWEEN CUTTING SPEED AND TOOL LIFE FOR VARIOUS TOOL MATERIALS AND CONDITIONS [1]

No.	Tool		Workpiece material	Size of Cut, in.		Cutting fluid	$VT^n = C$	
	Tool material	Shape		Depth	Feed		n	C
1	High carbon steel	8, 14, 6, 6, 6, 15, 3/64	Yellow brass (0.60 Cu, 0.40 Zn, 0.85 Ni, 0.006 Pb)	0.050	0.0255	Dry	0.081	242
2				0.100	0.0127	Dry	0.096	299
3	High carbon steel	8, 14, 6, 6, 6, 15, 3/64	Bronze (0.90 Cu, 0.10 Sn)	0.050	0.0255	Dry	0.086	190
4				0.100	0.0127	Dry	0.111	232
5	HSS-18-4-1	8, 14, 6, 6, 6, 15, 3/64	Cast iron 160 Bhn	0.050	0.0255	Dry	0.101	172
6			Cast iron, Nickel, 164 Bhn	0.050	0.0255	Dry	0.111	186
7			Cast iron, Ni-Cr, 207 Bhn	0.050	0.0255	Dry	0.088	102
8	HSS-18-4-1	8, 14, 6, 6, 6, 0, 0	Steel, SAE B1113 C.D.	0.050	0.0127	Dry	0.08	260
9			Steel, SAE B1112 C.D.	0.050	0.0127	Dry	0.105	225
10			Steel, SAE B1120 C.D.	0.050	0.0127	Dry	0.100	270
11			Steel, SAE B1120 + Pb C.D.	0.050	0.0127	Dry	0.060	290
12			Steel, SAE 1035 C.D.	0.050	0.0127	Dry	0.110	130
13			Steel, SAE 1035 + Pb C.D.	0.050	0.0127	Dry	0.110	147
14	HSS-18-4-1	8, 14, 6, 6, 6, 15, 3/64	Steel, SAE 1045 C.D.	0.100	0.0127	Dry	0.110	192
15		8, 22, 6, 6, 6, 13, 3/64	Steel, SAE 2340 185 Bhn	0.100	0.0125	Dry	0.147	143
16		8, 14, 6, 6, 6, 15, 3/64	Steel, SAE 2345 198 Bhn	0.050	0.0255	Dry	0.105	126
17		8, 14, 6, 6, 6, 15, 3/64	Steel, SAE 3140 190 Bhn	0.100	0.0125	Dry	0.160	178

#	Tool	Angles	Work material			Coolant		
18	HSS-18-4-1	8, 14, 6, 6, 6, 15, 3/64	Steel, SAE 4350 363 Bhn	0.0125	0.0127	Dry	0.080	181
19			Steel, SAE 4350 363 Bhn	0.0125	0.0255	Dry	0.125	146
20			Steel, SAE 4350 363 Bhn	0.0250	0.0255	Dry	0.125	95
21			Steel, SAE 4350 363 Bhn	0.100	0.0127	Dry	0.110	78
22			Steel, SAE 4350 363 Bhn	0.100	0.0255	Dry	0.110	46
23	HSS-18-4-1	8, 14, 6, 6, 6, 15, 3/64	Steel, SAE 4140 230 Bhn	0.050	0.0127	Dry	0.180	190
24			Steel, SAE 4140 271 Bhn	0.050	0.0127	Dry	0.180	159
25			Steel, SAE 6140 240 Bhn	0.050	0.0127	Dry	0.150	197
26	HSS-18-4-1	8, 22, 6, 6, 6, 15, 3/64	Monel metal 215 Bhn	0.100	0.0127	Dry	0.080	170
27				0.150	0.0255	Dry	0.074	127
28				0.100	0.0127	Em	0.080	185
29				0.100	0.0127	SMO	0.105	189
30	Stellite 2400	0, 0, 6, 6, 6, 0, 3/32	Steel, SAE 3240 annealed	0.187	0.031	Dry	0.190	215
31				0.125	0.031	Dry	0.190	240
32				0.062	0.031	Dry	0.190	270
33				0.031	0.031	Dry	0.190	310
34	Stellite No. 3	0, 0, 6, 6, 6, 0, 3/32	Cast iron 200 Bhn	0.062	0.031	Dry	0.150	205
35	Carbide (T 64)	6, 12, 5, 5, 10, 45	Steel, SAE 1040 annealed	0.062	0.025	Dry	0.156	800
36			Steel, SAE 1060 annealed	0.125	0.025	Dry	0.167	660
37			Steel, SAE 1060 annealed	0.187	0.025	Dry	0.167	615
38			Steel, SAE 1060 annealed	0.250	0.025	Dry	0.167	560
39			Steel, SAE 1060 annealed	0.062	0.021	Dry	0.167	880
40			Steel, SAE 1060 annealed	0.062	0.042	Dry	0.164	510
41			Steel, SAE 1060 annealed	0.062	0.062	Dry	0.162	400
42			Steel, SAE 2340 annealed	0.062	0.025	Dry	0.162	630

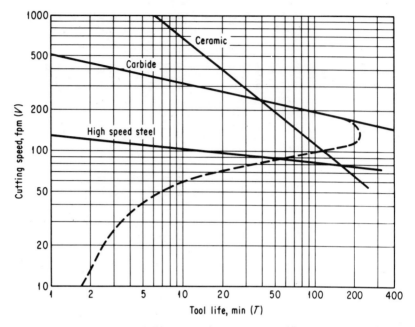

Fig. 1-29. Tool life vs. cutting speed for different tool materials.

Values of n and C for different work and tool materials are shown in Table 1-2 (see also Fig. 1-29). These values are for the particular feed, depth of cut, and tool geometry shown. Significant changes in the tool geometry, depth of cut, and feed will change the value of the constant C and may cause a slight change in the exponent n. In general, n is more a function of the cutting-tool material. The value of n for the common cutting-tool materials is as follows:

$$\begin{aligned} \text{H.S.S.:} &\quad n \cong 0.1 \text{ to } 0.15 \\ \text{Carbides:} &\quad n \cong 0.2 \text{ to } 0.25 \\ \text{Ceramics:} &\quad n \cong 0.6 \text{ to } 1.0 \end{aligned}$$

Equation (1-2) incorporates the effect of the size of cut:

$$K = VT^n f^{n_1} d^{n_2} \tag{1-2}$$

where K = constant of proportionality
f = feed, inches per revolution
d = depth of cut, inches
n_1 = exponent of feed (average value = 0.5 to 0.8)
n_2 = exponent of depth of cut (average value = 0.2 to 0.4)

The optimum cutting speed for a constant tool life is more sensitive to changes in the feed than to changes in the depth of cut. The tool life is most sensitive to changes in the cutting speed, less sensitive to changes in the feed, and least sensitive to changes in the depth of cut. This relationship is shown by Figs. 1-30 and 1-31.[1]

Example 2. The following equation has been obtained when machining AISI 2340 steel with high-speed steel cutting tools having a 8, 22, 6, 6, 6, 15, ³⁄₆₄ tool signature:

$$2.035 = VT^{0.13}f^{0.77}d^{0.37}$$

A 100-min tool life was obtained using the following cutting conditions:

$$V = 75 \text{ fpm}, \quad f = 0.0125 \text{ ipr}, \quad d = 0.100 \text{ in.}$$

Calculate the effect upon the tool life for a 20% increase in the cutting speed, feed, and depth of cut, taking each separately. Calculate the effect of a 20% increase in each of the above parameters taken together.

a. When $V = 1.2 \times 75 = 90$ fpm

$$T^{0.13} = \frac{K}{Vf^a d^b} = \frac{2.035}{90(0.0125)^{0.77}(0.100)^{0.37}}$$

$$= \frac{2.035}{90(0.034)(0.426)} = 1.56$$

$$T = 1.56^{1/0.13} = 1.56^{7.7}$$

$$= 31 \text{ min} \quad Answer$$

b. When $f = 0.0125 \times 1.2 = 0.015$ ipr

$$T^{0.13} = \frac{2.035}{75(0.015)^{0.77}(0.100)^{0.37}} = \frac{2.035}{75(0.038)(0.426)}$$

$$T = (1.67)^{7.7}$$

$$= 51 \text{ min} \quad Answer$$

c. When $d = 1.2 \times 0.100 = 0.120$ in.

$$T^{0.13} = \frac{2.035}{75(0.0125)^{0.77}(0.120)^{0.37}} = \frac{2.035}{75(0.034)(0.456)}$$

$$T = (1.745)^{7.7}$$

$$= 73 \text{ min} \quad Answer$$

d. When $V = 90$ fpm, $f = 0.014$ ipr, $d = 0.120$ in.

$$T^{0.13} = \frac{2.035}{90(0.038)(0.456)}$$

$$T = (1.3)^{7.7}$$

$$= 7.5 \text{ min} \quad Answer$$

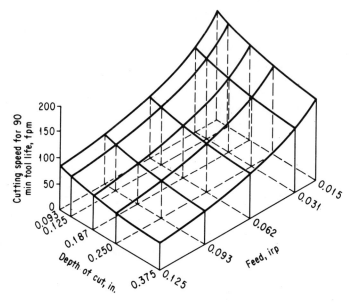

Fig. 1-30. Effect of feed and depth of cut on cutting speed for 90-min tool life.[1] Workpiece material—gray cast iron. Tool material—HSS.

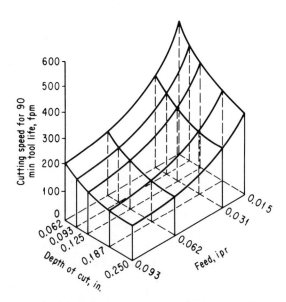

Fig. 1-31. Effect of feed and depth of cut on cutting speed for 90-min tool life.[1] Workpiece material—gray cast iron. Tool material—HSS.

Chap. 1 Design of Material-cutting Tools

Tool life is very sensitive to changes in cutting-tool geometry. However, tests varying the geometry do not generally yield curves that are consistent enough to interpret in general mathematical terms such as already given. Tool life is also very sensitive to the microstructure and the hardness of the workpiece. An approximate equation relating tool life to Brinell hardness number (Bhn) is

$$K = VT^n f^{n_1} d^{n_2} \text{Bhn}^{1.25}$$

The microstructure of the metal has a more pronounced effect on tool life than hardness alone. It is possible to have two pieces of steel at the same hardness but with different microstructures. The two pieces will yield a different tool life and surface finish when machined under the same cutting conditions.

Tool life is also sensitive to the tool material and the use of cutting fluids. The following general equation and Tables 1-7 and 1-8 take these factors into consideration.

$$V = \frac{K_1}{d^{0.37} f^{0.77}} \times \sqrt[6]{\frac{60}{T}} \times \text{C.F.}$$

Where

V = cutting speed, fpm
K_1 = proportionality constant

$\sqrt[6]{\dfrac{60}{T}}$ = a factor which will correct the cutting speed from that obtained for a basic 60-minute tool life to the cutting speed for the desired tool life.

T = tool life, minutes

C.F. = correction factor for the tool material. (18-4-1 HSS = 100)

Table 1-7 lists values for K_1. Table 1-8 lists correction factors (C.F.) for different tool materials. Table 1-9 lists values for $d^{0.37}$ and $f^{0.77}$.

Example 3. A 5-in. diam bar of SAE 1020 steel is to be turned to a 4-in. diam in one cut, using cutting fluid. A feed of 0.020 ipr will provide a satisfactory surface finish and a tool life of 180 minutes is desired.

Determine the cutting speed which will accomplish the cut if an 18-4-3 HSS tool is to be used.

$K_1 = 3.75$ (interpolate from Table 1-7, 60 min with cutting fluids, basic)

$d = 0.500$; $d^{0.37} = 0.774$ (from Table 1-9)

$f = 0.020$; $f^{0.77} = 0.049$ (interpolate from Table 1-9)

$T = 180$ min

C.F. $= 1.15$ (from Table 1-8)

$$V = \frac{3.75}{0.774 \times 0.049} \times \sqrt[6]{\frac{60}{180}} \times 1.15$$

$$= 99 \times 0.33^{.166} \times 1.15$$

$$= 99 \times 0.833 \times 1.15$$

$$= 95 \text{ fpm}$$

The recommended lathe rpm $= N = \dfrac{12V}{\pi D} = \dfrac{12 \times 95}{\pi \times 5}$

$$= 72.5 \text{ rpm}$$

TABLE 1-7. NUMERICAL VALUES FOR K_1

Metal to be cut	K_1 For 18-4-1 high-speed-steel tool and tool life of		
	60 min without cutting fluid, or 480 min with cutting fluid	60 min with cutting fluid	480 min without cutting fluid
Light alloys	25.0		
Brass (80–120 Bhn)	6.7		
Cast brass	4.2		
Cast steel	1.5	2.1	1.1
Carbon steel:			
SAE 1015	3.0	4.2	2.1
SAE 1025	2.4	3.3	1.7
SAE 1035	1.9	2.7	1.3
SAE 1045	1.5	2.1	1.1
SAE 1060	1.0	1.4	0.7
Chrome-nickel steel	1.6	2.3	1.1
Cast iron:			
100 Bhn	2.2	3.0	1.5
150 Bhn	1.4	1.9	1.0
200 Bhn	0.8	1.1	0.5

Chap. 1 Design of Material-cutting Tools

TABLE 1-8. CORRECTION FACTORS FOR COMPOSITIONS OF TOOL MATERIAL

Type	Approximate composition, %						C.F.
	W	Cr	V	C	Co	Mo	
14-4-1	14	4	1	0.7 –0.8	..	....	0.88
18-4-1	18	4	1	0.7 –0.75	..	....	1.00
18-4-2	18	4	2	0.8 –0.85	..	0.75	1.06
18-4-3	18	4	3	0.85–1.1	..	....	1.15
18-4-1 + 5% Co	18	4	1	0.7 –0.75	5	0.5	1.18
18-4-2 + 10% Co	18	4	2	0.8 –0.85	10	0.75	1.36
20-4-2 + 18% Co	20	4	2	0.8 –0.85	18	1.0	1.41
Sintered carbide	..	.	.	....	..	...	Up to 5

TABLE 1-9. NUMERICAL VALUES FOR $d^{0.37}$ AND $f^{0.77}$

d	$d^{0.37}$	d	$d^{0.37}$	f	$f^{0.77}$	f	$f^{0.77}$
0.01	0.182	0.25	0.598	0.001	0.004	0.025	0.059
0.02	0.235	0.30	0.640	0.002	0.008	0.030	0.067
0.04	0.305	0.35	0.678	0.004	0.017	0.035	0.075
0.06	0.353	0.40	0.712	0.006	0.019	0.040	0.084
0.08	0.393	0.45	0.744	0.008	0.024	0.045	0.092
0.10	0.427	0.50	0.774	0.010	0.029	0.050	0.099
0.14	0.482	0.75	0.899	0.014	0.037	0.075	0.135
0.18	0.530	1.00	1.000	0.018	0.045	0.100	0.170
0.22	0.571	....	.....	0.022	0.053		

Machining Economics

Although the equations already given will predict the tool life for a given cutting speed with reasonable accuracy, they do not answer the question of what tool life should be obtained for maximum production or for minimum cost per part. The following equations will provide the answers to these questions.

Tool life for maximum production:

$$T = \left(\frac{1}{n} - 1\right)K_2 \qquad (1\text{-}3)$$

Tool life for minimum cost per part:

$$T = \left(\frac{1}{n} - 1\right)\left(\frac{K_2 K_3 + K_4}{K_3}\right) \qquad (1\text{-}4)$$

where T = tool life, minutes
n = slope of the tool life curve
K_2 = tool changing time per tool, minutes
K_3 = machine labor plus burden cost, dollars per minute
K_4 = tool regrinding cost, dollars per cutting edge plus original cost of tool, divided by the number of available cutting edges per tool

Example 3. A 36-in. long cut is to be made on 4-in. diam steel bars in an engine lathe. The depth of cut is 0.250 in. and the feed rate is 0.020 in. per revolution. Tool-life tests conducted for this operation yield the following equation:

$$VT^{0.25} = 200$$

The following costs were determined for this operation:

Machine labor	$2.40 per hour
Machine burden rate	200% of labor
Grinding labor	$1.98 per hour
Grinding burden rate	250% of labor
Loading time	5 min/part

Three types of carbide tools were used, namely brazed inserts, solid inserts, and throw-away inserts.

For brazed inserts:

Original tool cost	$2.64
Grinding time	6 min (1 cutting edge per grind)
Tool changing time	2 min

5 grinds available per tool before discard plus original cutting edge.

For solid inserts:

Original tool cost	$3.60
Grinding time	10 min (4 edges per grind)
Tool changing time	0.5 min

5 grinds available per tool before discard plus original cutting edge.

For throw-away inserts:

Original tool cost	$1.60
Tools are not reground	
Tool changing time	0.5 min

Which tool is the least expensive to use?

Brazed inserts:

$$K_1 = 5 \text{ min}, \quad K_2 = 2 \text{ min}$$

$$K_3 = \frac{\$2.40 + 2 \times \$2.40}{60} = \$0.12/\text{min}$$

$$K_4 = \frac{\$2.64}{6} + \frac{6(\$1.98 + \$1.98 \times 2.5)}{60} = \$1.134/\text{edge}$$

$$T = \frac{1}{n-1} \times \left(\frac{K_2 K_3 + K_4}{K_3}\right) = \frac{2 \times 0.12 + 1.134}{0.12}$$

$$= 34.4 \text{ min}$$

$$V = \frac{C}{T^n} = \frac{200}{(34.4)^{0.25}} = 82.5 \text{ fpm}$$

$$N = \frac{12V}{\pi D} = \frac{12 \times 82.5}{\pi \times 4} = 78.5 \text{ rpm}$$

Chap. 1 Design of Material-cutting Tools 45

Loading cost = 5($0.12)	$0.60
Cutting cost = $0.12 $\dfrac{36}{0.020 \times 78.5}$	2.76
Tool changing cost = $\dfrac{\$0.12 \times 2 \times 36}{0.020 \times 78.5 \times 34.4}$	0.16
Tool cost = $\dfrac{\$1.134 \times 36}{0.020 \times 78.5 \times 34.4}$	0.757
Total machining cost per part	$4.277

Solid insert:

$$K_1 = 5 \text{ min}, \qquad K_2 = 0.5 \text{ min}, \qquad K_3 = \$0.12/\text{min}$$

$$K_4 = \frac{3.60}{24} + \frac{10 \times \$1.98 \times 3.5}{4 \times 60} = \$0.438/\text{edge}$$

$$T = 3\left(\frac{0.5 \times \$0.12 + \$0.438}{\$0.12}\right) = 12.4 \text{ min}$$

$$V = \frac{200}{(12.4)^{0.25}} = 106 \text{ fpm}, \qquad N = \frac{12 \times 106}{\pi \times 4} = 102 \text{ rpm}$$

Loading cost = 5($0.12)	$0.60
Cutting cost = $\dfrac{\$0.12 \times 36}{0.020 \times 102}$	2.12
Tool changing cost = $\dfrac{\$0.12 \times 5 \times 36}{0.020 \times 102 \times 12.4}$	0.087
Tool cost = $\dfrac{0.438 \times 36}{0.020 \times 102 \times 12.4}$	0.623
Total machining cost per part	$3.43

Throw-away inserts:

$$K_1 = 5, \qquad K_2 = 0.5, \qquad K_3 = \$0.12, \qquad K_4 = \frac{\$1.60}{8} = \$0.20$$

$$T = 3\left(\frac{0.5 \times \$0.12 + \$0.20}{\$0.12}\right) = 6.5 \text{ min}$$

$$V = \frac{200}{(6.5)^{0.25}} = 125 \text{ fpm}, \qquad N = \frac{12 \times 125}{\pi \times 4} = 120 \text{ rpm}$$

Loading cost = 0.5 × $0.12	$0.60
Cutting cost = $\dfrac{\$0.12 \times 36}{0.020 \times 120}$	1.80
Tool changing cost = $\dfrac{\$0.12 \times 0.5 \times 36}{0.020 \times 120 \times 6.5}$	0.139
Tool cost = $\dfrac{\$0.20 \times 36}{0.020 \times 120 \times 6.5}$	0.462
Tool machining cost per part	$3.001

Cutting Forces

Orthogonal cutting (Fig. 1-32) is defined as two-dimensional cutting in which the cutting edge is perpendicular to the direction of motion relative to the workpiece and the cutting edge is wider than the chip. The forces arising from this type of cutting action are shown in Fig. 1-33. F_c and F_N are the cutting and normal force, respectively. These components of the resultant force R can be measured by means of a dynamometer. F is the force required to overcome the friction between the chip and the face of the tool. If N is the coefficient of friction and a is the rake angle,

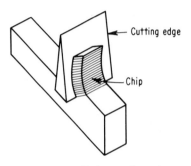

Fig. 1-32. Orthogonal cutting.

$$N = \tan \beta = \frac{F}{N} = \frac{F_N + F_c \tan \alpha}{F_c - F_N \tan \alpha}$$

$$F_\tau = F_c \cos \phi - F_N \sin \phi$$

F_τ cannot be calculated from the static mechanical properties of the material cut, since the strain rate encountered in metal cutting is much greater than that encountered in any conventional test. The shear angle ϕ must be determined before the shearing force and the shear flow stress can be determined from F_c and F_N. One method of approximating the shear angle is by measuring the thickness of the chip.

Let r = cutting ratio
t_1 = depth of cut or feed
t_2 = thickness of the chip

$$r = \frac{t_1}{t_2}$$

Then

$$\tan \phi = \frac{f \cos \alpha}{1 - r \sin \alpha}$$

The shearing force can be determined from F_c and F_N by the geometry of the force system in Fig. 1-33.

Figure 1-34 shows the resultant force acting on the cutting tool in a three-dimensional cut and the components of this force which can be measured with a three-component dynamometer. A typical relationship of the magnitude of these forces is shown in Fig. 1-35.[1]

Although some variation of tangential cutting force with respect to changes in speed may occur at low cutting speeds, the cutting force can be considered to be independent of cutting speed within the practical ranges of cutting speeds normally used. The effect of feed and the depth of cut on

Chap. 1 Design of Material-cutting Tools

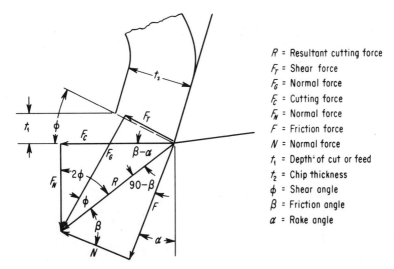

Fig. 1-33. Forces acting on a continuous chip in orthogonal cutting.

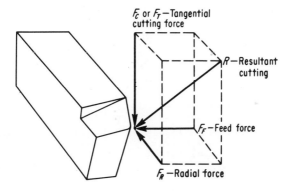

Fig. 1-34. Three components of cutting force.

the cutting force is illustrated in Fig. 1-36.[1] The following equation is obtained from Fig. 1-36.

$$F_T = C f^{n_3} d^{n_4} \tag{1-5}$$

where F_T = tangential cutting force, pounds
f = feed, inches per revolution
d = depth of cut, inches
c = constant of proportionality
n_3 = slope of F_T vs. f graph (typical values = 0.5–0.98)
n_4 = slope of F_T vs. d graph (typical values = 0.90–1.4)

The cutting-tool geometry has a considerable effect upon the cutting forces. The most pronounced effect is due to variations in the rake angle. This

48   Design of Material-cutting Tools   Chap. 1

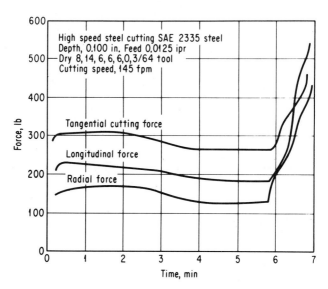

Fig. 1-35. Three components of cutting force through tool life.[1] Tool material—HSS. Work material—SAE 2335 steel. Tool geometry— 8, 14, 6, 6, 6, 0, 3/64.

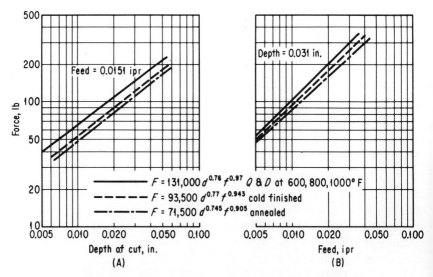

Fig. 1-36. (*A*) Cutting force vs. depth of cut.[1] Tool material—HSS. Tool geometry—6, 11, 6, 6, 6, 15, 0.01. Work material—SAE 1045 steel. (*B*) Cutting force vs. feed.[1] Depth—0.031 inches.

Chap. 1   Design of Material-cutting Tools

effect is shown in Fig. 1-37 for orthogonal cutting conditions. In three-dimensional cutting, the true rake angle is the particular rake angle on the face of the tool along which the chip is sliding.

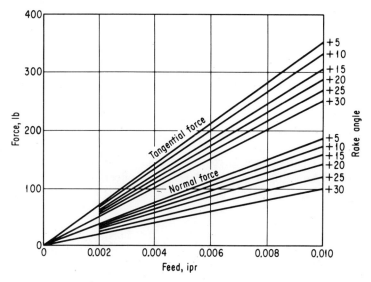

Fig. 1-37. Effect of feed on tangential and normal cutting forces for a range of rake angles when turning nickel steel.

The true rake angle can be closely approximated by the following equations (Fig. 1-38):

$$\tan \theta = \frac{d}{\text{NR} + (d - \text{NR}) \tan \text{SCEA}} \tag{1-6}$$

where θ = chip flow angle

NR = nose radius

SCEA = side cutting edge angle

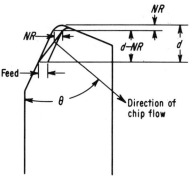

Fig. 1-38. Approximate chip flow direction.

This equation is correct for tools having a zero-degree rake angle. For most normal rake angles used it will be correct within a few degrees of error. For very large rake angles a significant error can be introduced. Equation (1-6) can be used for most practical applications.

Equation (1-7) expresses the true rake angle corresponding to the chip flow angle.

$$\tan \rho = \tan \gamma \sin \phi + \tan \theta \cos \phi \tag{1-7}$$

where ρ = true rake angle
γ = side rake angle
θ = chip flow angle
ϕ = shear angle

The location of the maximum rake angle is found from the following expression:

$$\tan \theta_{max} = \frac{\tan \gamma}{\tan \theta} \tag{1-7a}$$

Figure 1-39 is a graphical representation of the effect of feed on the several forces for various depths of cut. The effects of cutting speed on cutting force (Fig. 1-40) and tool forces (Fig. 1-41) are also illustrated.

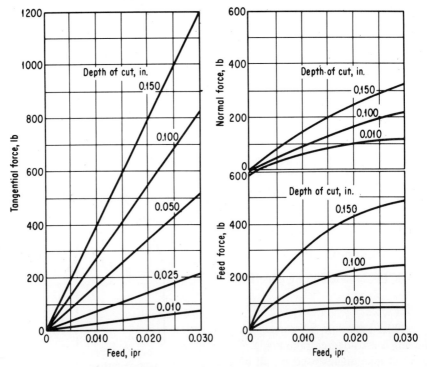

Fig. 1-39. Effect of feed on tangential feed and normal forces for various depths of cut. Tool material—HSS. Tool shape—8, 14, 6, 6, 6, 15, 3/64. Work material—SAE 3135 steel, annealed. Cutting speed—50 fpm (dry).

Chap. 1 Design of Material-cutting Tools

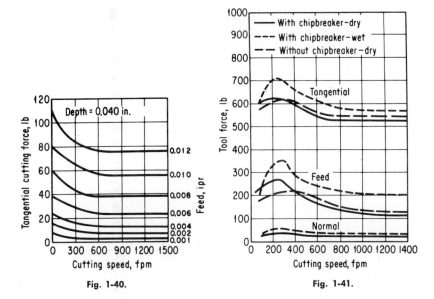

Fig. 1-40. Effect of cutting speed on cutting force for various feeds when cutting bronze with a tungsten carbide tool. Depth of cut—0.040 inches.

Fig. 1-41. Effect of cutting speed on tool forces. Tool material—tungsten carbide. Tool shape—10, 10, 8, 8, 7, 0, 0.015. Work material—AISI C 1118 steel. Depth of cut—0.150 inches.

Power Requirements

The horsepower at the cutting tool is defined by Eq. (1-8).

$$\text{hp}_c = \frac{F_T V}{33{,}000} \tag{1-8}$$

where hp_c = horsepower at the cutting tool
V = cutting speed, feet per minute
F_T = tangential cutting force component

The radial cutting force does not contribute to the horsepower. Although the feed force can be considerable in magnitude, the feeding velocity is generally so low that the horsepower required to feed the tool can be neglected.

Substituting Eq. (1-5) in (1-8),

$$\text{hp}_c = \frac{C f^{n_3} d^{n_4} V}{33{,}000} \tag{1-9}$$

There are losses in the machine which must be considered when estimating the size of the electric motor required.

$$\text{hp}_g = \frac{\text{hp}_c}{\text{eff}} + \text{hp}_t \tag{1-10}$$

where hp_g = gross or motor horsepower
hp_t = tare horsepower; the power required to run the machine at no-load conditions
eff = the mechanical efficiency of the machine

Specific Power Consumption or Unit Horsepower. The specific power consumption, W_p, or unit horsepower, is defined as the horsepower required to cut a material at a rate of one cubic inch per minute. Thus

$$W_p = \frac{\text{hp}_c}{\text{cubic inches per minute removed by cut}} \tag{1-11}$$

For turning,

$$W_p = \frac{\text{hp}_c}{12Vfd} \tag{1-11a}$$

$$= \frac{F_T}{396,000 fd} \tag{1-11b}$$

Note that W_p is independent of the cutting speed.

$$W_p = \frac{Cf^{n_3}d^{n_4}}{396,000 fd} = \frac{C}{396,000 f^{1-n_3} d^{1-n_4}} \tag{1-11c}$$

from which

$$\frac{C}{396,000} = C^1 = W_{p_1}(f_1)^{1-n_3}(d_1)^{1-n_4} = W_{p_2}(f_2)^{1-n_3}(d_2)^{1-n_4}$$

or

$$W_{p_2} = W_{p_1}\left(\frac{f_1}{f_2}\right)^{1-n_3}\left(\frac{d_1}{d_2}\right)^{1-n_4} \tag{1-12}$$

Equation (1-12) defines the effect of feed and depth of cut on the specific power consumption. This equation may be reduced to the following form since the value of n_4 is often nearly equal to one.

$$W_{p_2} = W_{p_1}\left(\frac{f_1}{f_2}\right)^{1-n_3}$$

Typical values of specific power consumption are given in Table 1-10.

TABLE 1-10. HP_c/CIM VALUES FOR VARIOUS MATERIALS

Dry Cutting; Depth, ⅛ in.; Feed, 1/64 ipr

Material cut	Tool shape	Brinell hardness no.	HP_c/cu in. per min
Plain carbon steel		126 179 262	0.59–0.66 0.70–0.79 0.85–0.95
Free-cutting steel	8, 14, 6, 6, 6, 0, 1/16	118 179 229	0.36–0.39 0.44–0.48 0.50–0.54
Alloy steel		131 179 269 429	0.46–0.57 0.55–0.68 0.67–0.83 1.10–1.90
Cast iron		140 179 256	0.22–0.32 0.45–0.68 0.85–1.30
Leaded brass		33 76 131	0.18–0.27 0.22–0.31 0.25–0.35
Unleaded brass	8, 14, 6, 6, 6, 15, 0	50.9	0.54
Pure copper		40.4	0.88
Magnesium alloys		32 49 68	0.084–0.10 0.094–0.11 0.10–0.12
Aluminum alloys	8, 14, 6, 6, 6, 15, 0	55 159	0.28 0.26
	20, 40, 10, 10, 10, 15, 0	32 94 115 153	0.12 0.15 0.17–0.21 0.20
Monel metal	8, 14, 6, 6, 6, 15, 3/64	147 160	0.58–0.75 1.35

Example 1. Alloy steel having a hardness of 250 Bhn is to be machined in a milling machine. The depth of cut is to be 0.250 in., the feed is 0.005 in. per tooth, and the cutting speed is 300 fpm. The milling cutter has 12 teeth and is of 10-in. diameter. The width of the cut is 5 in. The specific power consumption is 0.70 hp/cu in./min.

$$\text{hp}_c = W_p \times \text{cu in. per min}$$

$$= (W_p)\left(\frac{12V}{\pi D}\right)(n)(f)(w)(d)$$

$$= (0.70)\left[\frac{12(300)}{\pi(10)}\right](12)(0.005)(5)(0.250)$$

$$= 6.03$$

Assume

$$\text{hp}_t = 0.5 \text{ and eff} = 0.8.$$

$$\text{hp}_g = \frac{\text{hp}_c}{\text{eff}} + \text{hp}_t = \frac{6.03}{0.8} + 0.5$$

$$= 8.05$$

Use 10-hp motor.

Example 2. A steel part, shown in Fig. 1-42, is to be machined between centers in a lathe equipped with an air-operated tailstock spindle. The maximum depth of cut is to be 0.100 in., the feed is 0.010 ipr, and the cutting speed is 300 fpm. The cutting-tool geometry is 10, 10, 6, 6, 10, 15, 0.030. The feeding force is assumed to be two-thirds of the tangential cutting force. The following equation applies to this material.

$$F_T = Cf^{0.8}d$$

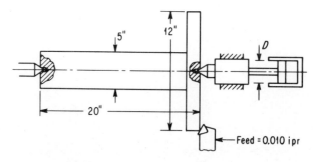

Fig. 1-42. Forces in a turning operation.

Chap. 1 Design of Material-cutting Tools

Calculate the diameter D of the air cylinder required to hold the work between centers against the cutting forces if the minimum pressure in the air cylinder may reach 60 psi. Both centers rotate; thus the friction between the work and the centers need not be considered.

$$W_p = 0.70 \text{ (from Table 1-10)}$$

$$W_{p_2} = W_{p_1}\left(\frac{f_1}{f_2}\right)^{1-0.8} = 0.70\left(\frac{0.0156}{0.010}\right)^{0.2}$$

$$= 0.765 \text{ hp}_c/\text{cu in.}/\text{min}$$

$$F_T = 396{,}000 W_p f d = 396{,}000 (0.765)(0.010)(0.100)$$

$$= 303 \text{ lb}$$

$$F_F = \tfrac{2}{3} \times 303 = 202 \text{ lb}$$

$$\tan \theta = \frac{d}{NR + (d - NR)\tan \text{SCEA}}$$

$$= \frac{0.100}{0.030 + (0.100 - 0.030)\tan 15}$$

$$\theta = 64°01'; \text{ say } 64°$$

$$F_R = F_F \tan(90 - \theta) = 202 \tan 26°$$

$$= 99 \text{ lb}$$

Taking moments about the headstock center (Fig. 1-43A),

$$\Sigma \quad M = 0 = 6F_F - 20F_R + 20F_X$$
$$F_X = 38.4 \text{ lb}$$

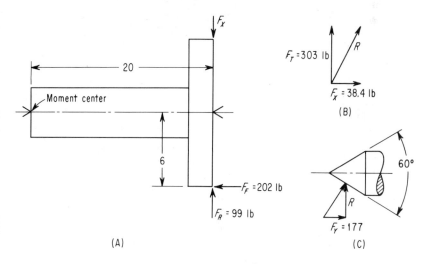

Fig. 1-43. Forces in turning operations.

Resolving the forces at the cutting tool in a vertical plane normal to the work axis (Fig. 1-43B),

$$R = (F_T{}^2 + F_X{}^2)^{1/2} = 307 \text{ lb}$$

Resolving R about the tailstock center (Fig. 1-43C),

$$F_Y = \frac{R}{\tan 60} = 177 \text{ lb}$$
$$D_{CYL} = \frac{4F_Y}{\pi 60} = 3.76 \quad Answer$$

BASIC PRINCIPLES OF MULTIPLE-POINT TOOLS

Multiple-point cutting tools are basically a series of single-point tools mounted in or integral with a holder or body and operated in such a manner that all the teeth (tools) follow essentially the same path across the workpiece. The cutting edges may be straight or may be in the form of various contours which are to be reproduced upon the workpiece. Multiple-point tools may be of either the linear-travel or the rotary type. With linear-travel tools the relative motion between the tool and the workpiece is along a straight-line path. The teeth of rotary cutting tools revolve about the tool axis. The relative motion between the workpiece and a rotary cutting tool may be either axial or in a plane normal to the tool axis. In some cases a combination of the two motions is used. Certain form-generating tools involve a combination of linear travel and rotary motions.

Figures 1-44 and 1-45 illustrate two types of milling cutters and indicate the differences in nomenclature between the angles of these cutting tools as compared to single-point tools. Figure 1-45 shows the peripheral cutting edge angle as zero degrees with positive and negative directions indicated.

Whether a cutting a tool is single-point or is one component of a milling cutter, the various angles are ground to provide the most efficient cutting

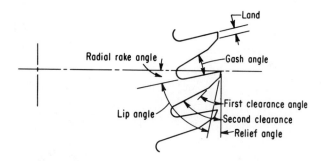

Fig. 1-44. A solid plain milling cutter showing the basic tooth angles and the tooth land.

Chap. 1 Design of Material-cutting Tools   57

action. Theoretical considerations may dictate larger angles, but actual cutting experience may dictate smaller angles for greater tool strength without chatter. Advantages from increasing any angle must always be considered together with the effect on tool strength.

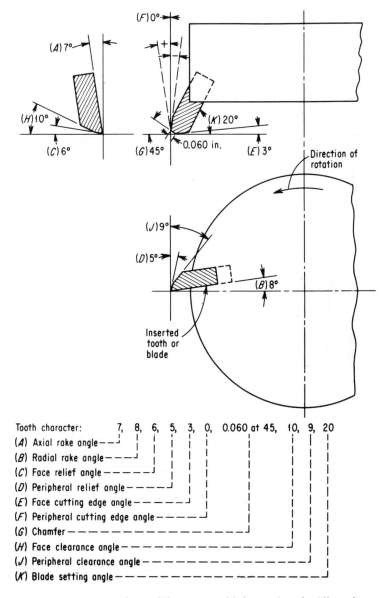

Tooth character: 7, 8, 6, 5, 3, 0, 0.060 at 45, 10, 9, 20
(A) Axial rake angle
(B) Radial rake angle
(C) Face relief angle
(D) Peripheral relief angle
(E) Face cutting edge angle
(F) Peripheral cutting edge angle
(G) Chamfer
(H) Face clearance angle
(J) Peripheral clearance angle
(K) Blade setting angle

Fig. 1-45. A face milling cutter with inserted teeth. All pertinent tooth angles are included.

Cutting Processes. The cutting processes for multiple-point tools are essentially similar to those for single-point tools. Linear-travel tools produce a series of chips that are similar to those produced by single-point tools on planing cuts. Milling cutters produce chips that vary in thickness because of the nature of the tooth path as illustrated in Fig. 1-46. The chips produced by axial-feed tools tend to be conical because the varying diameters across the cutting edge cause a difference in the distance travelled by different portions of the cutting edge. Aside from these differences, studies have shown that there is no fundamental difference between the metal-deformation process involved in forming chips with these tools and that involved in using single-point tools on turning or planing cuts.

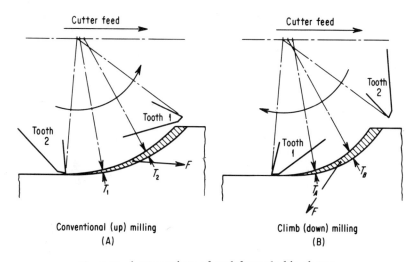

Fig. 1-46. A comparison of undeformed chip shapes.

The twist drill is unique in that it involves two different metal-deformation processes. The main cutting lips produce conventional chips as illustrated in the chip section shown in Fig. 1-47. Aside from some curvature of the finished surface, the chip is quite similar to the chip produced by a single-point tool. The metal deformation under the chisel edge is much more complex. At the exact center of the hole the only motion of the drill is axial, so the deformation resembles that produced by an indenting punch. As the radius increases, the rotation of the drill becomes important and the chisel-edge wedge appears to both cut and extrude the metal. An analysis of this region of the drill based upon such an assumption fits experimental data quite well. Figure 1-48 shows a diametral section normal to the chisel edge. The complex metal deformation shown is a major factor in the high thrust forces required by twist drills with conventional drill points.

Chap. 1 Design of Material-cutting Tools 59

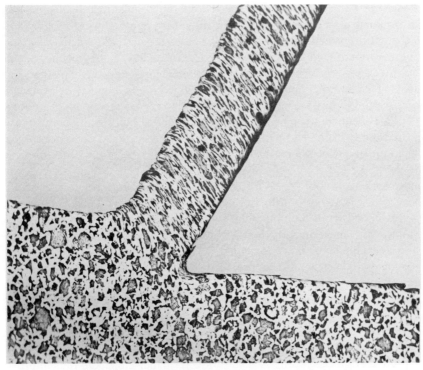

Fig. 1-47. Section showing conventional chip formation by a single-point tool.

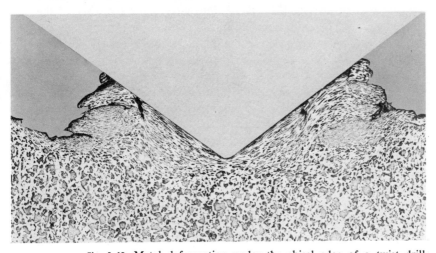

Fig. 1-48. Metal deformation under the chisel edge of a twist drill.

Design Considerations. The most important single factor affecting the performance of any cutting tool is the attainment of a high degree of rigidity in the entire machining system. This includes the cutting tool, the machine tool, the fixture, and the workpiece. A lack of rigidity in any of the system elements can largely nullify the benefits of high rigidity in the other elements. This interrelationship is all too often overlooked by fixture designers so that workpieces are not adequately supported at the point of cutting.

With adequate workpiece support and machine tool rigidity, an increase in the rigidity of the cutting tool can effect large improvements in tool life. Such improvements are particularly evident in the case of the super-strength alloys used in aircraft and missile production. When drilling such alloys improvements in tool life by a factor of 50 or more have been obtained by using short, heavy-web drills in place of drills of conventional design.

In the design of multiple-point tools, there must always be some compromise with maximum rigidity. Most multiple-point tools are required to carry the chips generated for some distance before they can be ejected. Adequate chip space must be provided to avoid jamming, which can cause tool breakage. The amount and shape of the chip space provided depend upon the material and the nature of the cut. If the chips are well broken or powdery, less chip room is required and closer tooth spacing can be used. More chip room and wider tooth spacing are required for loosely coiled or snarling chips. With some tools, it is possible to incorporate some form of chip breaker on the cutting face to produce smaller chips and thus improve the chip-conveying ability of the tool.

With tools such as milling cutters and broaches, tooth spacing can also affect smoothness of operation and the accuracy of the work. If too many teeth are in contact with the work at the same time, either the tool or the machine may be overloaded. Overloading may cause breakage of the tool or driving key. The deflection due to the increase in loading may be such that accuracy will suffer. With milling cutters it is generally desirable that at least one tooth be in contact with the work at all times. (See "Number of Teeth in a Milling Cutter," below.)

The rake and relief angles on multiple-point cutting tools affect both the tool performance and the strength of the teeth. High rake angles usually make cutting easier and more efficient. High relief angles reduce the rate of wear-land development. These angles, however, cannot be increased without limit. As either the rake angle or the relief angle is increased, cutting-edge strength is reduced. In addition, while high relief angles reduce the rate of wear-land generation, they do cause a greater change in size for the development of a wear land of given size. The selection of rake and relief angles therefore requires a compromise to suit the job being done.

Operating Considerations. In setting up the operating conditions for any multiple-point cutting tool, there are three important variables that can be adjusted. These are the cutting speed, the feed per cutting edge, and the cutting fluid. Of these, the feed is the most important and should be established first. With some tools, such as taps and broaches, the feed per tooth is determined by the design of the tool and can be modified only by regrinding. With most other tools the feed is determined by the machine settings. It has long been demonstrated that it is more efficient to remove metal in the form of thick chips than in thin ones, so the maximum possible feed per tooth should be used. The maximum feed per tooth is limited by the following factors: (1) cutting-edge strength, (2) rigidity and allowable deflection, (3) the surface finish required, and (4) tool chip space.

The feed cannot be increased without limit because some point will be reached at which either the tool or the machine will be overloaded. Overload can cause breakage either immediately or with accumulative increase in cutting forces due to dulling. As the feed is increased, the cutting force increase causes greater deflection between the tool and the workpiece. Deflection can become so large that it is impossible to hold the accuracy required. The surface finish produced on the workpiece usually deteriorates as the feed is increased. This may necessitate the use of light feeds. With some metals the greater volume of chips produced with heavy feeds may overload the chip space in the tool. In other cases, heavier feeds will produce broken chips which can be handled better than continuous chips resulting from light feeds.

In peripheral milling, the actual chip thickness is affected by the depth of cut. Maximum undeformed chip thickness equal to the feed per tooth is obtained when the depth of cut equals one-half the cutter diameter. With shallow finishing cuts the maximum undeformed chip thickness can be very much less than the cutter advance per tooth. However, it is common to use the expression "feed per tooth" to mean "cutter advance per tooth." As long as the depth of cut is constant, cutter advance per tooth is nearly proportional to maximum undeformed chip thickness, so no serious error is introduced into the analysis of the effects of feed changes.

After the maximum allowable feed has been established, the cutting speed should be considered. At a constant feed per tooth, there is relatively little change in cutting forces as the cutting speed is increased. In the normal operating range, speed has relatively little effect on surface finish, but sometimes a large increase of speed (usually made possible by a change of tool material) will yield an improvement in surface finish by elimination of built-up edge on the cutting edges. The principal effect of increasing cutting speed is to reduce the number of pieces that can be produced per sharpening. However, an increase in cutting speed does result in the production of parts at a more rapid rate. The optimum cutting speed is that which will permit production of parts at the lowest cost per piece. This requires economic analysis of all of the costs including machining cost, tool changing cost, and cutting-tool acquisition and maintenance costs.

The use of the correct cutting fluid can result in substantial improvements in a machining operation. The proper cutting fluid can permit higher feeds and increased speeds as well as attainment of better surface finishes. Cutting fluids should be directed to the exact point where the cutting is being done. No machining operation should be set up without some consideration of cutting fluids.

Forces and Power Requirements

Knowledge of cutting forces is essential to machine and fixture design, and in the determination of the power required. Force and power predictions for multiple-point tools are more complex than those for a single-point tool. Varying numbers of teeth may be in contact with the work; the chip size can vary in different parts of the cut, and the orientation of the cutting teeth may not be constant with respect to the workpiece. The best method to determine forces on such tools is by actual measurement on the machine to be used or on a simulated setup in a metal-cutting laboratory. Since such measurements are sometimes impossible or inconvenient to make, methods for making reasonable force and power estimates may be of value.

One approach is to consider the multiple-point tool equivalent to a series of single-point tools, then estimate the contribution of each tool and sum these to arrive at the resultant forces. The following methods of calculation are preferable.

Power Requirements for Drilling and Reaming. For rotary axial-feed tools, such as twist drills, core drills, and reamers, reasonably accurate estimates of forces and power can be made through the use of formulae developed experimentally and analytically by Shaw and Oxford.[8]

The torque and thrust for a twist drill operating in an alloy steel with a hardness of 200 Bhn may be represented by the following formulae:

$$M = 23{,}300 f^{0.8} d^{1.8} \left[\frac{1 - \left(\frac{c}{d}\right)^2}{\left(1 + \frac{c}{d}\right)^{0.2}} + 3.2\left(\frac{c}{d}\right)^{1.8} \right] \quad (1\text{-}13)$$

$$T = 42{,}600 f^{0.8} d^{1.8} \left[\frac{1 - \frac{c}{d}}{1 + \left(\frac{c}{d}\right)^{0.2}} + 2.2\left(\frac{c}{d}\right)^{0.8} \right] + 19{,}300 c^2 \quad (1\text{-}14)$$

where M = torque, inch-pounds
T = thrust, pounds
d = drill diameter, inches
c = chisel edge length, inches (approximately 1.15 times the web thickness for normal sharpening)
f = feed per revolution, inches

Chap. 1 Design of Material-cutting Tools

For drills of regular proportions the ratio c/d can be set equal to 0.18 and the equations simplified to:

$$M = 25{,}200 f^{0.8} d^{1.8} \tag{1-15}$$

$$T = 57{,}500 f^{0.8} d^{0.8} + 625 d^2 \tag{1-16}$$

For reamers or core drills, which are used for enlarging existing holes, the effects of the chisel-edge region can be eliminated and the equations reduced to:

$$M = 23{,}300 k f^{0.8} d^{1.8} \left[\frac{1 - \left(\frac{d_1}{d}\right)^2}{\left(1 + \frac{d_1}{d}\right)^{0.2}} \right] \tag{1-17}$$

$$T = 42{,}600 k f^{0.8} d^{0.8} \left[\frac{1 - \frac{d_1}{d}}{\left(1 + \frac{d_1}{d}\right)^{0.2}} \right] \tag{1-18}$$

where d = drill diameter, inches
  d_1 = diameter of hole to be enlarged, inches
  f = feed per revolution, inches
  k = a constant depending upon the number of flutes

The constant k is necessary since, for a given feed per revolution, the number of flutes affects the feed per tooth. It has been pointed out that it is more efficient to remove metal in the form of thick chips than thin ones, so tools with a large number of flutes will require proportionately more energy because of the thinner chips. Values of k for different numbers of flutes are tabulated below.

Number of flutes	Constant, k	Number of flutes	Constant, k
1	0.87	8	1.32
2	1.00	10	1.38
3	1.08	12	1.43
4	1.15	16	1.51
6	1.25	20	1.59

Whereas the thrust forces can be substantial and thus can have a large influence upon the required strength and rigidity, the power required in feeding the tool axially is very small (less than 2% of the total power requirements), and can usually be disregarded. The cutting power is a function of the torque and rotational speed and can be computed from the equation

$$\text{hp} = \frac{MN}{63{,}025} \qquad (1\text{-}19)$$

where hp = horsepower
N = speed, rpm
M = tool torque, inch-pounds

In using these formulae for estimating purposes, an allowance of at least 25% should be made for increases due to tool dulling, and further allowances should be made for the efficiency of the machine drive train. For other materials, the cutting forces vary in about the same ratio as noted for single-point tools.

When it is not possible to estimate forces and power directly, a fairly good estimate can be made by considering the cutting energy. It has been found that the removal of one cubic inch of an alloy steel with a hardness of about 200 Bhn at normal feeds requires the expenditure of about 500,000 inch-pounds of energy. In terms of engineering units this can be stated as 1.25 horsepower per cubic inch per minute. Thus, if the maximum rate of metal removal is computed in terms of cubic inches per minute, this value can be multiplied by 1.25 to give a reasonable estimate of the horsepower required at the cutting tool. If a rotary tool is involved, the tool torque M, in inch-pounds, can be estimated from the following equation:

$$M = \frac{63{,}025 \text{ hp}}{N} \qquad (1\text{-}20)$$

where hp = tool horsepower
N = tool rotational speed, rpm

In the case of linear-travel tools, the force in the cutting direction can be estimated from the following formula:

$$F = \frac{33{,}000 \text{ hp}}{V} \qquad (1\text{-}21)$$

where F = cutting force, foot-pounds
hp = tool horsepower
V = cutting speed, feet per minute

Equations 20 and 21 must be applied with caution since they will be useful in estimating torque or cutting force only if an accurate estimate of the maximum rate of metal removal is made. If they are applied to the average rate of metal removal they will indicate only average torque or average force, and this could be considerably below the peak forces in the case of intermittent cutting as might be encountered in milling or broaching.

Chap. 1 Design of Material-cutting Tools   65

Further, all of the equations presented here ignore the even higher peak forces resulting from impact or vibration during cutting. Because of these limitations, the equations should not be used as a basis for the design of critical elements in the machine or fixtures.

Power Requirements for Milling. Milling machines, like other machine tools, absorb part of the power exerted on them. The reasons are frictional losses, gear-train inefficiencies, spindle speeds that are too high for the particular machine, mechanical condition, etc. Consequently, the power for milling must include the machine power losses and the power actually used at the cutter. Efficient use of power at the cutter is influenced by cutter speed, design, and material, and by workpiece material.

The total horsepower required at the cutter (hp_c) is given by the equation

$$hp_c = \frac{cim}{K} \qquad (1\text{-}22)$$

where hp_c = horsepower at the cutter
  cim = metal removal rate, cubic inches per minute
  K = a factor reflecting the efficiency of the metal-cutting operation

The K factor varies with type and hardness of material; also for the same material it varies with the feed per tooth, increasing as the chip thickness increases. Time-consuming trials are required to determine the quantities involved, because in each case the K factor represents a particular rate of metal removal and not a general or average rate.

TABLE 1-11. MILLING-MACHINE SELECTOR TABLE

Rated hp of machine	3	5	7.5	10	15	20	25	30	40	50
Overall machine efficiency, per cent	40	48	52	52	52	60	65	70	75	80
Material	Max metal removal (cu in./min)									
Aluminum	2.7	5.5	8.7	12	18	27	37	48	69	91
Brass, soft	2.4	4.7	7.5	10	16	24	32	41	60	79
Bronze, hard	1.7	3.3	5.3	7.3	11	17	23	30	43	56
Bronze, very hard	0.78	1.6	2.5	3.4	5.3	7.8	11	15	20	26
Cast iron, soft	1.6	3.2	5.2	7.1	11	16	22	28	41	54
Cast iron, hard	1	2	3.3	4.6	7	10	14	18	26	35
Cast iron, chilled	0.78	1.6	2.5	3.4	5.3	7.8	10	13	19	26
Malleable iron	1	2.1	3.4	4.7	7.3	11	14	18	26	36
Steel, soft	1	2	3.3	4.6	7	10	14	18	26	35
Steel, medium	0.78	1.6	2.5	3.4	5.3	7.8	10	13	19	26
Steel, hard	0.56	1.1	1.8	2.5	3.9	5.7	7.7	10	14	19

Data courtesy of Kearney & Trecker Corp.

To make available a quick approximation of the total power requirements, a milling-machine selector table has been devised (Table 1-11), which estimates the metal removed in cubic inches per minute for various rated horsepowers of various machines under constant load conditions.

The metal-removal rates in Table 1-11 may be considered as the products of the K factor (Table 1-12) and over-all efficiencies listed in Table 1-11. All K constants given here are for dull cutters; hence no allowance for increase in horsepower due to dulling need be made. All values apply to average milling conditions for carbide cutter, i.e., milling speeds and rake angles recommended for the various materials, and 0.010 in. feed per tooth.[9]

TABLE 1-12. VALUE OF K FACTOR FOR VARIOUS MATERIALS

Aluminum and magnesium	2.5–4.0
Bronze and brass, soft	1.7–2.5
Bronze and brass, medium	1.0–1.4
Bronze and brass, hard	0.6–1.0
Cast iron, soft	1.5
Cast iron, medium	0.8–1.0
Cast iron, hard	0.6–0.8
Malleable iron and cold-drawn steel, SAE 6140	0.9
Cold-drawn steel, SAE 1112, 1120, and 1315	1.0
Forged and alloy steel, SAE 3120, 1020, 2320, and 2345, 150–300 Bhn	0.63–0.87
Alloy steel, 300–400 Bhn	0.5
Stainless steel, AISI 416, free-machining	1.1
Stainless steel, austenitic, AISI 303, free-machining	0.83
Stainless steel, austenitic, AISI 304	0.72
Monel metal	0.55
Copper, annealed	0.84
Tool-steel	0.505
Nickel	0.525
Titanium	0.75

LINEAR-TRAVEL TOOLS

Broaches

The most common multiple-point linear-travel tool is the broach. Broaches are used for producing either external or internal surfaces. The surfaces produced may be flat, circular, or of quite intricate profile, as viewed in a section normal to the tool travel. A broach is essentially a series of single-point tools following each other in the axial direction along a tool body or holder. Successive teeth vary in size or shape in such a manner that each following tooth will cut a chip of the proper thickness. The basic elements of broach construction are illustrated in Fig. 1-49.

Chap. 1 Design of Material-cutting Tools

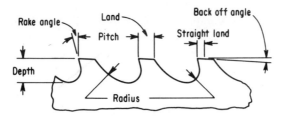

Fig. 1-49. Basic elements of broach construction. (*American Broach & Machine Division, Sundstrand Machine Tool Company.*)

The spacing and shape of broach teeth are determined by the length of the workpiece and the chip thickness per tooth as well as by the type of chips formed. The chip space between the broach teeth must be sufficient to take care of the volume of chips generated. Broach teeth are provided with rake and relief angles in the same manner as other cutting tools. Standard broaching nomenclature designates the rake angle as the *face angle* and the relief clearance as the *back-off angle*. The rake angles fall in the same range as used for other tools, but the back-off angles are normally quite low, in the range between $\frac{1}{2}$ degrees and $3\frac{1}{2}$ degrees. Low back-off angles are used on broaches to minimize the loss of size in resharpening. Final finishing teeth are often provided with unrelieved land behind the cutting edge to assure proper sizing of the workpiece. Sometimes noncutting burnishing teeth follow the final cutting teeth.

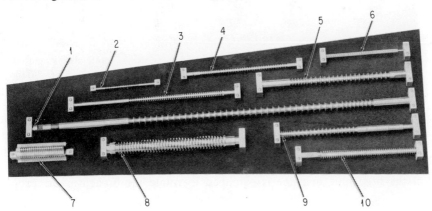

Fig. 1-50. Internal broaches: (1) round pull-type, (2) round push-type, (3) involute spline pull, (4) round push-type, (5) push-type with overlapping teeth, (6) two-spline type with round body, (7) shaving shell for pump rotor, (8) helical spline push-type, (9) round pull-type with overlapping teeth, (10) combination round and serration type. (*American Broach & Machine Division, Sundstrand Machine Tool Company.*)

Internal broaches (Fig. 1-50) are either pulled or pushed through the work. Strength considerations limit the design of such broaches. Surface broaches (Fig. 1-51) are ordinarily carried on a large guided ram; here strength is not so critical since the cutting load can be transferred to the ram at many points along the broach length.

Fig. 1-51. Surface broach sections showing tooth detail and chip breakers on taper teeth. (*American Broach & Machine Division, Sundstrand Machine Tool Company.*)

Broaches are commonly made of high-speed steel as solid units, but carbide-tipped and inserted-blade broaches are sometimes economical. This is particularly true in the case of surface broaches which are better adapted to such a tooth mounting. Broaches can be used to cut helical internal forms if the broaching machine is equipped to rotate the broach at the proper lead-rate as it passes through the work.

Gear Shaper Cutters

A gear shaper cutter is a tool that is basically a gear, the teeth of which are relieved to provide cutting edges. Typical gear shaper cutters are illustrated in Fig. 1-52. Gear shaper cutters are reciprocated in the axial direction, but at the same time are rotated in timed relationship to the gear being generated. The rotational speed is low compared to the speed of

Chap. 1 Design of Material-cutting Tools 69

reciprocation, so the shaper cutter is principally a linear-travel tool. The feed per tooth is determined by the rate at which the cutter is fed into depth in the gear blank and to some extent by the speed of rotation. Gear shaper cutters can, within limits, be designed for forms other than gears, such as splines, sprockets, etc.

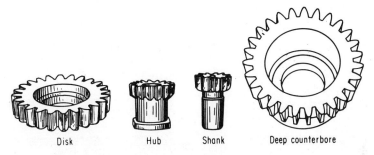

Fig. 1-52. Standard gear shaper cutters.

AXIAL-FEED ROTARY TOOLS

Twist Drills

In its most basic form a twist drill (Fig. 1-53) is made from a round bar of tool material. It has a pair of helical flutes which form the cutting surfaces and act as chip conveyors. Relief is provided behind the two cutting edges or lips. The intersection of the two relief surfaces across the web between the flutes is known as the *chisel edge*. The lands between the flutes are cut away to a narrow margin to reduce the area of contact between the lands and the wall of the hole and/or guide bushing. The metal cut away to form the margin is known as *body diameter clearance*.

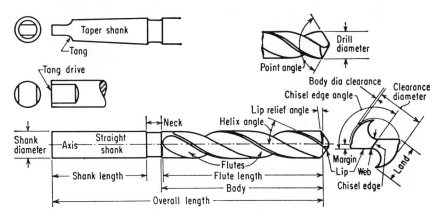

Fig. 1-53. Nomenclature for twist drills.

Twist drills can be provided with a variety of shanks, but the straight shank and the Morse taper shank are the most common. Twist drills are ordinarily made of high-speed steel, but certain sizes and types are available, made from solid carbide or with carbide cutting tips.

Standard twist drills are available in a wide variety of designs, of which a few are shown in Fig. 1-54, which adapt them to various materials and services. The most common variations are those of the helix angle and the web thickness. Tough materials usually require the use of rigid drills with heavy webs and reduced helix angles. Free-machining materials can be cut with drills having higher helix angles and lighter webs which provide for more efficient cutting and better chip ejection.

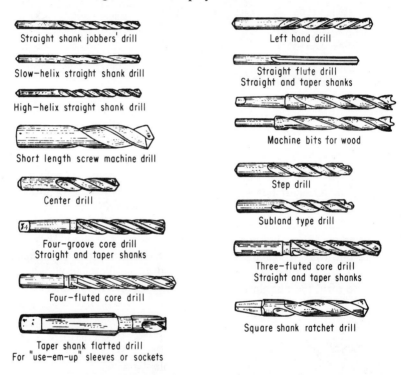

Fig. 1-54. Conventional and special-purpose drills.

Reamers and Core Drills

Twist drills produce holes; core drills and reamers are hole enlarging and finishing tools. The cutting ends of these tools have relieved chamfers which extend to a small enough diameter to permit their entry into the hole being enlarged or finished.

Core drills are designed principally to enlarge existing holes, and provide a greater degree of accuracy and better finish than a two-flute twist

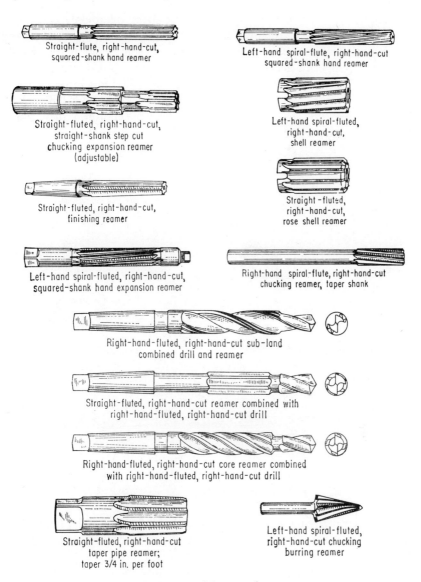

Fig. 1-55. Commercial types of reamers.

drill. These tools usually have three or four moderately deep helical flutes. The lands between the flutes are given body diameter clearance to produce a margin similar to that on a twist drill. Typical core drills are shown in Fig. 1-54. Core drills are usually made of high-speed steel but in some sizes they can be furnished with carbide tips. Sometimes they are made in a two-piece construction so that only the cutting end need be replaced.

Reamers are designed principally for hole sizing where only a moderate amount of stock is to be removed from the hole walls. The stock is removed by a larger number of flutes than are common with drills. Reamed holes usually have superior accuracy and finish. Because of the greater number of flutes and cutting edges, the margins on reamers are much narower than those on twist drills and core drills, which minimizes galling of the margins. Reamers have little to no starting taper, which improves their ability to accept guidance by bushings. The flutes on reamers may be either straight

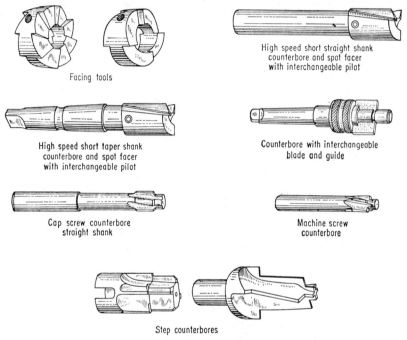

Fig. 1-56. Types of counterbores and spotfacers.

or helical depending upon the type of work to be finished. Solid high-speed-steel construction of reamers is most common, but carbide-tipped and solid carbide reamers are available in certain sizes and styles. Figure 1-55 shows various commercial types of reamers. Also shown are expansion and taper reamers.

Counterbores and Countersinks

Counterbores and countersinks are tools for modifying the ends of holes. They are often provided with pilots which engage the existing hole to improve alignment, and are commonly used to provide a seat in a plane normal to the hole axis. Where the flat seating surface is relatively shallow, the tools used are often called *spotfacers*. Deeper seats, such as those used for recessing the heads of socket-head cap screws, are called *counterbores*. Figure 1-56 illustrates various counterbores and spotfacers.

When a conical seat is required at the end of a hole, the operation is known as *countersinking* and the tools are called *countersinks*. Such seats are required to receive machine centers and to permit the use of flat-headed screws with the heads flush with a surface.

Except for the angle of the cutting edges, counterbores, spotfacers, and countersinks are of similar construction. They are commonly made with two or more flutes which usually have a right-hand helix. The flutes are usually shorter than those on drills because the seats produced are relatively shallow. When pilots are provided they may be either integral with the tool or of the removable type. The tools themselves are often made quite short

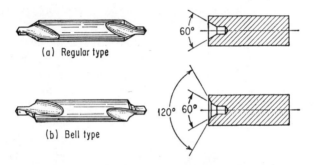

Fig. 1-57. Combined drills and countersinks.

and are used as readily replaceable tips in a holder which is mounted in the machine spindle. In some cases, one holder will drive a fairly wide range of sizes of cutters.

The combined drill and countersink is a specialized tool used for producing the center holes required for lathe and other machine centers. This tool combines a short two-flute drill with a two-fluted countersink so that the entire drilling and countersinking operation can be performed in one pass on a single setup. Such tools are usually made in double-end construction as shown in Fig. 1-57. When one end becomes dull the tool can be reversed in the chuck to provide another cutting end.

Multiple-Diameter Tools

It is sometimes desirable that multiple-diameter holes be produced or finished in a single operation. Common situations involve drilling and counterboring or drilling and chamfering a hole on the same setup. In high-production operations this can result in lower machining costs.

The simplest way to make a multiple-diameter tool is to start with a tool of the size that will produce the largest diameter and then grind down the cutting end of the tool to the size required for the small diameter. The cutting portions of the tool are provided with appropriate relief. This is known as *step construction*.

Another style of multiple-diameter tool employs what is known as *subland construction,* in which separate tool lands are provided for each cutting diameter. The small-diameter lands run between the large-diameter lands. Subland tools are more expensive because of the additional manufacturing operations required in producing the additional cutting surfaces.

Step and subland drills are included in Fig. 1-54. Multiple-diameter reamers (Fig. 1-55) and counterbores (Fig. 1-56) are also commercially available. Where the length of the small-diameter portion of the tool is relatively long in terms of the usable tool length, the simpler step construction is usually preferred. Besides lower acquisition cost, such tools provide more chip space in the flutes.

Where the small-diameter step is relatively short, the use of subland tools may be advantageous. In the step construction the usable sharpening life is limited to something less than the length of the small-diameter step. After this is used up it is necessary to cut off the tool and completely remanufacture the step end. In subland construction the two diameters can be sharpened for the entire length of the tool and the length of the step is not critical. The restricted flute space due to the presence of the additional lands sometimes limits the depth of holes to which they may be applied. With any multiple-diameter tool it is important that the difference in diameters be not too great. With a large difference in diameter it is not possible to maintain optimum cutting speeds on both diameters, and there is some difficulty in maintaining adequate strength in the tool.

Milling Cutters

Milling cutters are cylindrical cutting tools with cutting teeth spaced around the periphery (Fig. 1-58). Figure 1-44 shows the basic tooth angles of a solid plain milling cutter. A workpiece is traversed under the cutter in such a manner that the feed of the workpiece is measured in a plane perpendicular to the cutter axis. Sometimes the workpiece is plunged radially into the cutter and in rare cases there is also an axial feed of the cutter which results in a generated surface on the workpiece. Milling-cutter teeth intermittently engage the workpiece, and the chip thickness is determined by the motion of the workpiece, the number of teeth in the cutter, and the rotational speed of the cutter.

There are two modes of operation for milling cutters. In conventional (up) milling the workpiece motion opposes the rotation of the cutter (Fig.

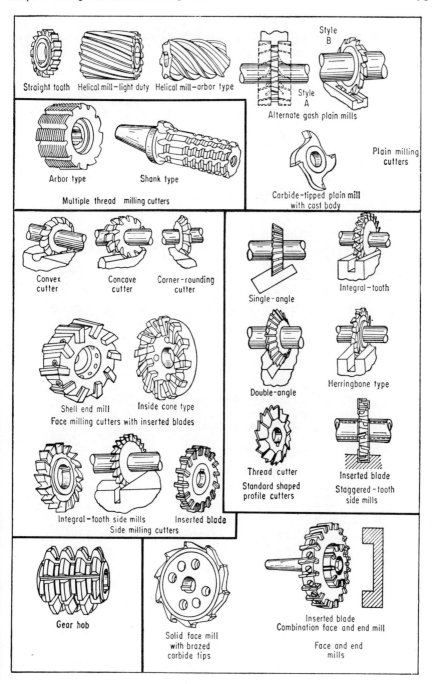

Fig. 1-58. Common types of milling cutters.

1-59A), while in climb (down) milling the rotational and feed motions are in the same direction Fig. 1-59B). Climb milling is to be preferred wherever it can be used since it provides a more favorable metal-cutting action and generally yields a better surface on the workpiece. Climb milling requires more rigid equipment and there must be no looseness in the workpiece feeding mechanism since the cutter will tend to pull the workpiece.

Milling cutters may have either profile-sharpened or form-relieved teeth. Profile-sharpened cutters are those which are sharpened on the relief surface, using a conventional cutter grinding machine. Form-relieved cutters are made with uniform radial relief behind the cutting edge. They are sharpened by grinding the face of the teeth. The profile style provides

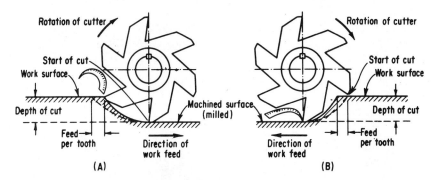

Fig. 1-59. (A) "Out-cut," "conventional," or "up" milling; also called "feeding against the cutter." (B) "In-cut," "climb," or "down" milling; also called "feeding with the cutter."

greater flexibility in adjusting relief angles for the job to be done, but it is necessary that any form on the cutter be reproduced during each resharpening. In the form-relieved style, the relief angle cannot be changed since it is fixed in the manufacture of the cutter. However, the form-relieved construction is well adapted to cutters with intricate profiles since the profile is not changed by resharpening.

Solid high-speed-steel construction is most common in milling cutters, but they are available in some styles with carbide-tipped teeth or with removable and replaceable tips of various tool materials. Replaceable-tip cutters are usually called "inserted blade" cutters.

Most large milling cutters are made in arbor-type construction. These are provided with an axial hole for mounting on an arbor, and usually have a keyway to receive a driving key. Certain small-diameter cutters and some cutters for specialized applications are made in shank-type construction where the cutting section is at the end of a straight or tapered shank which

Chap. 1   Design of Material-cutting Tools   77

fits into the machine tool spindle or adapter. In addition, some large facing cutters are designed to mount directly on the machine tool spindle nose.

Number of Teeth in a Milling Cutter. The number of teeth n may be expressed as [10]

$$n = \frac{F}{F_t N} \qquad (1\text{-}23)$$

where F = feed rate, in. per min
F_t = feed per tooth, in. (chip thickness)
N = cutter speed, rpm

Equation (1-24) has been found satisfactory for all ordinary cutters such as plain, side, and end mills, for both production and general-purpose use.

$$n = 19.5\sqrt{R} - 5.8 \qquad (1\text{-}24)$$

where R = cutter radius, in.

Example. For a 4-in. diam cutter,

$$n = 19.5\sqrt{2} - 5.8 = 21.8, \text{ or } 22 \text{ teeth}$$

Equation (1-25) [11] was developed for calculating the number of teeth n in high-speed milling when using carbide-tipped cutters. Since it involves power available at the cutter, it is helpful in permitting full utilization of available power without overloading the motor.

$$n = \frac{K \, hp_c}{F_t N d w} \qquad (1\text{-}25)$$

where K = a constant related to the workpiece material. It may be given the values of 0.650 for average steel, 1.5 for cast iron, and 2.5 for aluminum (conservative, taking into account the dulling of the cutter in service)
hp_c = horsepower available at the cutter
d = depth of cut, in.
w = width of cut, in.

End Mills. End mills are shank-type milling cutters which are usually designed with some form of relieved end teeth (Fig. 1-60). This construction, plus the fact that they are operated without an outboard support, makes them extremely versatile tools and permits their application to a wide variety of machining operations. End mills are usually considered in a separate category from other milling cutters.

Most end mills have helical flutes with a flute helix angle between 20 and 40 degrees. Flutes are usually several diameters long, but both longer and shorter designs are available. Tools with two and four flutes are most common, but the larger sizes are available with more flutes. The relieved end teeth permit their use in keywaying or pocketing operations with a minimum of contact with the surface at the bottom of the recess. Most two-flute end mills have end teeth which extend from the periphery to the center of the tool. This permits their use as drills to sink into depth before starting a cut. Some four-flute end mills are also made with two diametrically opposed teeth cutting to the tool center. The corners at the intersection of the peripheral and end teeth can be provided with radii to minimize the formation of stress risers in the work. When this radius becomes equal to the cutter radius, the end mill is said to have a *ball end*. Such tools are widely used in forming recesses and cavities in dies. An operation of this sort is often called *die sinking*.

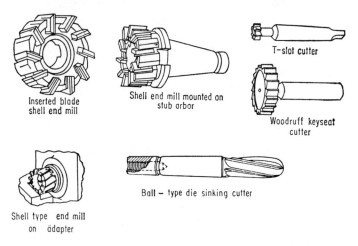

Fig. 1-60. End mills.

End mills are normally made of solid high-speed steel and usually have straight shanks. However, taper-shank tools can be furnished, and some styles are available with carbide-tipped cutting edges or are made from solid carbide blanks.

Taps

The tap is essentially a screw that has been fluted to form cutting edges. The cutting end of the tap has a relieved chamfer which forms the cutting edges and permits it to enter an untapped hole. If succeeding chamfered teeth are followed along the thread helix it will be seen that the effect of the chamfer is to make each tooth have a slightly larger major diameter than the preceding one. Thus, the major feed involved in tapping is a radial feed, although the feed is built into the tool rather than controlled by the machine

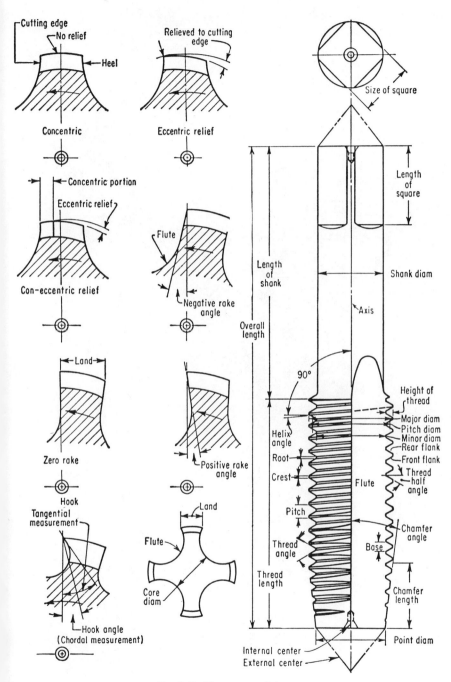

Fig. 1-61. Tap nomenclature.

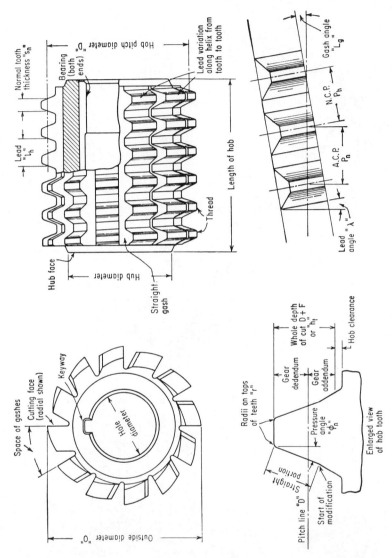

Fig. 1-62. Nomenclature of hobs.

as is the case with most other rotary tools. The actual feed per tooth depends on the chamfer angle and the number of flutes. A reduction in the chamfer angle or an increase in the number of flutes will reduce the feed per tooth.

The most common type of tap is the straight fluted hand tap (Fig. 1-61). This type of tap has a straight shank with a driving square at the shank end. In spite of the hand tap designation, these taps are generally used in tapping machines. Hand taps are usually available with three different chamfer lengths of nominally $1\frac{1}{4}$-, 4- and 8-thread pitches, which are designated respectively as *bottoming, plug,* and *taper* taps. Taps are also available with helical flutes and other modifications to suit them to specific applications. Most taps used in production operations are made of high-speed steel. Some solid carbide and carbide-tipped taps are used in work materials of high abrasiveness or hardness. However, these are more susceptible to breakage or chipping.

Hobs

A hob is a generating-type rotary cutting tool which has its teeth arranged along a helical thread (Fig. 1-62). The hob and the workpiece are rotated in timed relationship to each other, while the hob is fed axially or tangentially across or radially into the workpiece. The cutting edges of the hob teeth lie in a helicoid which is usually conjugate to the form produced on the workpiece. The teeth on a hob resemble those on a form-relieved milling cutter in that they are designed to be sharpened on the rake face. Most hobs are designed for generating gear teeth or splines. Certain other evenly spaced forms on the periphery of a cylindrical workpiece can also be hobbed. There are some limitations on the shape of forms that can be generated by hobs.

Most hobs are manufactured from high-speed steel. Carbide-tipped hobs are sometimes used for hobbing low-strength abrasive nonmetallic materials.

CONTROL OF THE CAUSES OF TOOL WEAR AND FAILURE

The design of cutting tools is not a pure science, involving only computations to be carried out in functional isolation. Of itself, a cutting tool is only a piece of metal of special shape and/or construction, although frequently a very expensive piece of metal.

Good tool design must be done in the light of many inseparably related factors: the composition, condition, and shape of the workpiece material; the rate and volume of the specified production; the type, motions, power, and speed of the machine tool to be used; the toolholders and workholders available or to be designed; the specified accuracy and surface of the finished workpiece; and many other factors, common or specific to the particular operation.

The following discussion is therefore very much the business of the good tool designer. He cannot escape responsibility by saying that some of the factors leading to undue tool wear, or to tool failure, are the responsibility of the process or methods functions. He should know, or anticipate, the possible applicational difficulties that lie ahead for the tool he is about to design, and then accommodate the difficulties by tool design, or consult with the other manufacturing functions as to changing certain conditions, or both.

Rigidity. Setup rigidity is vital to maintaining dimensional accuracy of the cut surface, since the tool shifts into or out of the cut with the accumulation of static deflections and take-up of loose fits. Rigidity also maintains surface-finish quality, avoiding the marks made by elastic vibration and free play of loose fits and backlash. In the control of vibration, rigidity is considered along with mass and damping contributed by each member and fit in the system.

Increasing the mass reduces the vibration amplitude and resonant frequency, while damping reduces amplitude by dissipating vibratory energy as frictional heat. Since each member can affect the mode and amount of vibration, most should be made oversize and broadly supported. This provides design latitude for those members having more severe cost and space limitations. Designers are generous with rigidity, anticipating fast, efficient cuts.

Strength. The strength of each member can be considered separately and related to the magnitude and application of the forces it will transmit. It should clearly be sufficient to prevent breakage or deformation beyond the elastic limit when the operation is performed correctly. The designer must also consider overloads and damage that may be encountered, providing abundant strength wherever economically possible. In particular, generous size and material specification should give good working life to areas subject to abrasive wear and work hardening under impact loads. But chatter, packed chips, or binding due to setup misalignment can multiply normal operating forces many times. Tool failure, mechanical malfunctions, and operating errors threaten destructive casualties even with costly overdesign.

Weak Links. A common practice is to protect the structural chain with weak links in anticipation of casualties and to confine or limit the possible damage. Suitable low strength with high rigidity is illustrated by the common soft shear pin. But these weak links must be strong enough to withstand normal operation and overload if possible. The permanent members are made unquestionably stronger by comparison and are protected by location. Identical design criteria make weak links an ideal combination with wearing details. They should be comparatively cheap, with duplicates widely stocked or at least readily produced, and of a form for easy, accurate replacement mounting. Mass-produced, delicate workpieces are of themselves natural weak links, though cutting-tool tips are the typical wear and breakaway members. Clamped-insert cutting tools, using replaceable backing shims, are ideal examples of wear and damage isolation.

Force Limitations. Operating forces may be obviously limited by a weak-link member, as in the case of a delicate workpiece. A machine tool such as a hydraulic-stroke planer or broaching machine may be rated in terms of force, with the ratings understated but subject to measurement and control. Some saws and grinders regulate the feeding force instead of the cutting force in the direction of cutting velocity. The conduction of heat generated in the cut is likely to be adequate in a cutting tool and holder of ample strength, though it may become an additional problem.

Speed, Feed, and Size. A machine tool's speed and feed ranges, its cutting-tool adaptor capacity, and its working clearances set up restrictions on tool design and production rate. The effect on forces is indirect but inevitable. A milling cutter with few teeth contacting a delicate workpiece exerts only a few times the cutting forces of one tooth, while high speed permits rapid completion of the cut. Limited cutter diameter and a low speed range would require more teeth and more force or a longer cutting time. Variable infeed rates can be used to speed rough stock removal and then minimize distortion while finishing, as in the case of the spark-out of a grinding wheel. The important thing is to design around limitations and take full advantage of flexibility.

Related Force Components. Total cutting force is usually resolved into three mutually perpendicular components. Force in the direction of feeding motion of a turning, boring, facing, plunge forming or parting tool corresponds to radial force on a peripheral-cutting milling cutter tooth or abrasive wheel or belt contact area. It is commonly taken as from one-fourth up to three-fourths of the thrust force F_T for sharp tools, the larger fraction being appropriate for light feeds. The straight-line cutting tools, namely teeth of saw blades or broaches and planer or shaper tools, develop this force component in the direction of feed into the work. The remaining force component is radial for turning or boring tools, normal to the finished work surface in facing, planing, and shaping, and axial for peripheral milling teeth. It may be negative, a pull into the work tending to cut undersize, with large back rake and/or work material like soft, ductile brass. However it commonly varies up to $\frac{1}{2} F_T$. That cutting forces are not perfectly steady is illustrated by the cutting of ductile work into segmented chips at lower speeds. And so energy can be fed into a resonant vibrating system in which the values of force given above become averages, with instantaneous force cycling from high peaks to possibly negative lows.

Tool Wear

Tool wear occurs in any satisfactory cutting operation as the cutting material is squeezed, rubbed, melted, or chemically eroded away. Bonded abrasive can be specified to match the grinding conditions for self-dressing action, in which dull grains continually break away and uncover sharp ones. This is the only common situation of a tool maintaining uniform cutting action without reconditioning. In general, the tool designer should foresee progressive tool wear changing forces and finish during the useful tool life. Two generalized forms of wear can be described in terms of a single-point tool as *face cratering* and *flank wear*.

Face cratering is characteristic of cutting steel or strong cast iron into hot curled or segmented chips. At moderate or high speeds and feeds, the strong, hot chip flowing across the tool face erodes a smooth, shallow scoop which can reduce cutting forces and assist chip curling. The scooped-out crater will start next to the built-up edge of the workpiece material, adhering to and protecting the sharp cutting edge if speed is not too high. Yellow to blue coloring indicates the moderate to high temperature reached by the chips. Increasing chip temperature and snarling or increasingly short chips give warning that the deepening crater is weakening support of the cutting edge and reconditioning is due. Eventually, substantial pieces of tool material will break from the edge, accurate and smooth work surface will end abruptly, and cutting forces will rise rapidly.

This cratering action is encouraged by selection of "abrasion-resistant" cast iron and nonferrous grades of cutting-tool material. If work material and aggressive cutting make cratering the dominant cause of tool failure, then the crater-resistant or steel-cutting grades will extend tool life.

Flank wear is characteristic of cutting nonferrous metal chips or cutting brittle cast iron or plastics into crumbly chips or powder. It also predominates when cutting steel into thin, cool chips at low speeds, as in reaming, threading, or broaching. A furrowed wear land grows down the tool flank which, in turn, tends to smear and burnish the work. Instead of remaining fairly constant during useful life, cutting forces rise progressively. This type of wear calls for safety factors of 2 on F_T and 4 on the other components to allow for increase during typical useful life. Adequate relief angles and a large side cutting edge angle and/or nose radius, respectively, will thin and widen the heat-generating wear land. Hence they postpone the eventual failure due to heat and force concentration, provided that blunting or chipping are not produced as a result. The selection of tool-material grade is the reverse of that in the cratering situation. Crater-resistant grades encourage flank wear instead. If work material and light cutting make flank wear the dominant cause of failure, then abrasion-resistant grades will extend tool life.

Blunting and chipping are inescapable machining factors. A basic criterion for tool-material selection is that it must maintain an advantage in hardness and strength enabling it to penetrate the work at the cutting temperature. The vulnerable tool tip must be well under any hard scale present, or any previous cuts, in materials of high work-hardening tendency. Even so, a cutting edge may quickly blunt or chip away, depending on its ductility, owing to poor chip control or misalignment or deflection of the setup. If not suspected or observed as it occurs, this will be disguised by the resulting flank wear. The remedy, however, is the opposite of that for flank wear. Converging chip flow across the tool face, particularly of strong, ductile work material, concentrates force; a wide, thinned-out chip promotes chatter. Large nose radius should be replaced by a narrow chamfer. Negative rake and honing the edge of carbide or oxide may help. Fundamentally, a tougher grade should be substituted for a harder grade of tool material which was supposed to give longer life.

Chip Disposal. One sure way to overload a cutting tooth is to block the path of the chip flowing across its face. Single-point tools cutting ductile work frequently employ a chip breaker to curl an otherwise stringy chip so that it will break up and fall away. If this breaker is set too close for the size of cut being taken, it can cause blunting, edge chipping, or breakage. Multiple-tooth cutters commonly have ample chip spaces, provided that the chips are thrown or washed out between successive passes through the cut. This may require strong streams of coolant or even brushing. It is difficult to remove work materials like soft steel or copper alloys and titanium, whose chips tend to weld onto the tool face. Chip disposal in milling slots or steps may demand high rake angles and climb milling instead of conventional cutter rotation. Involved selection and application methods have developed for tapping and deep hole drilling, where chip clogging, misalignment, and runout can readily break tools. Tool design should provide space for chip flow and means of disposal, which may well be the solution to problems of tool blunting and chipping.

Uneven Motions. Another sure way to overload a cutting tooth is to increase the feed rate drastically beyond its structural or chip-disposal capacity. Machine structural deflection accomplishes this in the example of a drill breaking as it breaks through the work. As the heavy thrust of the chisel edge is relieved, structural members spring back toward their unstressed shape, and the drill lips plunge into the work for an oversize bite. Feed mechanisms may employ air or hydraulic fluid whose compression is elastic; or gearing and a leadscrew nut fit may introduce blacklash. Machine way motion becomes jumpy at slow speeds ("slip-stick" motion), even with heavy lubrication. A milling cutter at slow feed may actually rub until pressure builds up. It then may dig into the work and surge ahead. Adding to the difficulty, the sudden change in cutting torque adds to the pounding caused by teeth entering the cut.

Torsional vibration and backlash tend to develop in a rotary drive train. Should cutter rotation become so erratic that it momentarily stops, carbide or oxide teeth will generally break at once by being bumped into the work. With some teeth gone, the entire cutter may fail progressively as each successive tooth is unable to carry the extra load left by the preceding damaged teeth. With slow cutting velocity and many teeth, tremendous forces develop owing to the flywheel energy of the motor and drive train as a failed cutter plows to a stop.

Chatter. The rapid, elastic vibration that sometimes appears between tool and work is easily detected by marks on the work surface and by the sound that gives it the name "chatter." It may be associated with chips breaking up and not adhering to the tool or wheel, or with free motion of machine ways. Chatter is a danger signal of the blunting and chipping or rapid abrasive loss that is a hazard to surface-finish quality. The remedy is to oppose uneven motion and loose fits. Chatter is less likely with few teeth moving at high velocity taking thick chip loads, and having high rake and ample relief angles. In grinding, softer action or smaller contact reduces chatter. Uneven motion calls for numerous, wide tooth contacts with the work to absorb backlash. Thin chip loads with moderate speeds may reduce shock. A negative rake angle may prevent pulling into the cut. In grinding, harder action or broader contact helps withstand bumping. As an extreme simplification, chatter can be combatted with lower cutting forces while looseness and backlash cannot.

PROBLEMS

1. What four elements interact in the machining process?
2. What factor will greatly influence the cutting tool material and geometry?
3. Can tool guidance be incorporated in the tool function?
4. What factors are influenced by the shape of a cutting tool?
5. What are the single-point tool angles?
6. What are the advantages of increasing the nose radius?
7. What normal tool angles should a single-point tool have for turning AISI 3120 steel?
8. With a tool holder angle of 10 degrees, what tool angles should be specified for a single-point tool for turning aluminum?
9. What is the primary function of a relief angle?
10. What is the rake angle requirement for ductile work materials; for brittle materials?
11. What are the machining factors used to evaluate machinability?

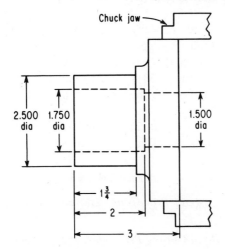

Fig. 1-63. Part for which a multiple tool is to be designed. The tool is to turn and bore the part on a turret lathe.

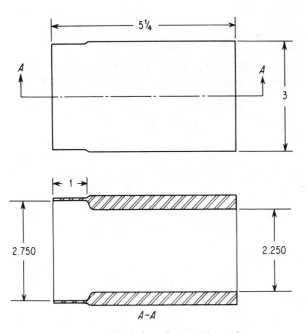

Fig. 1-64. Workpiece for Problem 18.

12. Name the three (3) most common types of chips.
13. What type of chip is generally produced by the machining of brittle materials?
14. Upon what does the capacity of a material for plastic flow depend?
15. What causes a built-up edge?
16. What effect does a smaller rake angle have on cutting forces?
17. The part shown in Fig. 1-63 is to be turned and bored in a turret lathe. Design a combination boring and turning tool to accomplish these operations.
18. Design a boring bar employing a standard adjustable boring tool cartridge for boring the 2.750-in. diam hole shown in Fig. 1-64.

REFERENCES

1. Boston, O. W., *Metal Processing,* 2nd ed., John Wiley & Sons, Inc., New York, 1951.
2. Shaw, M. C., *Metal Cutting Principles,* 3rd ed., Massachusetts Institute of Technology, 1954.
3. Ernst, H., "Physics of Metal Cutting," *Machining of Metals,* Am. Soc. Metals, 1938.
4. Ernst, H., and Merchant, M. E., *Chip Formation, Friction and Finish,* Cincinnati Milling Machine Company.
5. Opitz, I. H., "Present-Day Status of Chip-Formation Research," *Microtechnic,* No. 4 (1960).
6. Chao, B. T., and Trigger, K. J. "Temperature Distribution at the Tool-Chip Interface in Metal Cutting," *Trans. ASME,* **77** (1955).
7. Chao, B. T., and Trigger, K. J., "Temperature Distribution at the Tool-Chip and Tool-Work Interface in Metal Cutting," *Trans. ASME,* **80** (1958).
8. Shaw, M. C., and Oxford, C. J. Jr., (1) "On the Drilling of Metals," (2) "The Torque and Thrust in Milling," *Trans. ASME,* **79**:1 (Jan. 1957).
9. "Milling Cutters, Nomenclature, Principal Dimensions, etc.," American Standard ASA B5.3-1959, American Standards Association, New York.
10. *Tool Engineers Handbook,* 2nd ed., American Society of Tool and Manufacturing Engineers, McGraw-Hill Book Co., Inc., New York, 1959.
11. *A Treatise on Milling and Milling Machines,* 3rd ed., The Cincinnati Milling Machine Co., 1951.

2

WORKHOLDING DEVICES

The term *workholder* embraces all devices that hold, grip, or chuck a workpiece in a prescribed manner of firmness and location, to perform on it a manufacturing operation. The holding force may be applied mechanically, electrically, hydraulically, or pneumatically. This section considers workholders used in material-removing operations.

Figure 2-1 illustrates almost all the basic elements that are present in a material-removing operation intended to shape a workpiece. The right hand is the toolholder, the left hand is the workholder, the knife is the cutting tool, and the piece of wood is the workpiece. Both hands combine their motion to shape the piece of wood by removing material in the form of chips. The body of the person whose hands are shown may be considered as a machine that imparts power, motion, position, and control to the elements shown. Except for the element of force multiplication, these basic elements may be found in all of the forms of manufacturing setups where toolholders and workholders are used.

Figure 2-2 shows a pair of pliers or tongs used to hold a rod on which a point has to be ground or filed. This simple workholder illustrates the element of force multiplication by a lever action, and also shows serrations on the parts contacting the rod to increase the resistance against slippage.

Figure 2-3 shows a widely used workholder, the screw-operated vise. The screw pushes the movable jaw and multiplies the applied force. The vise remains locked by the self-locking characteristic of the screw, provides means of attachment to a machine, and permits location of the work in a more precise manner.

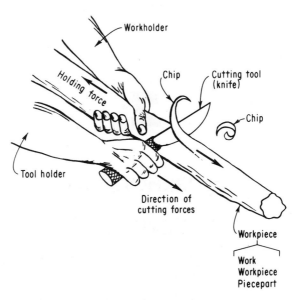

Fig. 2-1. Principles of workholders.

Figure 2-4 shows a vise with a number of refinements often used in workholders. The main holding force is supplied by hydraulic power, the screw being used only to bring the jaws in contact with a workpiece. The jaws may be replaceable inserts profiled to locate and fit a specific workpiece as shown. Other, more complicated jaw forms are used to match more complicated shapes of workpieces.

Another large group of workholders are the chucks which are attached to a variety of machine tools and used to hold a workpiece for turning, boring, drilling, grinding, and other rotary-type operations. Many types of chucks are available. Some are tightened manually with a wrench, others are power-operated by air or hydraulic means or by electric motors. On some chucks, each jaw is individually advanced and tightened, while others

Chap. 2 Workholding Devices

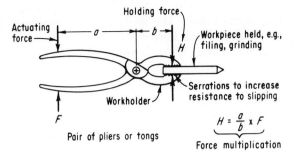

Fig. 2-2. Muliplication of holding force.

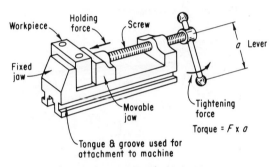

Fig. 2-3. Elementary workholder (vise).

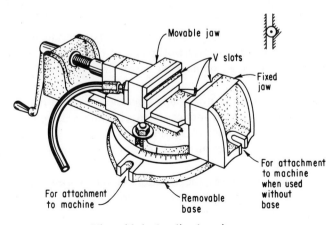

Fig. 2-4. Vise with hydraulic clamping.

have all jaws advance in unison. Figure 2-5 shows a workpiece clamped in a four-jaw independent chuck. The drill, which is removing material from the workpiece, is clamped in a universal chuck.

Purpose and Function of Workholders. A workholder must position or locate a workpiece in a definite relation to the cutting tool and must withstand holding and cutting forces while maintaining that precise location. A workholder is made up of several elements, each performing a certain function. The locating elements position the workpiece; the structure withstands the forces; brackets attach the workholder to the machine, and clamps, screws, and jaws apply holding forces. Elements may have manual or power actuation. All functions must be performed with the required firmness of holding, accuracy of positioning, and with a high degree of safety for the operator and the equipment.

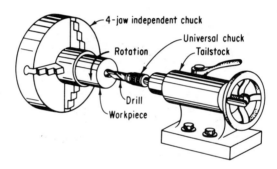

Fig. 2-5. Holding (chucking) a round workpiece.

The design or selection of a workholder is governed by many factors, the first being the physical characteristics of the workpiece. The workholder must be strong enough to support the workpiece, and to some extent will reflect its size, weight, and shape. The workholder material must be carefully selected with reference to the workpiece so that neither will be damaged by abrupt contact, e.g., damage to a soft copper workpiece by hard steel jaws.

Cutting forces imposed by machining operations vary in magnitude and direction. A drilling operation induces torque, while a shaping operation causes straight-line thrust. The workholder must support the workpiece in opposition to the cutting forces and will generally be designed for a specific machining operation.

The workholder establishes the location of the workpiece relative to the cutting tool. If the operation is to be performed at a precise location on the workpiece, locating between the workpiece and workholder must be equally precise. If the cutting tool must engage the workpiece at a specified distance from a workpiece feature such as a line or plane of the workpiece, then the workholder or workholding fixture must establish the line or plane at the specified distance. The degree of precision required in the workholder will usually exceed that of the workpiece because of cumulative error.

The strength and stiffness of the workpiece will determine to what extent it must be supported for the machining operation. If the workpiece design is such that it could be distorted or deflected by machining forces, the workholder must support the affected area. If the workpiece is sufficiently rigid to withstand the machining forces, workholder support at the edge of the workpiece may be adequate. The strength of the workholder is determined by the magnitude of the machining forces and the weight of the workpiece.

Production requirements will greatly influence workholder design. If a great number of workpieces are to be processed, the cost of an elaborate workholder might well be surpassed by savings due to increased hourly production made possible by the elaborate workholder, since the cost of the workholder will be prorated against the great number of workpieces. High production rates and volumes can therefore justify expensive fixturing. Conversely, if only one or two workpieces are to be machined, the operation will usually be performed with standard toolroom equipment and little to no fixturing costs can be justified. Production schedules may limit the time available for workholder acquisition and may compel the use of standard equipment.

Safety requirements must always dictate workholder design or selection. A workholder must not only withstand normal cutting forces and workpiece weight but may also have to withstand large momentary loads. In machining a cast workpiece, the cutting tool might strike an oxide inclusion causing instantaneous multiplication of force. The tool might cut through the inclusion. The tool might break, or the machine might stall. If the workholder broke, however, the tool might impart motion to the workpiece. A workpiece in uncontrolled motion is much like a missile. The workholder must also be designed to protect the workman against his own negligence. Where possible, a shield should be interposed between the workman and the tool.

A workholder should be designed to receive the workpiece in only one position. If a symmetric workpiece can be clamped in more than one position it is possible and probable that a percentage of workpieces will be incorrectly clamped and machined. To prevent this, workholders should be made foolproof.

It is advisable to use standard workholders and commercially available components whenever possible. Not only can these items be purchased for less than the cost of making them, but they are generally of adequate strength and dimensional accuracy.

There are many workholders used in industry that are not used on material-removing operations. Workholders may be used for the inspection of workpieces, for assembly, welding, and so on (see Chapters 5 and 6). There may be very little difference in their basic design and their appearance. Quite often a standard commercial design may be used in one application for a turning operation and for the same or another workpiece in an inspection operation.

Methods of Location

To insure successful operation of a workholding device, the workpiece must be accurately located to establish a definite relationship between the cutting tool and some points or surfaces of the workpiece. This relationship is established by locators in the workholding device, by which the workpiece can be positioned and restricted to prevent its moving from its predetermined location. The workholding device will then present the workpiece to the cutting tool in the required relationship. The locating device should be so designed that each successive workpiece, when loaded and clamped, will occupy the same position in the workholding device. Various methods have been devised to effectively restrict the movement of workpieces. The locating design selected for a given workholding device will depend on the nature of the workpiece, the requirements of the metal-removing operation to be performed, and other restrictive conditions inherent in the workholding device.

3-2-1 *Method of Location.* A workpiece in space, free to move in any direction, is designed around three mutually perpendicular planes and may be said to have twelve modes or degrees of freedom. It may move in either of two opposed directions along three mutually perpendicular axes, and may rotate in either of two opposed directions around each axis, clockwise and counterclockwise. Each direction of movement is considered one degree of freedom. The twelve degrees of freedom as they apply to a rectangular prism are shown in Fig. 2-6A. Figure 2-6B shows three views of the prism in orthographic projection, with all twelve degrees of freedom indicated in their respective positions. To accurately locate a workpiece, it must be confined to restrict it against movement in any of the twelve degrees of freedom except those called for by the operation. When this condition is satisfied, the workpiece is accurately and positively confined in the workholding device.

A workpiece may be positively located by means of six pins, so positioned that collectively they restrict the workpiece in nine of its degrees of freedom. This is known as the 3-2-1 method of location. Figure 2-7 shows the prism resting on three pins A, B, and C. The faces of the three pins supporting the prism form a plane parallel to the plane that contains the

Chap. 2 Workholding Devices   95

X and Y axes. The prism cannot rotate about the X and Y axes and it cannot move downward in the direction of freedom 5. Therefore, freedoms 1, 2, 3, 4, and 5 have been restricted.

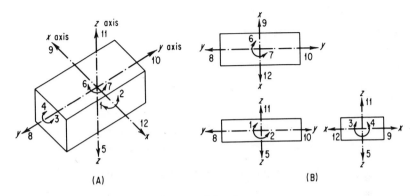

Fig. 2-6. Twelve degrees of freedom.

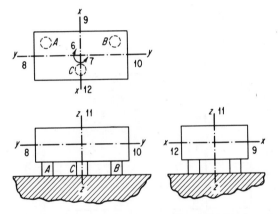

Fig. 2-7. Three pins arrest five degrees of freedom.

In Fig. 2-8, two additional pins D and E whose faces are in a plane parallel to the plane containing the X and Z axes prevent rotation of the prism about the Z axis. It is not free to move to the left in the direction of freedom 8. Therefore, freedoms 6, 7, and 8 have been restricted and the prism cannot rotate.

Finally, with the addition of pin F as shown in Fig. 2-9, freedom 9 is restricted. Thus by means of six locating points, three in a base plane, two in a vertical plane, and one in a plane perpendicular to the first two, nine degrees of freedom have been restricted.

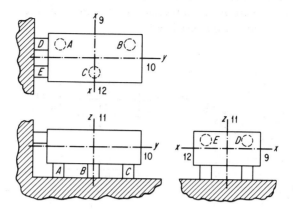

Fig. 2-8. Five pins arrest eight degrees of freedom.

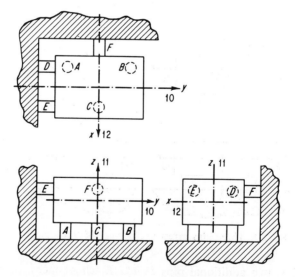

Fig. 2-9. Six pins arrest nine degrees of freedom.

Three degrees of freedom, 10, 11, and 12, still remain unrestricted. The addition of three more pins, one for each remaining freedom, would completely restrict movement of the prism. The pins would then entirely enclose the workpiece. This is not practical since it would prevent loading of the workpiece into the workholding device. The remaining three freedoms may be restricted by means of clamping devices, which also may serve

Chap. 2 Workholding Devices   97

to resist the forces generated by the metal-removing operation being performed on the workpiece. Any combination of three clamping devices and locating pins may be used if this is more suitable to the design of a particular workholding device.

Pins. The cylindrical locating pins shown in Figs. 2-7, 2-8, and 2-9 to illustrate the basic 3-2-1 locating concept are perhaps the simplest locating element. They are commonly made from steel drill rod. The making of a pin involves only cutting it to the required length. Locating a pin in its proper position in a workholding device requires only a round hole of a suitable size into which the pin may be pressed. The hole need only be

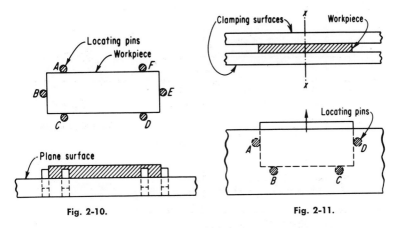

Fig. 2-10. Simple workholder made of plane surface and pins.

Fig. 2-11. Vertical workholder.

tight enough to prevent the pin from coming loose during normal usage. Because of its simplicity the round pin is the most commonly used locating element.

Considerations other than location of the workpiece will often affect the number of locating pins used. The workholding device must be designed to clamp the workpiece securely and to so support it as to resist the forces generated by the operation. In Fig. 2-7, pins A, B, and C are used to restrict six degrees of freedom. However, these pins support only a small portion of the workpiece. If the operation performed applies considerable force, the workpiece may spring out of shape. Hence, the locating elements must be designed to also provide adequate support for the workpiece against the forces acting upon it. Many workpieces are essentially flat in nature, or have a flat surface that can be used for locating purposes. These are commonly located by placing the workpiece on a plane surface to restrict it in five freedoms as shown in Fig. 2-10. The addition of six locating pins A, B, C, D, E, and F will restrict it in six more degrees of freedom. The workpiece can move only in an upward direction. This final degree of freedom is restrained by a suitable clamp having a plane surface parallel to the one on which the workpiece is placed.

Vertical Holding. If the workpiece must be held in a vertical position, the same principle of clamping and supporting between two plane surfaces may be used. This again will restrict motion in six degrees of freedom. Of the six remaining freedoms, only five must be restricted by locating pins as shown in Fig. 2-11. The four pins A, B, C, and D restrict freedom of

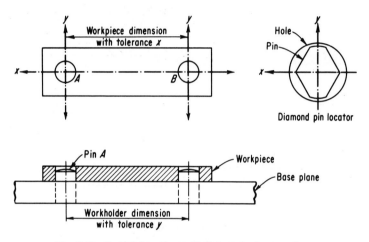

Fig. 2-12. Radial location with internal pins or plugs.

motion downward, to the left and right, and both clockwise and counterclockwise around axis X. Gravity may be used to locate the workpiece and restrict freedom of movement during the machining or other operation to be performed.

Radial Location. When the workpiece contains accurate holes that are perpendicular to a finished locating surface, adequate locating may be achieved with two pins or plugs extending from a base surface. The pins are made and positioned to fit the holes in the workpiece as shown in Fig. 2-12. A workpiece located from two pins secures its principal location from the base plane and pin A, and radial location from pin B. The base plane restricts five degrees of freedom and pin A eliminates movement in directions X and Y through its axis. The only need for pin B (the radial locator) is to prevent rotation of the workpiece about the axis A, and this is achieved by elimination of movement in the Y direction through the axis B. Thus pin B needs to fit its hole on only two points, these being the points where the Y axis intersects the periphery of the hole.

This may be thought of as the 3-2-1 method. The base contains three points to eliminate motion in five degrees of freedom, consisting of one vertical direction and both clockwise and counterclockwise rotation around two axes. One point on the cylindrical pin and one on the radial locator provide the two points in a vertical plane to eliminate movement in three additional freedoms, namely, clockwise and counterclockwise rotation around the third axis and in one horizontal direction. A second point on the cylindrical pin acts as the point in the third mutually perpendicular plane to restrict motion in one more horizontal freedom.

Diamond Pins. It is possible to accurately locate a workpiece with two round pins, but allowances must be made for the variations encountered in hole sizes and locations. For instance, the distance between holes A and B (Fig. 2-12) will vary to the extent of tolerance X. Similarly, the distance between pins A and B in the workholder has a tolerance Y. For accurate location there should be an allowance between pin A and hole A of only a few ten-thousandths of an inch. But if pin B is a complete cylinder (same as pin A) the allowance between pin B and hole B must be at least as great as the sum of tolerances X and Y. This is necessary for the pins to engage holes within the permissible tolerance X. Extreme cases occur when both hole and pin center-to-center dimensions are at maximum or minimum conditions. As a result, there will be a large allowance between the hole and pin at B in the Y direction. This will permit an undesirable amount of radial rotation around the axis A and defeats the purpose for which pin B is intended. To achieve more accurate radial location, B may be a diamond pin as shown in the inset in Fig. 2-12. It is relieved on two sides to allow for variations in the X direction and has two cylindrical portions to locate the hole in the Y direction. The minimum radial movement of the workpiece occurs when the diameter of the cylindrical portion of the pin is smaller than the diameter of the hole by the allowance necessary to slip the minimum size hole over the pin.

The combination of a locating pin engaging a hole and a radial locator can also be used when the workpiece contains only one hole. Figure 2-13 shows a workpiece so located. The principal location is secured from the base plate and pin A. Radial freedom in both directions at Y is restricted by two pins B confining the periphery of the workpiece.

The basic rule for radial location is that the position of radial locators shall be as far from the axis of rotation as possible to minimize deviations from true location. This is illustrated schematically in Fig. 2-14. Displacement D at distance X from the axis of rotation O results in angular deviation of AOA'. The same displacement D at the greater distance Y results in a smaller angular deviation BOB'.

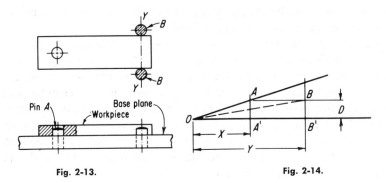

Fig. 2-13.   Fig. 2-14.

Fig. 2-13. Radial location by external pins.

Fig. 2-14. Displacement as a factor in minimizing deviation from true radial location. (*L. E. Doyle, "Tool Engineering," Prentice-Hall, Inc., 1950*).

Principles of Pin Location. Three basic principles apply to the use of locating pins.

1. The principle of *minimum locating points*. No more points than necessary should be used to secure location in any one plane. The 3-2-1 principle determines the minimum number required. More can be used, but the additional points should be used only if they serve a useful purpose, and care must be taken that they do not impair the location function.
2. The principle of *extreme positions*. Locating points should be chosen as far apart as possible on any one workpiece surface. Thus, for a given displacement of any locating point from another, the resulting deviation decreases as the distance between the points increases.
3. The principle of *mutually perpendicular planes*. The most satisfactory locating points are those in mutually perpendicular planes. Other arrangements are possible but not desirable. Two disadvantages result from locating from other than perpendicular surfaces: (1) the consequent wedging action tends to lift the workpiece, (2) the displacement of a locating point or a particle (chip or dirt) adhering to it introduces a correspondingly larger error. In Fig. 2-15 the introduced error T is projected to become the resulting error E. The projection factor F is zero when the locating surfaces are perpendicular, and increases as the angle between them becomes more acute.

V Locators for Cylindrical Workpieces. A cylinder, like the prism, also has twelve degrees of freedom. The cylinder in Fig. 2-16 is free to move in two opposed directions along each axis, and to rotate both clockwise and counterclockwise around each axis. To accurately locate a cylindrical workpiece, it must be confined to restrict its motion in each of its twelve freedoms.

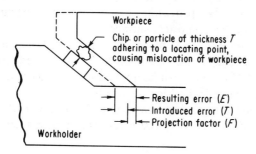

Fig. 2-15. Magnification and projection of error.

Figure 2-17 shows a cylinder placed in the intersection of two perpendicular planes. The base plane is parallel to the X and Z axes, and the vertical plane is parallel to the Y and Z axes. The horizontal plane restricts movement in the two rotational freedoms around the X axis and the downward freedom along the Y axis. The vertical plane restricts the two rotational freedoms around the Y axis and the leftward movement along the X axis. The pin that forms the end stop restricts one freedom, i.e., forward movement along the Z axis. This corresponds to the basic 3-2-1 method of location used for the prism, but it restricts movement only in seven freedoms. The cylinder can move backward along the Z axis, to the right along the X axis, and upward along the Y axis; in addition, it is free to rotate clockwise and counterclockwise around the Z axis.

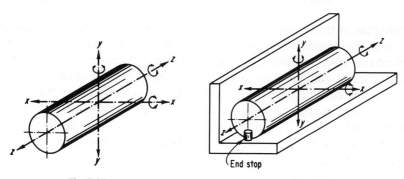

Fig. 2-16. Fig. 2-17.

Fig. 2-16. Twelve degrees of freedom of a cylindrical workpiece.

Fig. 2-17. Seven degrees of freedom arrested by V locator with stop pin.

Rotation around the Z axis can be restrained by clamping friction applied against the V formed by the two planes. This does not locate in a definite angular position about the Z axis and therefore cannot be considered true locating. No provision has been made to accurately locate a particular point on the cylindrical surface.

Locating a cylinder in a V places its longitudinal axis in true location. This is often sufficient for the operation to be performed. In addition, the basic principle of V location can be applied to workpieces that are not true cylinders but do contain cylindrical segments.

A single V locator provides two points for locating, the points where the cylindrical end of the workpiece is tangent to both sides of the V. The

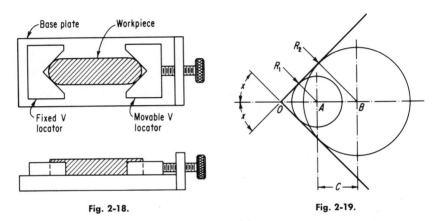

Fig. 2-18.   Fig. 2-19.

Fig. 2-18. Workholder with multiple V locators.

Fig. 2-19. Positions of circular sections of varying diameter in a V locator.

equivalent of three points in a base plane and a radial locator are required for complete location of the workpiece. In Fig. 2-18, a workpiece with two cylindrical ends is confined by means of two V locators. The movable V locator serves only to locate one point, the center of the cylindrical portion.

The included angle between the two surfaces of a V locator governs the positions of circular sections of varying diaemters (Fig. 2-19). A V with an included angle of 2 X locates a circle of radius R_1 with center A and a circle of radius R_2 with center B. The distance between centers $= C$. By similar triangles

$$\frac{R_1}{OA} = \frac{R_2}{OB} = \frac{R_2}{OA + C}$$

but $OA = R_1$ cosecant X
  $C = $ cosecant $X(R_2 - R_1)$
and $D_2 = 2R_2$; $D_1 = 2R_1$

therefore
$$C = \frac{\text{cosecant } X(D_2 - D_1)}{2}$$

Consequently, the distance between positions of any two diameters in a V varies as ½ of the cosecant of ½ the included angle of the V. The smallest variation occurs when $X = 90°$, where cosecant $\frac{X}{2} = .5$. However, as X approaches 90°, there is less inclination for the circular section to seat positively in the V and more difficulty in retaining it. Note that with $X = 90°$, $2X = 180°$ or a straight line. The best compromise is achieved when $X = 45°$ and the included angle of the V is 90°.

Irregularities in the circular section of a workpiece, or a chip lodged between the workpiece and the V locating surface can introduce errors of location. The included angle of the V has a definite influence on the effect of such displacement. In Fig. 2-20, E is the displacement caused by a rough

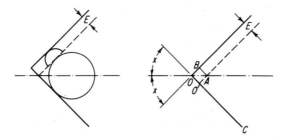

Fig. 2-20. Influence of the included angle or errors of location.

surface or chip. The circular section of the workpiece may be considered to rest on another side of the V indicated by a dotted line. The displaced side of the V, shown by the solid line, forms a new V with the opposite side to define the displaced position of the circular section. The change in the position of the center of the workpiece is identical to the shift in the apex from the original to the new V. The original V is BOC and the new V is AO′C. The axis of the workpiece is displaced by the distance OO′. For a constant displacement E, OO′ is a minimum when $2X = 90°$.

Consideration must be given to the axis of the rotating tool and its relationship with the position of the V locator. The V locates the longitudinal axis of the cylindrical workpiece. When work is done perpendicular to this axis, the position of the V locator should be arranged to keep displacement of the workpiece to a minimum. In Fig. 2-21A, a cylindrical workpiece is placed in a V locator so that a hole can be drilled perpendicular to the longitudinal axis. Any variation in diameter of the workpiece will cause a displacement in the location of the vertical axis. The drill bushing, however, remains in its original position and the drilled hole will deviate from its required position by the amount of the displacement. In Fig. 2-21B, the V locator is so positioned that its axis is parallel to that of the drill bushing. Variation in diameter of the workpiece will cause no displacement of the vertical axis and the drilled hole will not deviate from its required position.

Nesting or Cavity Locating. The nesting method of locating features a cavity in the workholding device into which the workpiece is placed and located. If the cavity is the same size and shape as the workpiece this is an effective means of locating. Figure 2-22 illustrates a nest which encloses

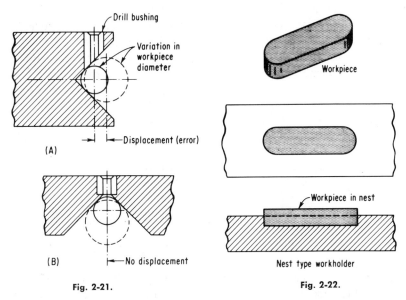

Fig. 2-21. Minimizing error by proper placement of a V locator.

Fig. 2-22. Nest-type workholder.

the workpiece on its bottom surface and around the entire periphery. The only freedom remaining is in an upward direction. A similar nest can be used to locate cylindrical workpieces. Nests of the cavity type are used to locate a wide variety of workpieces regardless of the complexity of their shape. All that is necessary to provide a cavity of the required size and shape. No supplementary locating devices such as pins are normally required.

The cavity type of nest possesses some disadvantages. Since the workpiece is completely surrounded, it is often difficult to lift it out of the nest. This is particularly true when no portion of the workpiece projects out of the nest to afford a good grip for unloading. The workholding device can, of course, be turned over, and the workpiece shaken out. When the work-

Chap. 2 Workholding Devices   105

piece tends to stick, an ejecting device can be incorporated in the workholder. This, however, introduces an additional time element into the operation. Another disadvantage is that the operation performed may produce burrs on the workpiece which tend to lock it into the nest. In this case, the workpiece must be pried out or an ejector must be provided. Chips from the cutting operation may lodge in the nest and must be removed before loading the next workpiece. Any chips remaining may interfere with the proper positioning of the next workpiece.

To avoid the disadvantages of the cavity type of nest, partial nests are often used for locating. Flat members, shaped to fit portions of the workpiece, are fastened to the workholding device to confine the workpiece between them. Figure 2-23 shows two partial nests, each confining one end of the arcuate workpiece. Each nest is fastened to the flat supporting surface of the workholder by means of two screws. Accurate positioning of the nests is ensured by dowels which prevent each nest from shifting from its required position.

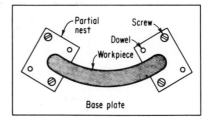

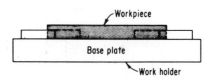

Fig. 2-23. Workholder with partial nests.

Partial nests eliminate many of the disadvantages of the cavity type of nesting method. Since they do not have the entire contour of the workpiece they require less time to make. Since they do not completely confine the workpiece it is easily lifted out of the workholder. Normally, an ejecting device is not required. The cavity in each nest is open at one end to permit easy removal of chips.

With the development of plastic casting materials, the making of cavities for locating workpieces has been simplified. The casting material can be poured around a workpiece prototype. After solidification and cure, the workpiece is removed. The resulting cavity becomes the nest for locating workpieces during the production operation. This method is much simpler than machining the cavity from a solid piece of material and requires considerably less time, particularly when the workpiece is of complex shape.

Many varieties of plastic materials are available, some with fillers of solid material, including steel or aluminum, to increase strength. Those with metallic fillers may be sawed, milled, drilled, and tapped after curing to a solid state. This machinability facilitates fastening the poured plastic cavity to the workholding device, and alteration to accommodate future changes in the workpiece. Selection of the proper plastic depends on the forces generated by the operation to be performed, the quantity of parts to be made, and the effects of shrinkage of the material as it solidifies and cures.

Since plastic materials do not possess the strength of steels, plastic cavities are often reinforced with steel members. They increase resistance to tool forces but do not materially increase the wear resistance of the plastic nesting surface. Consequently, plastic cavities are not generally used when large quantities of parts must be made. Also, if the plastic material used has a high degree of shrinkage, the resulting cavity may be too small to adequately locate the work. This, however, is not too critical a factor since continuing improvements in the formulation of plastic compounds have consistently reduced the inherent shrinkage.

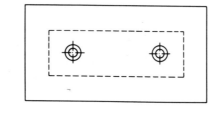

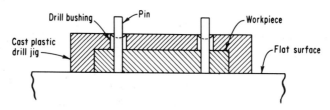

Fig. 2-24. Cast plastic plate-type drill jig.

Since the workpiece itself is required for the preparation of a nesting cavity, this type of workholder cannot be completed in advance of the production process. Of course, a prototype workpiece may be used, but making the prototype costs time and money so that the primary advantage of the cast locating cavity is lost. Therefore, the cavity is not usually made until production parts are available. This, however, does not cause a serious delay, since a cast cavity can be poured and cured in a few hours and be ready for use the following day.

A minimum-size workpiece must not be used for making the cavity since it will not accept parts that vary toward the maximum dimension. When the contours of the workpiece permit, its surfaces can be built up to increase vital locating dimensions to insure casting a cavity large enough to adequately locate a maximum-size part. Often this can be done by applying strips of thin masking tape to the proper surfaces of the workpiece.

Chap. 2 Workholding Devices   107

Experience has developed certain sophistications of the casting process of making locating cavities to the extent that a complete workholding device can be poured in one piece. Figure 2-24 shows a plate-type drill jig made this way. The workpiece containing the required holes is laid on a flat surface. Round pins are pressed into the holes and project upward to locate the drill bushings whose outer periphery is serrated to insure a firm grip in the cast plastic. A dam may be made from four rectangular bars placed around the workpiece and the plastic is poured to the height of the drill bushings. After solidification and curing, the workpiece and pins are removed and the drill jig is ready for use.

Tool Forces

A clear understanding of the direction and magnitude of cutting forces may eliminate the need to restrain all twelve degrees of freedom of a workpiece. Figure 2-25 shows how two pins and a table absorb the torque and thrust of a drilling operation. Although the workpiece is free to turn in a direction opposite to the torque, this freedom is insignificant until and unless a force is applied in that direction. No such force will be encountered in the planned drilling operation, and the remaining freedom need not be restrained. If, however, as part of the drilling operation the spindle rotation is reversed for tool removal, such force will be encountered, and the freedom must be restrained.

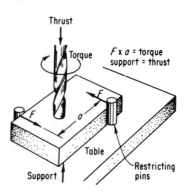

Fig. 2-25. Pin-type drill fixture resisting torque and thrust.

Theoretically there is no need to hold the workpiece down, as this is accomplished by the thrust of the drill. When the drill breaks through the thickness of the workpiece, an upward force may be created by interaction between the drill flutes and material remaining around the periphery of the hole. If there is no restraint in this upward direction, the workpiece may be lifted above the pins, creating a very dangerous condition. An upward force may also be produced when a drill or reamer gets lodged in a workpiece and the tool is to be withdrawn.

Figure 2-26 shows how a workpiece must be restrained for tapping. Torque in both directions must be absorbed and the lifting pull of the tap must be resisted by some form of a holddown such as a clamp (not shown). For leadscrew tapping, torque resistance is still required but no holddown is needed, because the leadscrew, having the same lead as the tap, eliminates all thrust. When two holes are tapped simultaneously on a two-spindle setup, each tap prevents workpiece rotation by the torque exerted by the other tap. Without a leadscrew, a holddown would still be needed to prevent accidental lifting of the workpiece by the spindles.

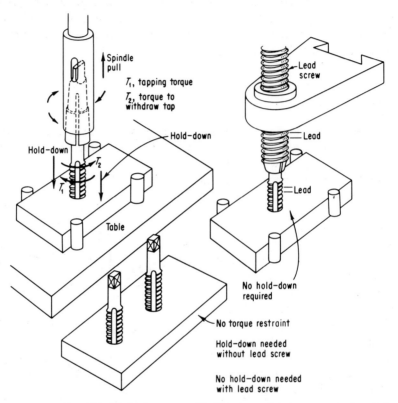

Fig. 2-26. Designing a workholder to resist torque and thrust in a tapping operation.

Figure 2-27 shows another instance where the cutting force holds the workpiece, in this case against the support plate of a broaching machine. The broach is guided in the support plate and to some extent in the workpiece. The broach in turn also holds the workpiece. The cutting tool and cutting force both contribute to the workholding operation.

Once the designer of a workholder has identified the possible direction and magnitude of the forces, he has two ways to restrain the workpiece to counteract these forces. One utilizes the strength and rigidity of some part of the workholder against which the workpiece rests or is forced by a clamp, screw, or wedge. The other way utilizes friction between the workpiece surface contacting under pressure a surface of the workholder.

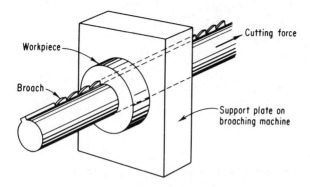

Fig. 2-27. Workholder for broaching operation.

Figure 2-28 shows a workpiece held in a vise. The horizontal component of the cutting force is absorbed by the solid jaw of the vise. The

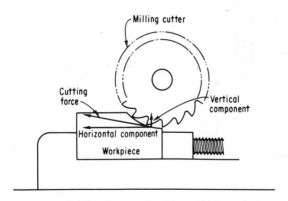

Fig. 2-28. Cutting force resisted by solid jaw of vise.

vertical component is resisted by friction between the workpiece and the jaws. Figure 2-29 shows the cutting force absorbed only by the friction between the jaws and the workpiece. Wherever possible, cutting forces should be opposed by the structure of the workholder and preferably by the strongest and most rigid parts of it. If necessary, a movable element may be used to absorb cutting forces, but only if it is properly designed for strength and rigidity.

110 Workholding Devices Chap. 2

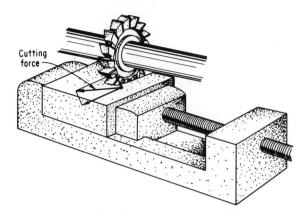

Fig. 2-29. Cutting force resisted only by friction.

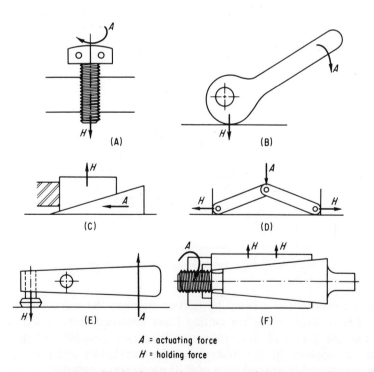

A = actuating force
H = holding force

Fig. 2-30. Mechanical methods of transmitting and multiplying force: (A) screw, (B) cam, (C) wedge, (D) toggle linkage, (E) lever, (F) combined screw and wedge.

Clamping Forces

Complete analysis of the tool forces in a proposed operation will disclose which of the twelve degrees of freedom must be restrained, and to what extent. Quite often tool forces are of such magnitude and direction that a workpiece may be dislodged or moved from its required location. If the locating elements of a fixture cannot assure adequate restraint, it may be necessary to clamp the workpiece against them.

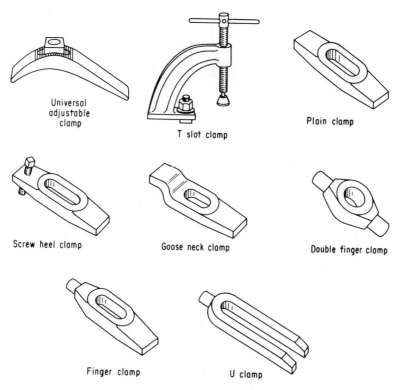

Fig. 2-31. Commercially available fixture components. (*Armstrong Bros. Co.*)

Clamping Forces. Clamps are used to hold a workpiece against a locator. Perhaps the most common application is the bench vise, where a movable jaw exerts pressure on a workpiece, thereby holding it in a precise location determined by a fixed jaw. The bench vise uses a screw to convert actuating force into holding force. Figure 2-30 shows a number of commonly used mechanical methods for transmitting and multiplying force.

The clamping forces applied against the workpiece must counteract the tool forces. Having accomplished this, further force is unnecessary and may be detrimental. The physical characteristics of the workpiece greatly influence clamping pressure. Hard vise jaws can crush a soft, fragile workpiece. The clamping pressure must hold, but not damage, deform, or impose too great a load on the workpiece.

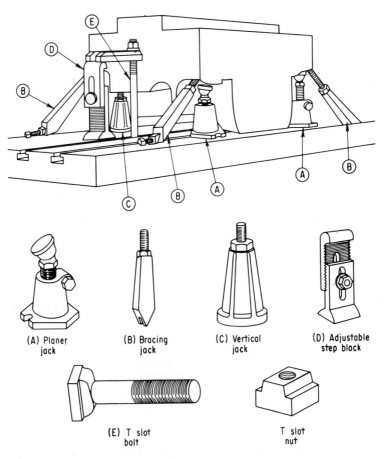

Fig. 2-32. Commercial components as used to hold a large workpiece. (*Armstrong Bros. Co.*)

The direction and magnitude of clamping pressure must be consistent with the purpose of the operation. An example is the boring of a precise round hole, with the workpiece clamped in a heavy vise. Excess clamping pressure can compress the workpiece. The bored hole may be perfect in size and roundness while the workpiece is compressed. The release of clamping pressure might permit the workpiece to return to its normal rather than compressed condition, and the hole might then be offsize and elongated. Another example of excess or misdirected clamping pressure may be found in a cutoff operation. If a workpiece clamped in a vise is to be cut between the jaws, in a direction parallel to the jaws, the removal of metal

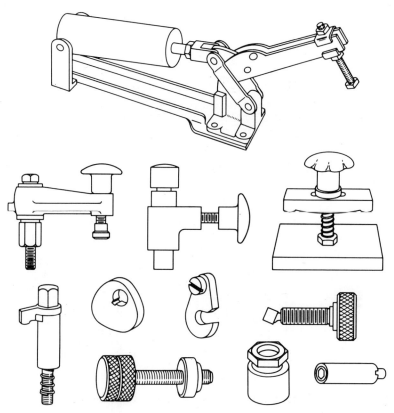

Fig. 2-33. Commercially available fixtures components. (*Universal Engrg. Co.*)

by the saw blade permits the remaining metal to flow into the cut area. This metal movement can effectively wedge and stop the saw blade.

Clamping pressure should not be directed toward a cutting operation, but should wherever possible be directed parallel to it. Clamping pressure should never be great enough to change any dimension of the workpiece.

Clamping Elements. In addition to complete clamps, a number of elements such as straps, T bolts, brackets, toggles, cams, and wedges are offered by many suppliers. These components are commercially available in an almost endless variety of shapes and sizes. They are comparatively low priced and as a rule have undergone a long history of testing and usage by industry. They are reliable devices and the designer should use them unless special requirements make it advisable to go to special equipment. Figures

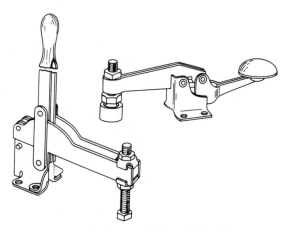

Fig. 2-34. Toggle clamps. (*Lapeer Mfg. Co.*)

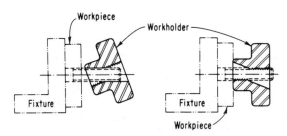

Fig. 2-35. Quick-acting screw used as workholder. (*Northwestern Tools, Inc.*)

2-31 through 2-35 show an array of workholder components. These are often used as furnished by the supplier; many attach temporarily and directly to the T slots of machine tables or are bolted to the machine. They can be used, therefore, without being assembled with a base into a self-contained workholding unit. They are used for single workpieces and temporary setups and also for large workpieces, where the cost of locating and holding is only a minor part of the total machining cost.

All these devices, when properly made and used, are self-locking and cannot be loosened by the work forces. Toggles and quick-acting screws (Fig. 2-34, 2-35) permit rapid approach to contact the workpiece, and then exert a large holding force during the final tightening stroke. These elements can be combined with bases and locating elements to form self-contained workholders which can be attached to a machine and, after use, be removed for storage. All these basic workholding elements multiply the actuating force into a larger workholding force. The holding force is maintained, even if no actuating force is subsequently exerted. There are exceptions where a holding force is exerted only as long as an actuating force is applied (direct-acting air or hydraulic-operated piston plungers). In such cases it may be wise to have the machine stop automatically if the actuating pressure drops below a predetermined value.

ELEMENTS AND TYPES OF FIXTURE DESIGN

Workholders may be generally divided into two groups depending on the shape and surface of the workpiece being held.

The first group includes workholders for workpieces that are bounded by flat and irregular surfaces upon which the holding forces act.

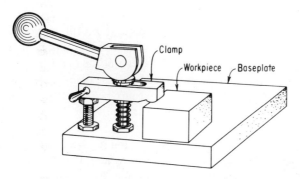

Fig. 2-36. Single-clamp workholder for rectangular workpiece.

The second group includes workholders for workpieces that are located and/or held in reference to round surfaces. The workpiece as a whole need not be round; it may be bounded by irregular and flat surfaces, provided the holding forces act on or in, and locating is accomplished on or in, the round surface. The meaning of "round" is here extended to include surfaces consisting of elements arranged in a circular form such as gears, splines, squares, or even hexagonal bar stock.

Workholding Principles for Flat or Irregular Surfaces. Figure 2-36 shows a simple workholder having a single clamp holding a rectangular workpiece. Both the location and the holding are in reference to flat surfaces. Figure 2-37 shows a milling fixture in which a workpiece is located by reference to a flat base, and the flat surfaces of a keyway. The holding force is directed against the irregular upper surface of the workpiece.

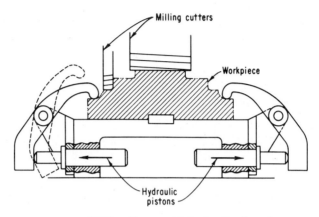

Fig. 2-37. Milling fixture with hydraulic clamping.

Figure 2-38 shows a round part clamped with a strap and T bolts while a cross hole is being drilled. The workpiece, the drill, or both may revolve. This example fits the definition of the second group if that location is accomplished on a round surface. However, it is listed with the first group because the roundness of the surface is not directly used for fixturing.

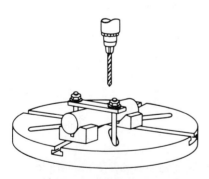

Fig. 2-38. Round workpiece held with strap clamp and T bolts.

Figure 2-39 shows another exception where a workpiece is apparently located by a round dowel. During the actual milling the dowel resists the thrust of the cutters in the direction indicated by the arrows, and does not hold all around the circumference of the hole. Location is accomplished by a flat shoulder on the dowel or arbor, a jack point, a screw clamp, and the unidirectional load on the dowel. This workholder is considered as belonging to the first group.

Workholding Principles for Round Surfaces. Most of the workholders in this group hold a workpiece for an operation that requires rotation. Examples may be found in turning, boring, grinding, or intermittent rotation for indexing. The workpiece may be held and located on an external or internal diameter for a milling operation requiring indexing, which means performing and repeating the operation in a certain angular relation. Other examples include drilling holes on a certain bolt circle, and grinding gear teeth.

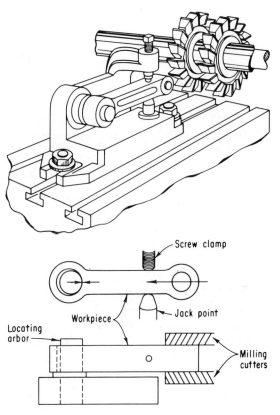

Fig. 2-39. Fixture for straddle milling a connecting rod.

Figure 2-40 shows an independent lathe chuck used to hold a round piece of tubing for a boring operation. Figure 2-41 shows a nut arbor. The workpiece is centered on the arbor within an accuracy that is the result of tolerances of the actual bore size and the size of arbor diameter. The resulting eccentricity could at the most be equal to the amount shown.

Means of Actuation. Workholders of both types may be actuated manually or with power. Force may be applied directly to the workpiece, or may be transmitted through levers, toggles, cams, wedges, or screws, to obtain secure and strong holding. In the transmission, the mechanical principles of force multiplication, the friction laws, and the strength and stresses of materials are all applicable. Figure 2-30 shows examples of force application and transfer.

Rigid vs. Elastic Workholding. Workholders may be either rigid or elastic. Since there are no absolutely rigid bodies and materials, rigid will here mean that the holding elements are preset to fixed position. Figure 2-42 shows a screw holding a workpiece as an example of rigid workholding. There is some elasticity in the screw and nut but it is not intentionally provided. Figure 2-43 shows a pressure-supported piston which bears down on and holds the workpiece as an example of elastic workholding.

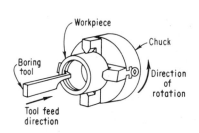

Fig. 2-40. Workpiece held in an independent-type chuck for a boring operation.

In Fig. 2-42 a shift sideways will bring the screw out of contact with a workpiece that is not of uniform thickness, causing complete loosening. The workpiece, however, very well resists any upward force against the screw, because the screw is a self-locking,

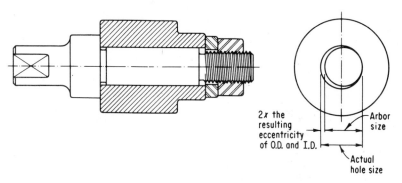

Fig. 2-41. Nut arbor.

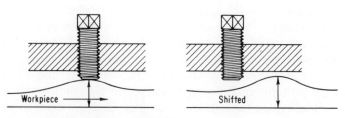

Fig. 2-42. Rigid workholding.

mechanically irreversible element. In Fig. 2-43 the piston clamp will continue to exert holding force in case of a shift sideways. An upward force is resisted only by hydraulic pressure exerted on the piston.

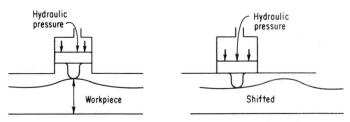

Fig. 2-43. Elastic workholding.

Figure 2-44 shows examples of elastic workholding, using self-contained hydraulically operated clamping cylinders. The workholder may be clamped by hydraulic pressure and released by spring pressure (Fig. 2-44A), or may be held by spring pressure and released by hydraulic pressure (Fig. 2-44B).

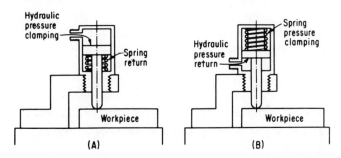

Fig. 2-44. Elastic workholding with self-contained hydraulically operated clamp cylinders.

Air pressure may be used but usually requires considerably larger piston areas to obtain a sufficient holding force. The size of the cylinders may be reduced by the use of force-multiplying mechanical elements. Before elastic workholding devices can be used safely, the work forces and their direction must be determined.

Figure 2-45 compares a hydraulic mandrel for elastic workholding with a mechanical mandrel for rigid workholding. Hydraulic pressure is produced by a screw-piston arrangement. Line A–A traces a path through a solid workpiece, a solid expansible shell, and an elastic layer of a hydraulic compound. A similar path A–A on the split-collet expanding mandrel passes only through consecutive layers of rigid metal. This does not mean that the elastic hydraulic mandrel is not as positive as the rigid mandrel. The hydraulic mandrel may be more positive and rigid than the mechanical mandrel. Its torsional stiffness may be greater and it may possess many other desirable features such as higher inherent precision and less vulnerability to dirt.

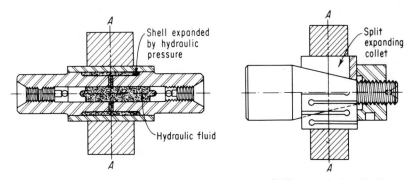

Fig. 2-45. Elastic and rigid mandrels.

Accuracy of Work Location. In workholders for flat or irregular workpieces, the holding of accurate tolerances on dimensions is determined primarily by the fixed locating and positioning means used in the workholder and, to a very small degree, by such movable holding elements as screws, cams, levers, and toggles. Figure 2-46 illustrates the relative importance of the fixed and moving elements in workpiece location accuracy. In workholders for round workpieces accuracy usually depends on concen-

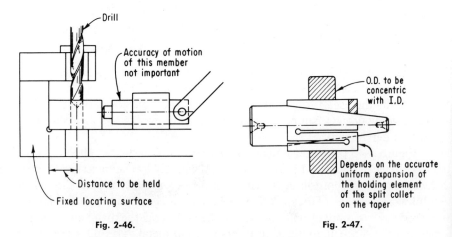

Fig. 2-46. Accuracy of work location.

Fig. 2-47. Concentricity as affected by collet accuracy.

tricity and is primarily determined by the holding elements such as the collets, the jaws, and the pins. Figure 2-47 illustrates a typical case where the concentricity between the workpiece internal diameter and the external diameter being ground depends very much on the care with which the expanding split collet and the tapered actuating cone are made.

Workholders for Flat or Irregular Workpieces

Pump Jigs. Pump jigs are so named because the handle often used with them for actuation resembles a pump handle. They are called jigs because tool guide bushings are frequently used in the movable plate of this workholder. The general appearance of pump jigs made by various manufac-

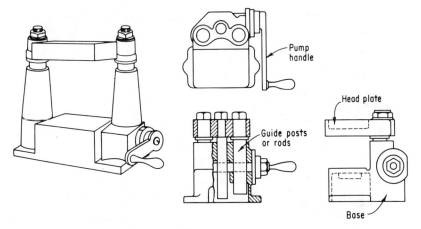

Fig. 2-48. Pump-type jigs.

turers shows much similarity. Some variation is found in the design of the actuating and locking mechanisms. The pump jig may be called a basic workholding element but of a more complete and complex nature.

Pump jigs (Fig. 2-48) have a fixed base which can be attached to the machine table and a movable head plate usually guided and moved by cylindrical rods. These rods and, with them, the head plate are moved up and down by a mechanism, actuated by the pump handle. This mechanism meets the basic requirement for a workholder: it cannot be loosened except by operating the handle. Drill bushings are located in the head plate according to the required hole pattern. Positioning and locating elements and supports may be attached to the fixture plate and to the head plate as required. The pump jig could easily be visualized as a modified quick-acting vise, put on end by making the solid jaw the base.

The workpiece is located on the stationary plate of the pump jig; the movable plate then locks the workpiece in place. With the addition of miscellaneous bushings, locating pins, and adjusting screws, the basic pump jig again becomes an efficient self-contained workholder. Sometimes need justifies the conversion of a manually operated pump jig to a power-operated unit. This is done by the use of air, hydraulic, or electric actuators.

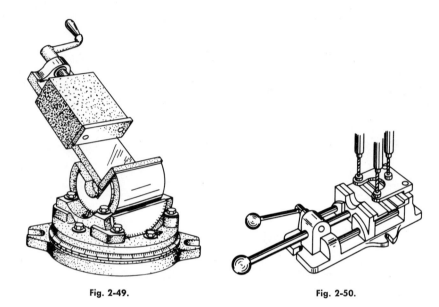

Fig. 2-49. Fig. 2-50.

Fig. 2-49. Universal-type vise.

Fig. 2-50. Drill jig with quick-acting vise as component. (*National Machine Tool Co.*)

Pump jigs are offered by various manufacturers in a variety of sizes and configurations. At an economic price they offer the tool designer a readily available building block around which he can readily develop efficient workholders for many requirements.

Vises. Vises are perhaps the most widely used and best-known workholders. All vises have in common one fixed and one movable jaw that grip or hold a workpiece between them. In all other respects, such as configuration of the jaws to grip particular workpieces, means of actuation, means of mounting, ability to position the workpiece, and sizes, an endless variety is commercially available. A vise is a good basic workholder element. By reworking the jaws or making special jaws, and adding such details as locating pins, bushings, and plates, vises can be easily converted into very efficient workholders.

Figure 2-3 shows a heavy plain vise for holding flat workpieces. With

Chap. 2 Workholding Devices

V-shaped grooves in the jaws, it can be used for locating round workpieces such as bars and tubing. Figure 2-4 shows a vise so equipped, that is, mounted on a rotary base permitting angular rotation, positioning, and locking as may be required. Figure 2-49 shows a universal vise that can position a workpiece at any angle relative to the cutting tool.

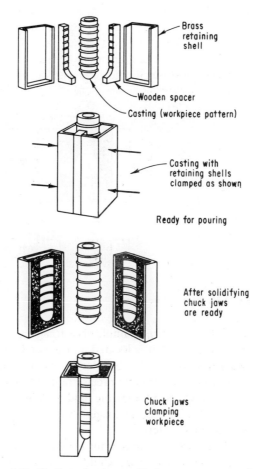

Fig. 2-51. Casting vise jaws with low-temperature alloy. (*Arro Sales Corp.*)

Figure 2-50 shows a workholder using a vise with a quick-acting jaw which is moved rapidly into contact with the workpiece. A handle-operated cam then locks the movable jaw to produce a large holding force. A special plate attached to the vise precisely positions three guide bushings in relation to the fixed (locator) jaw of the vise.

Special Vise Jaws. The usefulness of viselike workholders can be greatly extended by making special cast and molded jaws, which adapt it to the holding and locating of workpieces of rather complex shapes. Two methods are used to make such special jaws.

Figure 2-51 shows the method of casting jaws for holding a nipple by pouring a low-temperature alloy around the workpiece which is used as the pattern. Two wooden spacers are used to locate the pattern and to separate the cast jaw halves. The pattern is coated with a releasing agent for easy removal from the cast jaws. These jaws are then attached to the vise jaws.

In the second method (Fig. 2-52) a plastic material is used. The material has a metal filler to give it more wear resistance and strength. The plastic has a puttylike consistency, and is placed on each jaw of the vise.

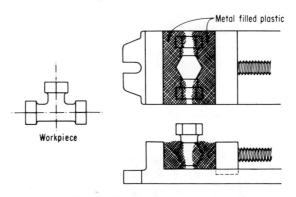

Fig. 2-52. Plastic vise jaws.

The workpiece, a T fitting in the example shown, again acts as the pattern. Coated with a releasing agent, it is located and pressed into the plastic material on the two sides of the vise by closing the vise jaws. The plastic hardens within two hours to form two precise half impressions of the workpiece which make an excellent locating and holding arrangement. The plastic material may also be used to locate and secure pins, bushings, and other details used in workholders.

V Blocks. V blocks are very important elements often used in building workholders. Figure 2-53 shows a workholder using a V block. Used alone, the V block restrains the round workpiece only in one direction and can be used only for very light work forces acting downward and centrally. The addition of straps and clamps can make the V block an effective workholder. Figure 2-54 shows another example of a workholder with a V block centering a round workpiece. Holding is accomplished with a quick-acting clamp. This workholder also has a tool guiding device (drill bushing), and therefore should be termed a jig.

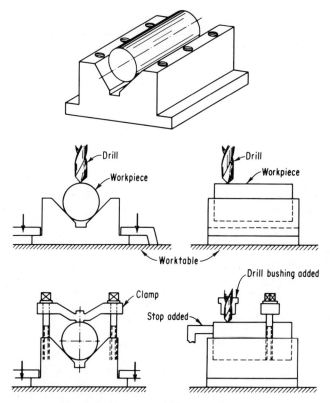

Fig. 2-53. V block as workholder component.

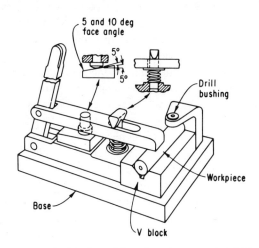

Fig. 2-54. V block, quick-acting clamp, and drill bushing combined to form a drill jig.

Angle Plates. Occasionally it is inconvenient to attach a workpiece directly to the table of a machine tool. Owing to the shape of the workpiece and the machining operation to be performed, turning the workpiece through an angle of 90° or some other angle may be desirable. Angle plates serve this purpose for light-duty work. The workpiece may be clamped to the angle plate by simple C clamps. For heavy work, large angle plates with T slots similar to the ones in the machine table are used. All workholding elements can be used in conjunction with angle plates. As more elements are interposed between the point where the cutting forces are exerted and the points where they are finally absorbed, and the larger the distance between these points becomes, the less rigid will be the whole workholding setup and the harder it will be to take heavy cuts or obtain good surface finishes and accuracy.

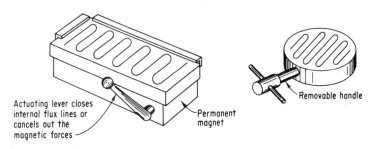

Fig. 2-55. Magnetic chucks.

Nonmechanical Chucking Methods. It is sometimes not practical to hold a workpiece by direct clamping pressure because of possible distortion, or perhaps because of the size of the workpiece. Magnetic, vacuum, and electrostatic workholders may be of value in such cases.

Magnetic chucks are available in a variety of shapes. They can hold only ferrous workpieces unless intermediate mechanical workholders permit the holding of workpieces made of nonmagnetic material. Magnetic chucks are suitable for light machining operations such as grinding. More strongly magnetic materials and better utilization of magnetic force permit their use also for heavier operations such as light milling and turning. Magnetic chucks can be operated by permanent magnets or by electromagnets powered by direct current. The gripping power obtainable depends on the strength of the magnets and the amount of magnetic flux that can be directed through the workpiece. Figures 2-55 and 2-56 illustrate various magnetic chucks.

A magnetic chuck is fast-acting and, by holding a large surface of the workpiece, causes a minimum of distortion. Magnetic chucks are available in rectangular shape, in circular shape as a rotary chuck, and also as a V

Chap. 2 Workholding Devices

block. Magnetic chucks impart some residual magnetism to workpieces. This must be removed by demagnetizing if it would interfere with proper functioning of the workpiece.

Vacuum Chucking. Quite often workpieces of nonmagnetic materials and/or of special shapes and dimensions must be held flat and securely without any mechanical clamping. An example might be a large flat plate. In such cases, vacuum chucking may be the only practical holding method.

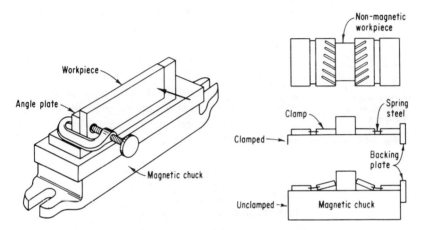

Fig. 2-56. Magnetic chuck and angle plate used as workholder.

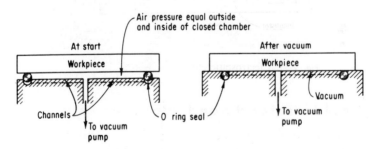

Fig. 2-57. Vacuum-chucking principle.

The basic principle of vaccum chucking is very simple. On each square inch of a surface a pressure of approximately 14.7 lbs is exerted by the atmosphere. This represents approximately $144 \times 14.7 = 2116.8$ lb or close to 1 ton per square foot. Part of this pressure is utilized by creating a vacuum in a closed chamber made up of the locating surface of the workpiece and the mating surface of the workholder. At first, the pressures on the outside and inside of the chamber are equal. As a vacuum is produced inside the chamber, the outside pressure holds the workpiece against the locating surface (Fig. 2-57). Figure 2-58 shows a typical vacuum chuck. An O ring seal laid in a groove around the chucking area creates the closed chamber to be evacuated. Orifices and a grid of connecting small channels in the chucking surface assist in the speedy creation of the needed vacuum.

Electrostatic Chucks. The attraction of opposite electrically charged parts may be used to hold flat or flatsided workpieces that cannot be magnetized. Electrostatic chucks (Fig. 2-59) can hold any electrically conductive material. Glass, ceramics, and plastics also may be held by flash metal plating on one flat side to provide a suitable electrical contact.

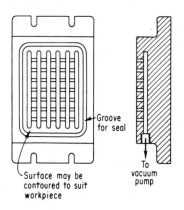

Fig. 2-58. Vacuum chuck.

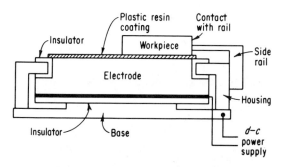

Fig. 2-59. Electrostatic chuck.

Workholders for Round Workpieces

Workholders of this type are used to hold, drive, and locate workpieces on their circular or round surfaces. Figure 2-60 shows in schematic form workpieces with round surfaces as they may be held when located and clamped to be concentric with their centerlines. The operation of a workholder is often called *chucking*. Chucking is the attachment of a workpiece to a machine tool by means of a workholder (chuck) in a definite relationship and with a firmness required to perform the desired machining operations.

Chucking operations can be distinguished by the location at which they are performed on the workpiece. Common methods shown in Fig. 2-60 include external chucking (*a*), internal chucking (*b*), endwise chucking

(c, d, k, l), holding and/or driving on centers (e, f, g, h), and combinations of these methods (j). Although different workholders may be used for the various methods, the differences are primarily in name and appearance rather than in function or mechanical principles employed.

Motion is imparted to the workpiece either by friction or by positive means. In chucking, the jaw elements of the workholder bear on but do not positively engage the workpiece. The friction between the jaws and the workpiece is used to rotate or drive the workpiece. Positive means for

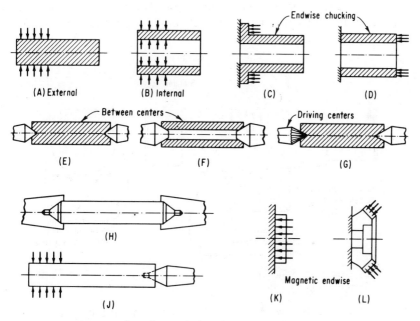

Fig. 2-60. Location and clamping of round workpieces.

driving the workpiece can be provided by making use of the structure of the workpiece. Cutouts, keyways, gear teeth, or splines on a workpiece can be mated with matching elements on the workholder to give a positive drive. One distinguishing element can be found should failure occur: in a positive drive something breaks, while in a friction drive something slips.

Nomenclature. A *chuck* is a workholder generally used for gripping the outside or end of a workpiece, and is usually attached to a machine tool spindle. An *arbor* is a workholder generally used for internal chucking, holding, or gripping, and is usually attached to a machine tool spindle. A *mandrel* is a workholder that, like an arbor, is used for internal chucking, holding, or gripping. A mandrel is not generally as precise as an arbor, and is usually held between centers instead of being attached to the spindle.

Range is the amount of variation in workpiece diameter that can be accommodated by workholder expansion and contraction. *Interference* is the amount by which the chucking diameter in its holding condition (without the workpiece) differs from the diameter of the workpiece to be held. *Chucking area* is the area of a workholder that can be used for holding. The holding elements of workholders are called by various names depending on the type of workholder. They are the jaws on some chucks; the fingers or prongs on some collets; the expanding wedges or inserts on some arbors; the expanding and/or contracting cylinders or sleeves on some special arbors and mandrels, and so on. These elements of the workholders are made to deflect by the actuating mechanism, which transmits the force

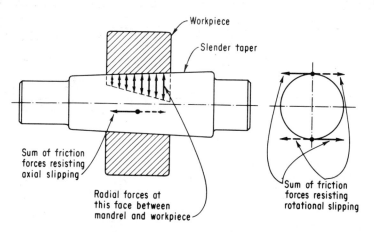

Fig. 2-61. Solid mandrel.

needed to operate the workholder. This force in turn may be applied manually, or by power (pneumatic, hydraulic, electric). The holding elements are backed up by the actuating elements; these again are backed up by the operating mechanisms and forces.

Solid Arbors and Mandrels. The solid, slender taper mandrel (Fig. 2-61) is about the simplest possible workholder for round workpieces. Its main characteristic is a slightly tapered chucking surface with a 0.004 to 0.006-in. taper per foot. The workpiece diameter must be smaller than the largest diameter of the mandrel. The workpiece is forcibly pushed endwise onto the mandrel. This produces a gripping force all around the hole in the workpiece, decreasing axially in relation to the interference produced between the outer diameter of the mandrel and the inner diameter of the workpiece. The driving torque that can be transmitted depends on the radial gripping forces and the tangential friction forces produced. The resistance against axial slipping depends on the axial friction forces produced.

It is not always easy to obtain the same driving power or to position the workpiece to a definite stop when trying to control the resulting interference between workpiece and mandrel. Pressing the workpiece on the mandrel requires an arbor press, is slow, and may damage the finish of the workpiece bore and score the mandrel. Workpieces with accurate round and straight bores are held with great accuracy on this simple mandrel. If the bore is not round and straight, the workpiece and mandrel will mutually distort under the forces used to press on the workpiece.

Straight Mandrel. This mandrel resembles the previous mandrel but has a straight (untapered) chucking area. The outer diameter of this mandrel is made larger than the bore of the workpiece by an amount called the *interference,* to produce the required pressfit. The amount of permissible interference is determined by the wall thickness, the diameter, and the material of the workpiece. It must be controlled in order not to exceed the elastic limits of the workpiece and the mandrel. Exceeding such limits could produce a change in bore diameter, especially in materials of low strength and low modulus of elasticity.

The obtained driving torque together with the axial resistance to slipping depends on the interference and is not easy to control. Possible mandrel wear and damage to the workpiece offset the advantages of the simple low-cost design. The range of interference-fit mandrels is very limited.

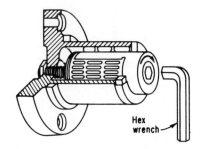

Fig. 2-62. Roll-Lock type expanding solid mandrel. (*Scully Jones & Co.*)

Combination of Slender Taper and Straight Mandrel. This mandrel has a short tapered chucking length followed by a straight length. The straight area fits snugly without interference into and helps prealign the bore of the workpiece. The tapered surface provides the driving area. To obtain reliable results for torque, accuracy, and axial location, hole tolerances must be carefully held. Like the taper and straight mandrels, this type also requires a pressfit condition between the workpiece and the chucking surface. The larger diameter of the mandrel is forced into the smaller bore of the workpiece. This axial pressing often damages the workpiece and workholder.

Solid Expanding Mandrels. The hydraulic mandrel (Fig. 2-45), and the roll-lock type of arbor (Fig. 2-62), produce a shrinkfit by expanding tubular shells which are not split lengthwise. The internal actuating mechanism expands the shell into the workpiece bore, and the amount of pressfit produced can be controlled by stops limiting the amount of expansion. The hydraulic type expands by hydraulic pressure; the roll-lock type expands by the gradual rolling and wedging action of straight rollers between the tapered inner diameter of the shell and a tapered plug which is turned by a wrench.

Mandrels having a chucking area that is not weakened by axial slots generally have only small ranges. The solid mandrel will permit a resulting interference of 0.001 to 0.002 in. per inch of diameter. The hydraulic and roll-lock workholders expand from 0.002 to 0.003 in. per inch of diameter. The obtainable accuracy is approximately 10 per cent to 20 per cent of the range of the workholder. A mandrel expanding 0.002 in. may then hold a round piece within a 0.0002 to 0.0004-in. total indicator reading or 0.0001 to 0.0002-in. eccentricity.

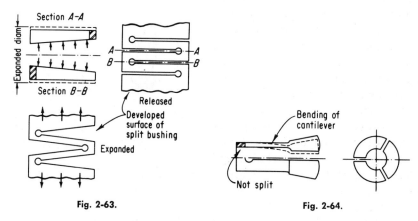

Fig. 2-63. Expansion of a split bushing.

Fig. 2-64. Split collet.

Split-Collet and Bushing Type Workholders. The solid type expanding mandrel has a very small range. To increase the range and to hold workpieces with larger diameter variations, split collets and bushings are used. These are basically slotted shells of various shapes. The slots permit greater flexibility to increase for internal chucking and to decrease for external chucking. The more flexibility that is provided, the greater the range will be. Figure 2-63 shows how the shell is split, and acts like a spring. Figure 2-64 shows a very popular collet in which the range is obtained by cantilever deflection. The cantilevers are produced by splitting the collet from one end only and leaving a solid ring on the other end. Figure 2-65 shows a high-range collet where large flexibility is obtained by imbedding loose individual collet jaws in a suitable rubber compound. Many varieties of these chucking elements (Figs. 2-63, 2-64, and 2-65) are used to form the

Chap. 2 Workholding Devices   133

basis of many workholders. They are used for internal and external chucking, and are actuated in most cases by cones or tapers acting on corresponding surfaces of the chucking elements. Some workholders use small tapers which are self-locking in that a force must be applied to disengage the mating tapers. A taper of less than 10 degrees is usually self-locking (Fig.

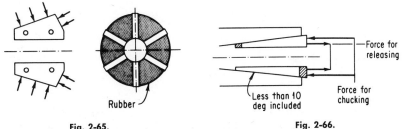

Fig. 2-65.

Fig. 2-66.

Fig. 2-65. High-range type collet.

Fig. 2-66. Collet with self-locking taper.

2-66). Above 12 to 16 degrees, the taper becomes self-releasing and requires force only to engage the mating tapers (Fig. 2-67). Split collets and bushings, made with care, give satisfactory results for most applications. They are hard to maintain where extreme accuracies are required, and are vulnerable to the entry of dirt and to wear on the sliding surfaces.

Axial Location. Axial location of a workpiece is effected by a moving collet. Figure 2-68A shows a collet chuck, the diameter of which contracts upon axial motion. Figure 2-68B shows a collet mandrel which expands upon axial motion. At first, the clearance between the workpiece diameter and the chucking diameter is closed; upon sufficient axial motion a shrink-

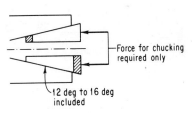

Fig. 2-67. Collet with self-releasing taper.

fit is created between the workholder and the workpiece. As soon as contact is made at the chucking surface the workpiece has a tendency to move together with the collet relative to the workpiece holder body. This movement produces an axial shift of the workpiece and prevents precise axial location. Figure 2-69 shows a method of obtaining precise location by providing a stop surface. The collet moves the workpiece toward the stop and pushes it firmly against the stop. Some slipping occurs between the collet and workpiece during the chucking operation. The slippage tends to reduce the driving power of the workholder by absorbing part of the actuating force; the mechanical efficiency of the chucking operation is thus lowered. To obtain accuracy, the stop surfaces of the workholder and faces of the workpiece must be square with the centerline. Lack of squareness may cause distortion of the arbor by one-sided loading against the stop surfaces (Fig. 2-70).

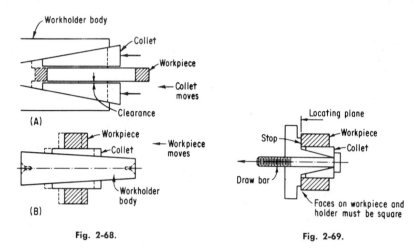

Fig. 2-68. Axial location of workpieces as affected by collets.

Fig. 2-69. Collet with stop plate to assure correct axial location.

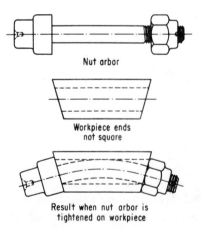

Fig. 2-70. Distortion caused by lack of squareness.

Figure 2-71 shows a collet arrangement designed to eliminate axial shift by not moving the collet. An intermediate bushing is used to interact with the tapered surfaces of the collet. An additional element with additional mating surfaces between the two cylinders is thus introduced. The need to produce additional surfaces to a high dimensional and relational accuracy will lower the resultant final accuracy of the workholder. The more mating surfaces required in any workholder, the lower the accuracy tends to be.

Some frequently used workholders have sliding elements or inserts actuated by movement on inclined planes (Fig. 2-72 A, B). They are similar to a lathe chuck in that their jaws or driving keys are held, guided, and moved in the workholder body. The actuating mechanism consists of a bar with inclined wedge surfaces. The

Chap. 2 Workholding Devices

body, jaws, and bar are three different elements working together. The accuracy depends on the excellence of workmanship and careful use. A desirable simplification is the combination of the actuating mechanism with the workholder body (Fig. 2-72B). These workholders are usually built as arbors and mandrels and have good range and gripping power. They are very useful tools for even the largest tubular workpieces.

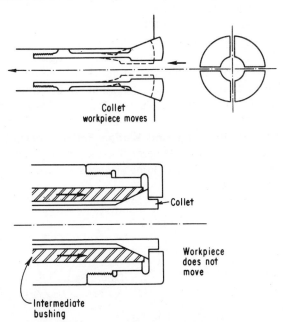

Fig. 2-71. Collet with intermediate bushing to eliminate axial shift.

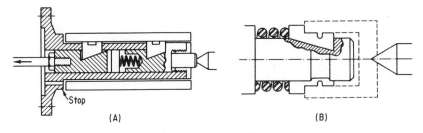

Fig. 2-72. Collets for internal chucking.

Lathe Chucks. Lathe chucks consist of a body with inserted workholding jaws that slide radially in slots and are actuated by various mechanisms, such as screws, scrolls, levers, and cams, alone and in a variety of combinations. The number of jaws varies. Chucks in which all the jaws move together are self-centering and are used primarily for round work. Two-jaw chucks operate somewhat like a vise and may be used for round and for irregular-shaped workpieces by the use of suitably shaped jaws or jaw inserts.

The accuracy of a new chuck deteriorates with usage owing to wear, dirt, and deformation caused by excessive tightening. Independent-jaw chucks permit each jaw to move independently to chuck irregular-shaped workpieces or to center a round workpiece.

The jaws of most lathe chucks can be reversed to switch from external to internal chucking. Jaws may be adapted to fit workpiece shapes that are not round. The means of attaching a lathe chuck to different machine tools have been well standardized, so that chucks made by different manufacturers can be easily interchanged. Figures 2-5 and 2-40 illustrate lathe chucks.

Self-Actuating Wedge Cam and Wedge Roller Workholders. These workholders are tightened by the tangential work forces of the cutting tools. They actuate themselves once the holding elements are brought even slightly in contact with the workpiece.

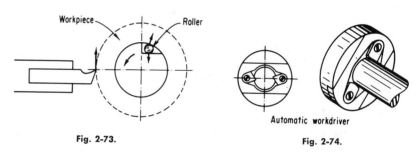

Fig. 2-73.   Fig. 2-74.

Fig. 2-73. Roller-clutch type arbor.
Fig. 2-74. Wedge-cam type chuck.

Figure 2-73 shows an arbor using the principle of the roller clutch. One or more rollers are nested in cutouts in the body of the workholder, retained by wire clips. Turning the workpiece in the direction shown relative to the workholder wedges the rolls in between the workpiece and the workholder. An increase in the applied cutting force increases the wedging action almost proportionally.

The wedge-cam chuck (Fig. 2-74), has two inwardly spring-loaded cams. The jaws are lifted by a ring (not shown) and then released till they touch the workpiece. The cam surfaces are usually serrated to obtain a better grip on the workpiece. The tangential cutting forces wedge the cam jaws tightly against the workpiece.

Chap. 2 Workholding Devices

General Considerations in Workholder Design and Selection

Workholder design and selection start, in general, with an analysis of the workpiece and the manufacturing operation to be performed. The shape, material, and tolerances of the workpiece for the manufacturing operation are specified by the product engineer. Certain functional and appearance requirements must be met. The manufacturing engineer specifies the most economical steps in the manufacturing process.

Physical Characteristics. The first consideration is the condition of the workpiece to be held by the proposed workholder for the next operation to be performed. This includes the physical characteristics of the workpiece, i.e., whether it is round, irregular, large, heavy, of weak or strong sections,

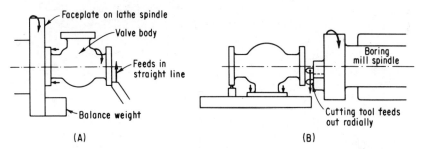

Fig. 2-75. Machining a valve body by either (*A*) workpiece rotation or (*B*) tool rotation.

and the like, and also the operation to be performed. This will indicate whether the workpiece has to remain stationary or be moved along a definite path relative to the cutting tool. Different machine tools have means to provide the needed motions for the workpiece and the cutting tools. Most of these motions are straight-line or rotary, or combinations of both. There are operations that to all appearance seem to be producing and requiring an irregular path of the cutting tool on the workpiece. This seemingly irregular path is often the resultant of the combination of several straight-line and rotary motions. Except for speed consideration and the degree of simplicity obtainable, only the relative motion of the cutting tool to the workpiece is of importance. For instance, the turning of flanges on a valve body or a flanged T pipe fitting could be performed with either the workpiece or the tool revolving (Fig. 2-75).

Whether the workpiece or the cutting tool moves in a straight line, revolves or moves in some combination of both, design requires careful coordination of workholder to the workpiece, and the workholder to the machine tool. Operations in which the workpiece revolves require great care in the attachment of the workholder to the machine tool and the means of actuation of the workholder. Unbalanced masses in the workholder and the workpiece must be minimized by proper balancing. This is particularly true in high-speed applications such as turning with tungsten carbide, diamond, and ceramic cutting tools.

Cutting Forces. On all operations, the magnitude and the direction of the forces produced by the material-removing operation determine the necessary holding forces. Cutting forces must be held within limits, so that the part itself cannot be distorted to an amount that would affect the required accuracy. Rigidity and strength of the workpiece limit the appli-

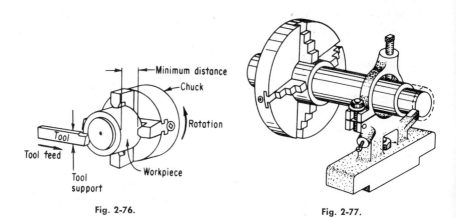

Fig. 2-76. Fig. 2-77.

Fig. 2-76. Minimizing cutting force by applying holding force as near as possible to point of tool application.

Fig. 2-77. Steady rest used to support workpiece in area of cutting-force application.

cable holding forces and the speed and amount of metal removal per unit of time. A thin-walled part may not be able to sustain heavy cutting and holding forces without distortion or damage.

Machine Selection. Workpiece weight and size influence what type and size of machine tool can or should be used for a particular operation. The combined weight of the workholder and the part must be carefully matched to the capacity of the ways, bed, table, and spindles of the machine tools. Excessive weight may cause distortion in the machine tool and produce inaccurate work.

Mounting of the Workholder to the Machine Tool. The mounting or attachment of the workholder or the workpiece to the machine tool should be so arranged that the forces produced in the material-removing operation are absorbed by the strongest and most rigid parts of the machine tool.

The cutting forces should tend to hold the workholder down against the bed of the machine rather than lift it away. Projection between the point at which the cutting force is applied and the nearest support should be minimized. Cutting forces that act parallel to the bed, table, and face plate should be applied as close to it as possible. Cutting forces should never be permitted the advantage of a large lever arm which could increase the tendency to loosen or pry away the workpiece and the workholder from their attachment. This is in contrast to the holding forces, where the effect of a large lever arm or mechanical advantage is always desirable.

Figure 2-76 shows a lathe chuck and a short workpiece to be turned. The cut is relatively close to the supporting spindle bearing and no difficulty is to be expected. But increasing the workpiece length will give the cutting tool larger leverage. Higher side thrust will consequently be pro-

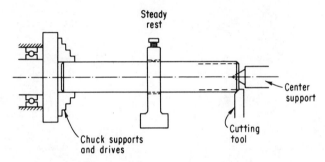

Fig. 2-78. Steady rest and center used to support workpiece in area of cutting-force application.

duced and may cause difficulties in the spindle bearing. The workpiece will deflect considerably more if the same size of chip is removed as on the short piece. The result may be inefficient material removal, low accuracy, and poor tool life. To reduce the effect of the large lever arm a center support can be used. If this does not improve performance, then a steady rest must be used. The center and the steady rest may be called workholders. They hold, locate, and support. Figure 2-77 shows a turning operation with the workpiece held by a lathe chuck. The steady rest supports the bar to be turned while three adjustable shoes center and support the bar. This support absorbs most of the bending forces. Figure 2-78 shows a turning operation with a steady rest and center support. The cutting tool is directed against the workpiece near the center support. Cutting forces should be absorbed as near as possible to the point where the force is created. The supports for the workpiece should be placed near the cutting forces. The cutting forces should be applied as close as possible to the bed, table, face plate, and the spindle bearings.

Standard Mounting Methods. Quite often only standard accessories such as clamps, straps, T slot bolts, T slot nuts, and jacks are needed to hold a workpiece. This is especially the case where only a few pieces have to be produced and where economic considerations do not justify more elaborate workholders. Most machine tools have provisions to receive such equipment. The beds and tables of machine tools such as drill presses, boring mills, and jig-borers have T slot milled into their work tables. The spindles of machine tools such as lathes and grinders have spindles to which, directly or by means of suitable adapters, the various types of workholders, chucks, arbors, and collets may be attached. There are standards for the sizes and the spacing of T slots of tables, beds, and other equip-

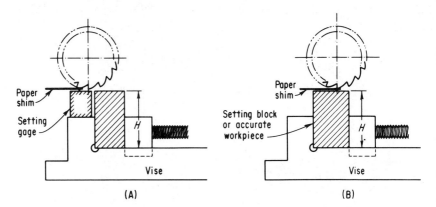

Fig. 2-79. Use of a gage block in setting up a milling operation.

ment to which workholders are to be attached. Standards for spindles are also established. Adherence to standards can result in economic interchangeability and multiple sources of tooling. Tool costs will be lower because the supplier can produce standard components in larger quantities at lower cost. Figure 2-32 shows a large workpiece held by an array of simple standard workholding devices.

Relationship Between Workholder and Cutting Tool. The direction and magnitude of the forces created at the cutting tool-workpiece area must be known. This knowledge should be used to minimize the size of force moments (force times distance of force application) by reducing the moment arm. Excessive projection should be avoided. Additional supports should be provided where needed.

The relative motions between the workpiece and the cutting tool may change the tool geometry during a cutting cycle. Rake and clearance angles may change from a selected optimum condition to a bad one.

Every material-removing operation has as its final objective the removal

Chap. 2 Workholding Devices   141

of a certain amount of material per unit of time, to certain specifications of depth, thickness, diameter, contour, and other related factors. These quantities can be obtained by control of machine motions with stops, gages, or tape controls. Workholders may be equipped with stop surfaces such as the top of a drill bushing against which a drill stop abuts. Setting gages may be placed between a milling cutter and a reference plane on a workholder to obtain the right thickness of a workpiece after milling (Fig. 2-79). Swing stops on lathe fixtures and arbors may be used to locate workpieces for removing stock evenly from both sides (Fig. 2-80).

It is necessary to check for and control interference between any cutting edge of the cutting tool and any part of the workholder during any possible relative position of workholder and cutting tool. It is advisable to simultaneously check for, minimize, and control excessive nonproductive approaches of cutting tools to the workpiece. Planning should be done to avoid "cutting air."

Space must be provided to remove and load workpieces easily without danger to the operator or damage to the workpiece and equipment. Space is also required for the insertion and removal of the cutting tools. This includes space for the application of any wrenches, keys, and other tools used for change of cutting tools. Such change should be possible without removing the workpiece.

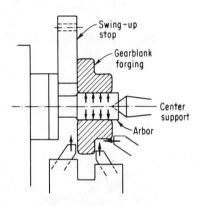

Fig. 2-80. Lathe fixture with swing stop.

Tool Positioning. Tool positioning refers specifically to locating the tool in respect to the work, or vice versa.

Preparatory to setting up the workpiece, the blueprint and/or workpiece are studied to determine primary and secondary locating points or surfaces. Once these are determined, it is then necessary to visualize how these points or surfaces may be accurately located in relation to the locating means.

Relationship to Locating Means. The locating means are the alignment or gaging surfaces of any angle, plate, bar, V block, vise, or the like that is secured to or is a part of the work table or fixture for properly positioning the work relative to the tool (Fig. 2-81). A T slot may be considered a locating means but, generally, such slots are only accurate enough for rough machining.

The most practical procedure for establishing the relationship of the tool to the work will be governed by the type and size of the machine, type and size of work, production rate, and specified dimensions and tolerances of cut. The end results of correct tool positioning are proper depth and location of the finished cut. Regardless of the type of machine involved, there are several different techniques for locating the work in relation to the tool, the exact technique being determined by the specific job requirement. Mass production may dictate great expenditure of money to minimize the time required for locating the part, compared with less expenditure and more time for location of a single piece part.

Setting Blocks. Gage or setup blocks are common means of reference for cutter setting (Fig. 2-82). In many cases the reference may be a designated surface on a locator (locating means). In its correct cutter position, the cutter should clear the setting surface by at least $\frac{1}{32}$ inch. Usual shop practice standardizes on one feeler thickness to avoid use of the wrong size on any particular operation. The thickness of the feeler to be used should be stamped on the fixture base near the setup block.

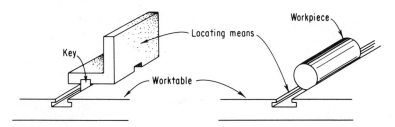

Fig. 2-81. Positioning a workpiece relative to locating means.

Keys. Keys are used under the base of an angle plate, vise, or workholding fixture. They provide an easy and accurate method of aligning a workholding fixture to a T slot to the same degree of accuracy as the T slot itself. Before using the T slots as a basic locating reference, their accuracy should be established in relation to table or cutter movement. Removable keys are used extensively, especially in vises with slots at right angles to each other for lengthwise or crosswise mounting. Both the key and the T slots should be periodically inspected for wear to assure proper dimensions and accuracy between the key and locating means.

Optics. Optical methods are used extensively for gaging the accuracy of tool position in relation to the work and also to the locating means. The multiplicity of optical tooling equipment design makes possible broad applications in tool-to-work locations. One typical application is the establishment of an exact drill center location. An optical instrument is inserted in the chuck. Through high magnification, the eyepiece and cross hairs are used in determining the exact center. The work is clamped securely to the

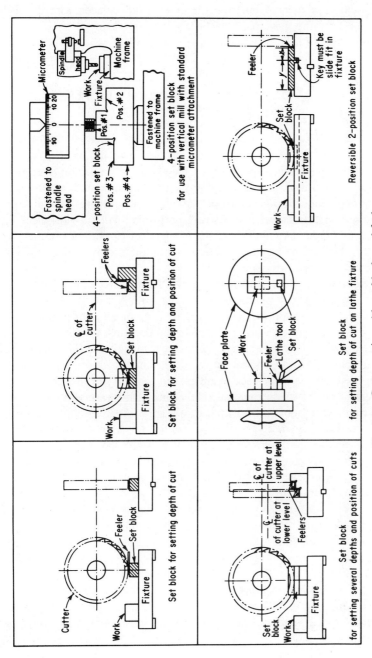

Fig. 2-82. Cutter setting with set blocks and feelers.

locating means and rechecked for proper tool center alignment. The optical instrument is then removed from the chuck and replaced with the drill for drilling. The same principle may be applied for gaging the indexing accuracy of a rotary fixture or index plate onto which work is positioned for machining.

EVOLUTION OF WORKHOLDERS

Direction of Forces. The application of any metal removal process to a specific workpiece will result in a distinctive combination of forces. It is possible to list the many processes and anticipate forces to some extent. The torque of a drilling operation or the thrust of a shaping operation can readily be visualized. It is also quite possible to anticipate or visualize the direction of the cutting forces in circumstances where several different machining operations are successively performed while a workpiece is in a single workholder. An experienced designer can carefully examine a workpiece or blueprint and quickly form a mental picture of all necessary fixturing elements to be considered.

Magnitude of Forces. Although the various forces and their direction can be foreseen, the magnitudes of the forces remain to be determined. The workholder must support the workpiece in a precise location while it is subjected to the cutting forces. The workholder must therefore be designed to withstand forces of specific direction and magnitude. The magnitude of any cutting force is dependent on many variables such as the hardness and homogeneity of the workpiece material, the geometry of the workpiece, tool geometry, and the metal removal rate. Some of the variables change during the cutting cycle (tool and workpiece geometry), therefore the magnitude of the force will vary. It is possible to predict for fixture design purposes a maximum value that any magnitude will probably not exceed.

The designer can use two methods to determine whether a workholder will withstand a specific force. The first is the cut-and-try method, the second is an analytical approach:

(1) The *cut-and-try method* involves actually building the workholder and trying out the proposed operation. If failure occurs immediately, the type and point of failure are noted; the workholder is redesigned accordingly, and another trial is made. If failure is progressive, the workholder can be reinforced as required and placed in service.

(2) The *analytical approach* involves determining the force required to perform an operation. The shear strength of the workpiece material may be obtained from a handbook or table. The proposed operation may require the removal of metal by shearing a given cross-sectional area. Ignoring a possible breakout factor, the cross-sectional area multiplied by the shear strength per square inch will give the required force with reasonable ac-

curacy. For cases where a more precise determination of the cutting forces is required, refer to the detailed analytical determination of "Cutting Forces" discussed in Chapter 1. The force must be absorbed by some detail of the workholder. This detail must, if also loaded in shear, be of sufficient cross-sectional area to withstand the force. Assuming the detail is fastened to the workholder by bolts and/or dowels, the combined cross-sectional area of the bolts and dowels must be sufficient to withstand the force.

In essence, the analytical approach requires determining the direction and probable maximum magnitude of each force, followed by a step-by-step determination that each part of the workholder can withstand the force. In practice the analytical approach is usually applied mentally, without recourse to reference data or mathematical computations. The designer mentally computes the cross-sectional area of the workpiece to which the cutting force is applied, considers the workpiece material, and often finds that an available commercial workholder such as a machine vise far exceeds the strength requirements. In many cases where special workholders are required the designer finds that preliminary design based on location factors and convenience of loading is functionally adequate. Often the smallest commercial components that are available far exceed strength requirements.

Combined Methods. The cut-and-try approach in itself is neither efficient nor safe. The resulting workholder may be either much stronger than required or may be on the brink of failure with potential danger to personnel. At best, the finished workholder is an example of material and manpower wasted in obtaining more strength than that required.

The analytical approach in itself is not practical. Complete determination of the magnitude and direction of all forces coupled with virgin design of all fixture components would in most cases be economically impossible. A workholder so designed would probably have fasteners of different diameter at each attachment point to match the anticipated load, and therefore would be impractical from a maintenance standpoint.

Workholder selection is usually a compromise of the two methods (cut-and-try and analytical). The designer is first governed by established standards. If a threaded fastener is required, he will specify a standard bolt of more than adequate strength. If a workholder is to be attached to a machine table, the T slots in the table establish the mounting bolt size. If a workpiece can be held in a vise already mounted on the machine, the designer will specify its use, and not be concerned by the fact that the vise can absorb the applied force many times over.

The designer will further be governed by a pyramid of knowledge and the established practices of his company. If an existing workholder can with slight modification be satisfactorily adapted to a new workpiece, the design and acquisition of a new workholder might not be economical.

The analytical approach must, however, always predominate. Although standard bolts may be used, the designer must select the size and quantity. He may for structural reasons select the larger of two standard machine vises available. He must be certain that modification will not prejudice the structural integrity of a fixture.

Cut-and-try aspects will always be present. A designer might specify conventional attachment of a workholder to a machine table, knowing that an undetermined thickness of shim stock must be interposed to raise the workpiece to a convenient height. A standard machine vise may be specified with the notation that special jaws will be provided should try-out prove their necessity.

Evolution of Drill Jigs and Fixtures

In the removal of material by drilling, two primary forces are directed against the workpiece. A *thrust force* is directed along the axis of the drill and is focused on the workpiece by the drill point. *Torque* is directed against the workpiece by the cutting edges of the drill as they revolve about the drill axis.

Thrust. Figure 1-48 shows the deformation of material under the chisel edge of a twist drill. The point of the drill removes metal only by extruding it into the path of the cutting edges. This displacement is caused by and continues only as long as the thrust force exists. A twist drill cannot pull itself into a solid workpiece. If the thrust ceases, the drill will not move axially and continued revolution will only generate heat, which can cause failure of the cutting edges.

Torque. With a constant infeed rate, each cutting edge of a drill will shear a continuous chip of constant thickness. Each cutting edge acts as a single-point tool. Figure 1-47 shows conventional chip formation by a single-point tool. The cutting edges of a drill impart torque in the direction of drill rotation. The edges will either shear a secured workpiece or rotate a free workpiece.

Secondary Forces. As the drill breaks through the workpiece, a momentary force may occur in the direction opposite to the initial thrust. This is characteristic of a high infeed rate. As the drill breaks through, the shearing action is incomplete and the remaining workpiece material in the hole's area engages and tends to ride upward on the drill flutes.

Drill removal may cause a lifting force. If the drill is reversed for removal, the lifting force may be accompanied by reverse torque.

Drilling Process. As the drill point is brought to bear on the workpiece, the thrust will cause deflection of the drill axis. As the drill bends, the point will wander on the workpiece surface until the thrust force is sufficient to cause displacement of the workpiece metal by the chisel edge and cutting can begin. The actual point at which the cutting begins can be anywhere in the area traversed by the wandering drill point. Accurate location under these circumstances is impossible. If the point does not promptly engage and displace metal, the drill might break as the bending increases.

Centerpunching. To assure accuracy of hole location and to minimize drill bending and breakage, a pointed metal punch (centerpunch) may be driven into the work piece at the intended hole location. The indentation thus formed receives the drill point and limits the amount of wandering. Thrust is thereby concentrated at one point, cutting action begins quickly, and drill bending is minimized.

Drill Guide Bushings. Another way to insure accuracy of hole location and minimize drill breakage is to mount a hardened collar or bushing directly above the intended hole location. The drill passes through the bushing to engage the workpiece. The bushing supports the drill as thrust causes the bending moment. As the thrust cannot be dissipated by bending, it is concentrated on the chisel point. Without bending, wandering is minimized and accuracy is assured.

By definition, a workholder having an integral device for tool guidance is called a *jig*. A workholder without tool guidance is called a *fixture*.

Drill Fixtures. Figure 2-25 shows a simple pin-type drill fixture consisting of a table and two pins. The table is part of the drillpress and absorbs the thrust load. The pins are set into the table and absorb the torque. If the number of workpieces to be processed is very small, disfigurement of the machine table might be objectionable, in which case parallels can be placed along the workpiece edges instead of the pins, and can be attached to the table with C clamps.

Although the fixture (Fig. 2-25), absorbs the cutting forces, it does not locate the workpiece relative to the tool. Location could be achieved by using six pins as shown in Fig. 2-10. If the workpiece has an irregular shape that cannot be conveniently supported by pins, a nest-type workholder (Fig. 2-22) might be required. If a small number of workpieces of varying size and shape are to be drilled, a simple vise (Fig. 2-3) can be temporarily attached to the table. If great numbers of a workpiece of complex shape are to be drilled, special vise jaws (Fig. 2-52) may be provided.

Fixture Limitations. Simple drill fixtures and standard vises can be advantageously used for both low- and high-volume production of a great variety of workpieces. Limitations are, however, imposed by the nature of the drilling process. Accuracy of hole location will not be precise owing to drill-point wander and drill bending, so workpieces normally processed in this manner will be those that do not require precise hole location. To minimize drill bending and breakage, the workpiece surface must always be perpendicular to the drill axis. The diameter-to-length ratio of the drill should be as large as possible. The drill point should be as acute as possible, consistent with the tool material, the workpiece material, and the drill manufacturer's recommendations. The ratio of hole depth to hole diameter will also be limited by process factors.

Accuracy can be improved somewhat by several methods. The hole location can be centerpunched, or the operator can manually position the drill point against the workpiece and twirl the drill chuck to start the cutting action before imposing the full thrust and torque loads. The hole location may first be established with a center drill or a larger-diameter twist drill, or may be established at a prior operation. A cast workpiece may have a subdiameter cored hole at the required location. A stamped workpiece may have a subdiameter pierced hole at the required location. Existing indentations or holes will allow the drill to cut without initial bending and wander. Accuracy will still be limited. An unguided drill will follow and duplicate rather than correct the location of a pilot hole. With careful location of the pilot hole in the prior process, the final location will be accurate enough for most purposes and applications.

Drill Jigs. If the required accuracy of hole location is more precise than that attainable without tool guidance, guidance must in some way be provided. Figure 2-50 shows a quick-acting vise which can be used as a simple drill fixture. The vise has been converted into a precise drilling jig by the addition of a plate with three drill guide bushings. Each guide bushing precisely establishes the location of a hole relative to the workpiece locators and relative to the other bushings. The accuracy obtainable is limited only by the accuracy of the jig. Figure 2-54 shows another example of a simple drill jig. Pump jigs (Fig. 2-48) can be converted into precise drill jigs by adding workpiece locators to the base and by placing drill guide bushings in the head plate.

In the examples of drill jigs already cited, all holes are drilled perpendicular to a designated base reference plane. The designer prescribes locators that will support the workpiece so that its primary base plane is perpendicular to the drill axis. He provides a method of retaining the workpiece in position on the locators. He constructs a plane parallel to the reference plane in which the drill bushings are to be located, and locates the bushings with the reference plane X and Y dimensions. The design is predicated on the analysis of the cutting forces and is further complicated by workpiece tolerance specifications. A hole location tolerance that might appear liberal seems to become very tight when successive portions are consumed by tolerances for the dimensions of the locating elements, tolerances for the attachment of the locators to the jig structure, tolerances for the jig structure, and tolerances for the location of the bushing in the jig structure. All of the design problems are compounded when the holes to be drilled are not parallel to each other.

Chap. 2 Workholding Devices   149

Box-Type Jigs. The shape of any workpiece is defined by reference to a base plane and two other planes that are perpendicular to the base plane and to each other. Quite often hole patterns must be drilled in several of the primary planes or faces of a single workpiece. Such a workpiece may be placed in a rectangular enclosure called a *tumble jig*. When the workpiece is clamped in the tumble jig, each surface of the rectangle is parallel to one of the three primary reference planes. Any hole to be drilled in the workpiece can then be precisely located by providing a drill bushing in the

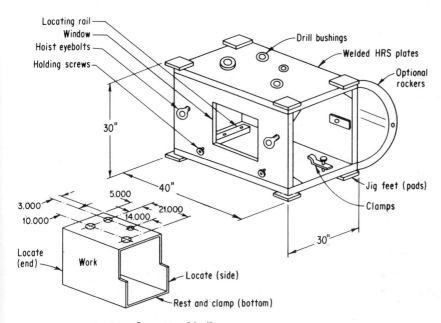

Fig. 2-83. Large tumble jig.

parallel surface of the jig. A tumble jig simultaneously locates a hole relative to other workpiece features; locates a hole relative to other holes in the same reference plane, and coordinates the location of holes in each of the three planes.

The tumble jig locates a workpiece by direct physical reference to features in all three planes. This superior location method is often utilized even though the holes to be drilled are all in a single plane. Figure 2-83 shows a large tumble jig so designed, together with the workpiece processed therein.

Leaf-Type Jigs. Box drilling jigs of the tumble and trunion types usually have one or more open sides for loading. A leaf-type jig is box-like, the cover of which has a hinge and clasp. All six sides of the jig can be equipped with drill bushings, and all holes perpendicular to any of the three reference planes can be drilled in one jig. The workpiece remains locked in the jig, which is moved from machine to machine to complete all drilling. The locations of all holes in the workpiece are determined and coordinated by one jig.

Leaf-type box jigs are commercially available in a wide variety of sizes. Jig feet and drill bushings may be added to any or all of the sides. Figure 2-84 shows a commercially available leaf-type tumble box jig which can be purchased in many sizes. The entire unit is squared to a tolerance of ±0.001 inch. The handle can be attached to any of the six surfaces as needed.

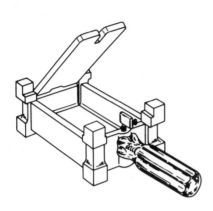

Fig. 2-84. Leaf-type tumble box jig. (*Standard Parts Co.*)

Templates. Templates are often used to coordinate the location of holes on large workpiece surfaces. Figure 2-85 shows a template used for locating and drilling a hole pattern in a rectangular metal sheet. The workpiece is located against pads and is held by toggle clamps while the holes are drilled with portable drilling units (hand drills).

If a workpiece is too large, or cannot be conveniently moved or nested, a template may be placed on or against the workpiece. The template is positioned relative to several workpiece features, and held in position while the hole pattern is drilled. Large templates are often made of lightweight plastic materials for portability and ease of handling.

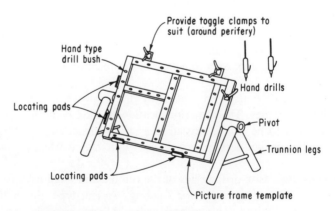

Fig. 2-85. Hole location template.

Templates are sometimes used in series. The first template is used to drill a hole pattern. A second template, identical to the first except for bushing size, is positioned on the workpiece while the holes are counterbored. Both templates must in turn be positioned relative to the same workpiece features.

Other Rotating Tool Processes

Jigs and fixtures are often required for other rotating-tool processes such as reaming, tapping, and boring. The principles and practices of drill jig and fixture design can be directly applied.

The directions of the tool forces incurred in reaming and boring operations are the same as those of the drilling process. The magnitude of the torque will depend directly on the workpiece material and the rate of metal removal. Thrust magnitude will be governed by the infeed rate.

Problems of hole location may vary. A reamer will tend to follow and enlarge an existing hole, while a boring tool will generate a new hole. If a close-tolerance hole is required at a precise location, a reaming jig may be required. This jig may be similar in appearance and function to the first-operation drill jig, but will be much more precise dimensionally. A properly guided reamer will usually bring the hole to its final diameter, and can also correct minute mislocation.

The boring process usually generates a very precise hole. Machines normally used for the process have a high degree of accuracy. A boring tool does not displace or extrude metal and consequently does not bend or wander. Hole location is determined by the accuracy of the fixture and proper adjustment of the machine. Tool guidance is not normally required. The mounting details of a boring fixture will be determined by the design of the machine. The workpiece location and details of support and retention should be determined by the analytical method. To maximize the advantage of the inherent accuracy of the process, boring fixtures are more precise dimensionally than drilling fixtures.

Tapping may be done by either of two methods (Fig. 2-26). In the first method, the infeed rate is determined by the tap, the spindle not being restricted in its axial motion. Thrust is used to engage the tap in the workpiece. Torque results from the cutting action of the tap. As the tap passes through the workpiece, lifting action may occur. Tap removal may cause further lifting action. The workpiece must be supported against thrust, torque, and reverse thrust (lifting).

In the second tapping method, the axial motion of the spindle (infeed rate) is controlled by a leadscrew. Torque results from the cutting action of the tap, but no thrust or lifting action occurs. The workpiece need not be clamped down.

A tap cannot generate a hole but can correct minor location error of a workpiece by the tap's resistance to bending. Guidance is rarely used. The fixture must position the workpiece so that the center of the hole coincides with the axis of the tap. The fixture must also support the workpiece against the tooling forces described above.

Evolution of Milling Fixtures

Whenever the structure or form of a material is changed, internal stresses may result. As a result, a bent workpiece may partially straighten. A perfect casting might warp to a useless condition. The removal of a large amount of material or the removal of material from a large area of a workpiece can cause internal stresses or can permit existing stresses to relieve themselves. It is quite possible to machine a perfectly flat surface, only to find upon subsequent examination that it has warped to an unacceptable condition. External stresses caused by poor fixturing or excessive clamping pressure can also cause later deformation.

First-Step Milling Operations. First-step milling operations are usually the hardest to perform on a casting. If the first milling cut is not made so that the workpiece will remain flat afterward, the entire part may become worthless. All subsequent operations may rely on the flatness of the first cut and its relation to boss centerlines, webs, and rib sections. A rough casting cannot merely be clamped flat on a machine table or held by vise pressure against its rough sides. The workpiece must be held firmly in its existing condition and must not be deformed by or subjected to more pressure than the minimum required to support it against the tool forces.

Deformation. Deformation caused by clamping and machining a workpiece in a poorly designed fixture may never be subsequently removed. The stresses imposed might be relieved at a very slow rate. The workpiece may creep for days, a few thousandths of an inch at a time.

Stresses may also be caused by dull cutters, lack of coolant, or removing too much metal at one time. The probability of success can be greatly enhanced by using a well-aged casting, taking a series of light cuts with sharp tools and ample coolant, and, if possible, spreading the cuts over a period of time.

Conventional Milling. Figure 1-46A shows how a milling cutter passes through a workpiece during conventional milling. The cutting edge tends to push the workpiece away. Constant thrust (infeed rate) is required to continue the process. The cutting force may tend to lift the workpiece out of the fixture. Any backlash present in the feed method employed is taken up as the workpiece approaches the cutter.

Climb Milling. Figure 1-46B shows how a milling cutter passes through a workpiece during climb milling. The cutting edges tend to pull the workpiece into the cutter. If the workpiece is not properly restrained, the cutter will embed itself in the workpiece and stall, with possible tool or machine breakage. Any backlash in the table leadscrew will permit the cutter to pull the workpiece forward suddenly. In this manner a planned material-removal rate of perhaps 0.003 in. per tooth, if tried on a machine having 0.025 in. of table travel due to backlash, could result in a sudden load of 0.028 in. per tooth. Even if breakage did not occur, the workpiece finish would be unacceptable.

Climb milling is high-production material-removal which is very satisfactory when performed on a good, tight machine. The overriding motion of the cutter helps propel the workpiece along its intended path and also tends to force the workpiece into its fixture. Properly conducted, this method will result in less machine wear than conventional milling.

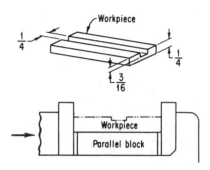

Fig. 2-86. Workpiece held in a machine vise for milling a shallow slot.

Milling Machine Vise. The most commonly used workholder for material removal by milling is a standard machine vise. Figure 2-86 shows a rectangular workpiece in which a slot is to be milled, together with a side view of the workpiece as supported by a parallel block between the vise jaws. The method is satisfactory because of the relative shallowness of the slot. Figure 2-87 shows the probable result of using the same method when

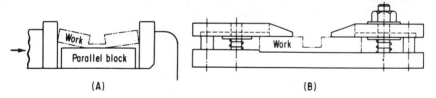

Fig. 2-87. Workpieces being held while a deep slot is milled: (A) probable distortion caused by conventional vise; (B) simple fixture that prevents distortion.

milling a deep slot. Clamping pressure may cause distortion when the supporting material is milled from the center of the workpiece. Shown also is a simple fixture which can be used in place of a vise when deep slots must be milled.

Special vise jaws (Figs. 2-51, 2-52) may be used to hold workpieces for milling operations. If the vise jaws duplicate the contour of the workpiece, chip removal may present a problem. Chips may collect in cavities and necessitate cleaning the fixture after every workpiece. Chips may be brushed aside or removed with a vacuum cleaner. Compressed air guns are sometimes used to blast away chips and debris, but are not generally acceptable because of safety reasons. An air blast can carry chips and debris back into the face of the workman.

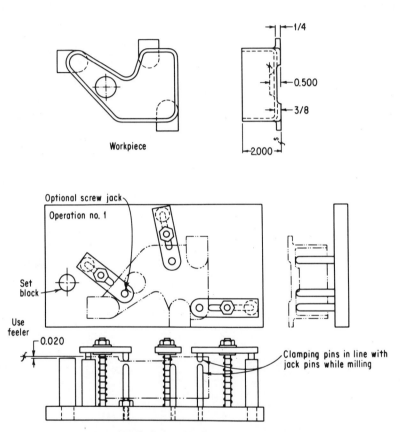

Fig. 2-88. Workpiece with first-operation milling fixture.

Design Evolution for Milling a Casting. Castings of irregular shape are often difficult to hold for a milling operation. The workpiece must be firmly held, but not distorted. The surface being milled must be precisely located in reference to other workpiece features.

Figure 2-88 shows an aluminum casting with the fixture for its first milling operation. Three bottom pads of the casting must be milled smooth and flat. The finished surface must be $3/8$ in. from the inside wall of the casting. The wall thereby becomes the reference surface for the first operation. This surface is located by three jack pins. Three clamping pins hold the workpiece against the jack pins during milling. A set block mounted on the fixture base is used with a feeler gage to establish the required path of the cutter (⅜ in. above the jack pins).

The casting is supported near but not at the points where cutting forces will be applied. Heavy cutting forces could deflect the unsupported areas of the workpiece. To avoid this, the operation consists of a series of light cuts rather than one heavy cut.

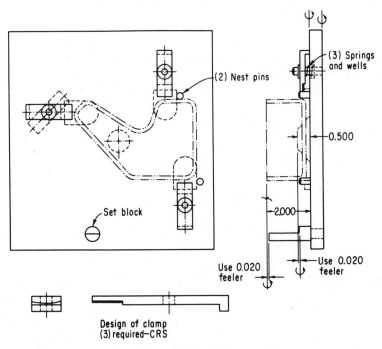

Fig. 2-89. Fixture for vertical milling operation.

Figure 2-89 shows a fixture for the second milling operation on the same workpiece (Fig. 2-88). The workpiece is located by the surface generated in the first operation. Three strap clamps hold the workpiece firmly against the fixture base during milling. A set block mounted on the fixture base is used with a feeler gage to establish the required paths of the cutter (one pass at 2.000 in. above the base and one pass at 0.500 in. above the base). This operation is performed on a vertical mill. Because of possible workpiece deflection under load, heavy cutting forces are avoided. Figure 2-90 shows how the workpiece may be attached directly to the machine table if no fixture is provided.

Figure 2-91 shows the same workpiece with the fixture for the third milling operation, i.e., the vertical milling or profiling of the outside edges of the three mounting pads. The top plate of the fixture is a hardened steel template with ground edges. A roller (spacer) on the cutter engages the template to keep the cutter edges at a constant distance from the template. The workpiece is held so that the edges of the mounting pads protrude and are in the path of the cutter. The roller diameter is 0.250 in. greater than the cutter diameter so the edge of the template is displaced inward 0.125 in. from the intended cutter path. The workpiece is located by stop pins and is held by clamping pressure as close as possible to the surfaces being milled.

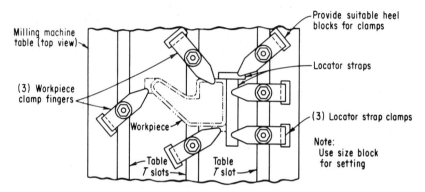

Fig. 2-90. Workpiece attached directly to machine table for vertical milling operation.

Figure 2-92 shows an alternate fixture for the same operation. The template is mounted separately and the workpiece can be rapidly loaded and unloaded. The template may have straight or tapered edges to match the chosen follower. The relative positions of the cutter and the follower are coordinated by the pantograph-type machine used for the process.

Various Milling Methods and Fixtures. A workpiece width or thickness dimension can often be established by mounting two cutters on a single milling machine arbor, with the distance between them equal to the dimension, and then feeding the workpiece between them (straddle milling). Figure 2-39 shows a workpiece, a fixture, and a machine so arranged. The surfaces being cut will always be parallel and will be at the correct fixed dimension as established by the relative position of the cutters. The fixture must hold the workpiece so that the surfaces are at the correct angular relationship to the primary reference plane of the workpiece. This is accomplished in the fixture shown (Fig. 2-39) by the locating arbor. Figure 2-91 shows a workpiece with a 6.000-in. width dimension that could have been achieved by straddle milling.

Chap. 2 Workholding Devices

Key slots must often be milled in shafts. Figure 2-93 shows how a workpiece may be presented to a cutter in either a vertical or a horizontal machine. Shown also is a simple fixture for use with a vertical milling machine.

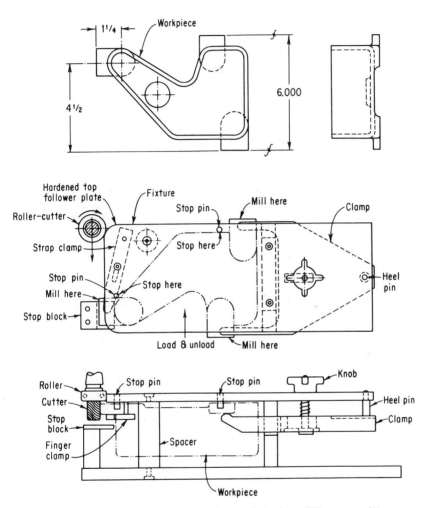

Fig. 2-91. Fixture for template-guided edge milling or profiling.

A workpiece having a number of surfaces to be milled may be successively positioned by a single fixture. If the various surfaces are parallel to an axis of the workpiece it may be possible to design a fixture that will rotate or index the workpiece around that axis. Figure 2-94 shows a fixture that rotates a workpiece to mill four flats. An index plate determines the angular relation between the milled surfaces. Within the size limits of the fixture, a great variety of workpieces can be held and positioned. Each, of course, will require an index plate to determine the locations of the surfaces to be milled.

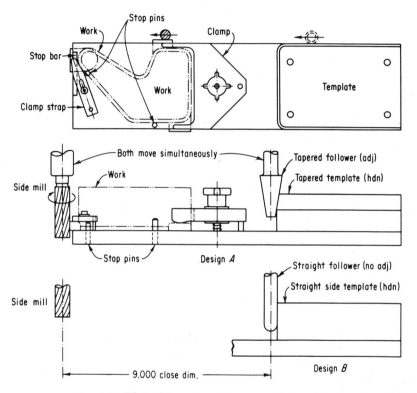

Fig. 2-92. Fixture for edge milling or profiling with a pantograph-type machine.

A well-designed fixture may be used on several machines of different types. Figure 2-95 shows a workpiece and its fixture as it might be processed in several different ways. The workpiece is a magnesium casting; the base and primary reference plane are established at the first milling operation. For this, the second milling operation, the workpiece is clamped to the fixture and is repeatedly indexed to present the four surfaces to the milling cutters. The rotating plate of the fixture has hardened bushings at 90-deg. intervals, into which an index pin plunger can nest.

Chap. 2 Workholding Devices

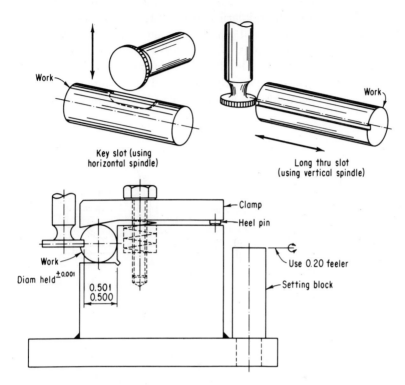

Fig. 2-93. Methods of milling key slots and vertical milling fixture.

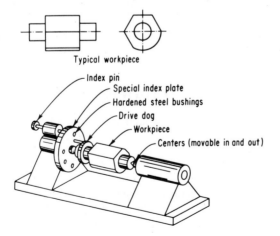

Fig. 2-94. Indexing fixture for milling flats on shafts.

Other Material-Removal Processes. Other material-removal processes in which the tool force is primarily thrust instead of torque are shaping, sawing, planing, broaching, and grinding. Torque may be imposed by some milling and grinding operations. The fixtures required for these processes are similar to milling fixtures in shape and function. Planing and shaping operations generally have high material-removal rates with greater thrust, so fixtures for these operations will generally be of heavier construction.

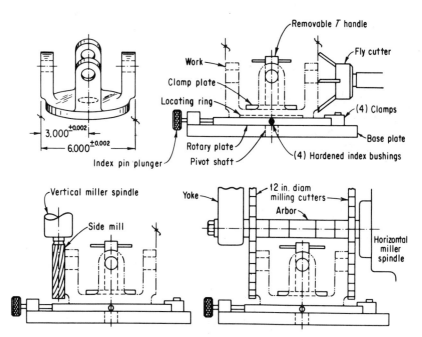

Fig. 2-95. Fixtured workpiece processed by several different methods.

FIXTURE DESIGN SUMMARY

The following step-by-step approach will prove of value for the design of all jigs and fixtures.

1. *View the workpiece as a whole.* A study of the complete workpiece, including its intended function, will disclose the relationship between the various workpiece features. With this insight, the fixture designer can either establish or better understand a proposed sequence of operations. This knowledge may enable him to combine operations and thus minimize fixturing.

2. *Gather all necessary data.* All information that may affect jig and fixture design should be readily available. The designer must know the physical characteristics of the workpiece, such as composition of the material, condition (hardness), and rough and finished weight. If the workpiece is to be made directly from raw material (mill extrusions or sheet stock) the designer must know the shape, size, and tolerances of the mill stock. All production data, including total pieces, rate of production, tooling budget allotted, and proposed production sequence must be available.

3. *Consider standard workholders.* Many operations can be performed by using available commercial workholders such as machine vises, T slots and bolts, jacks and clamps. The design of a special workholding jig or fixture must be economically justifiable. If the planned operations are similar to present operations, the rework of present fixtures may be considered.

4. *Determine what special workholders will be required.* Every proposed operation should be carefully examined to see whether it can be performed economically with commercial workholders (machine vises and so on), or whether available fixtures can be used. The designer should also consider whether minor alterations can accomplish this. After assigning as many operations as possible to commercial or available workholders, a small number of operations will remain for which special workholders must be provided. The number of special workholders may be further reduced by combining operations within a single fixture.

5. *Study fixtures for similar operations.* Every operation for which a special workholder is required should be considered individually. The designer should seek out—within his own plant, in other plants, in technical journals, and so on—similar operations for which fixtures were provided. By examining a number of existing fixtures, the designer can combine the best features of each.

6. *Review the fixturing plan.* The designer should consider in turn every operation of the production sequence and review his fixturing decisions. At every step he should confirm that the proposed workholder will be structurally adequate (cutting forces) and will be as precise as required (location aspects). The preliminary design of the required special workholders should be completed. Before calling any fixture design decisions final the designer will do well to check for compliance with the "Rules for Good Design" in Chapter 8 of this text.

7. *Execute the fixturing plan.* The designer should remain available during the execution of the fixturing plan. No plan can be regarded as final because of possible changes in workpiece dimensions and the typical cut-and-try aspects of much fixture planning. After the line has been used for production, further process improvement may suggest fixturing changes.

PROBLEMS

1. What is the purpose and function of a workholder?
2. What are the basic elements which make up a workholder?
3. Describe the degrees of freedom of a workpiece located in space.
4. Draw a simple sketch to show the 3-2-1 locating principle as applied to the workpiece shown in Fig. 2-96.

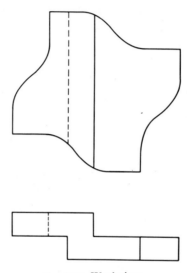

Fig. 2-96. Workpiece.

5. Does the 3-2-1 locating principle apply to radial location of workpieces?
6. Dimension the pin diameters for the fixture shown in Fig. 2-97.
7. What error is caused by the improper orientation of a "V" locator?
8. What are the disadvantages of a nesting type locator?
9. What forces does a drill exert on a workpiece?
10. List some basic mechanical elements which can be used to apply holding forces.
11. What other means are available for applying clamping forces?
12. What is a pump jig?
13. What other basic type of workholders can be used to hold flat or irregular workpieces?
14. List the basic workholding devices which can be used to hold round workpieces.
15. What is the first step in the design of any workholding device?

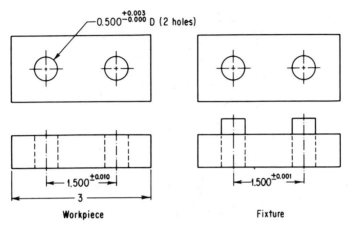

Fig. 2-97. Workpiece and pin type fixture.

16. What force factors must be counteracted in the design of a workholding fixture?
17. What is the difference between a drill jig and a drill fixture?
18. What are the limitations of a drill fixture?
19. List three basic types of drill jigs.
20. How can a drill jig be adapted to a reaming operation?
21. Name three types of drill bushings.
22. What is the primary function of a liner bushing?
23. Are there standardized throat dimensions for "T" slots on milling machine tables?
24. Are quarter-turn screws used for holding workpieces in milling fixtures?
25. How can an operator establish the depth of cut on a milling machine operation employing a fixture?
26. Figure 2-98 shows a spacer block made of SAE 1010 cold rolled steel. The workpiece is cut to rough length from 0.250 x 1.000 in. bar stock, and then ground to a manufacturing tolerance of 0.994-0.997 x 1.619-1.622 inches. A jig must be provided for the drilling of the two 0.249-0.257 in. holes and for a third hole which produces the required 5/32 in. radius. The production volume is low (400 pieces per month), and the duration is uncertain.

    (a) Design an inexpensive jig that will provide the required accuracy.

    (b) Since the parts will be ground to a 0.003 in. tolerance on the width and length, how can the part be nested to hold the positional tolerance of the holes?

    (c) How shall the part be held or clamped?

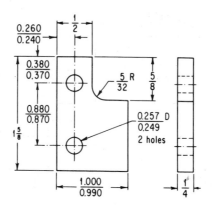

Fig. 2-98. Spacer block.

27. Figure 2-99*A* shows a latch bolt head made of brass rod. The rough workpiece (Fig. 2-99*B*) is made on a screw machine. The workpiece is then milled to the required shape (Fig. 2-99*C*). Yearly production is approximately 1,000,000 pieces.

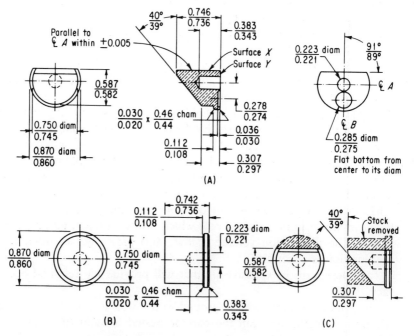

Fig. 2-99. Latch bolt head: (*A*) finished part, (*B*) rough workpiece, (*C*) milled workpiece.

The last operation drills a 0.275-0.285 in. diam. hole, 0.030-0.036 in. deep. Hole depth is measured from surface Y. A center distance tolerance of 0.274-0.278 in. must be held. The bottom of the hole must be flat.

(a) Design a jig to drill the 0.275-0.285 in. diam hole.
(b) What kind of tool can be used to produce a flat bottom hole?
(c) Will chip disposal be a problem?
(d) What kind of mechanism can be made or purchased which will force or elevate the Y surface of the workpiece against a fixed plane?
(e) How shall the jig members be arranged to facilitate loading and unloading?

3

DESIGN OF PRESSWORKING TOOLS

Characteristic of the pressworking process is the application of large forces by press tools for a short time interval which results in the cutting (shearing) or deformation of the work material.

A pressworking operation, generally completed by a single application of pressure, often results in the production of a finished part in less than one second.

Pressworking forces are set up, guided, and controlled in a machine referred to as a *press*.

POWER PRESSES

Essentially, a press is comprised of a frame, a bed or bolster plate, and a reciprocating member called a ram or slide which exerts force upon work material through special tools mounted on the ram and bed.

Chap. 3 Design of Pressworking Tools

Energy stored in the rotating flywheel of a mechanical press or supplied by a hydraulic system in a hydraulic press is transferred to the ram for its linear movements.

An *open-back inclinable (OBI) press* (Fig. 3-1), widely used, has a C-shaped frame which allows access to its working space (between the bed and the ram). The frame can be inclined at an angle to the base, allowing for the disposal of finished parts by gravity. The open back allows the feeding and unloading of stock, workpieces, and finished parts through it from front to back.

Major components of a press are:

1. A rectangular *bed,* part of the frame, generally open in its center, which supports a bolster plate.
2. A *bolster plate,* a flat steel plate, from 2 to 5 in. thick, upon which press tools and accessories are mounted. Bolsters having standardized dimensions and openings are available from press manufacturers. The JIC standard dimensions for open-back inclinable (gap-frame) presses are shown in Fig. 3-2.

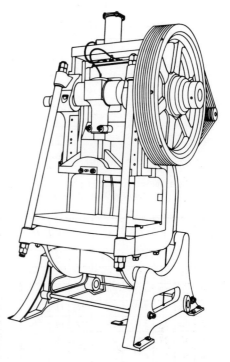

Fig. 3-1. Open-back inclinable (OBI) press.

3. A *ram,* or slide, that moves through its stroke, a distance depending upon the size and design of a press. The position of the ram, but not its stroke, can be adjusted.

The distance from the top of the bed (or bolster) to the bottom of the slide, with its stroke down and adjustment up, is the *shut height* of a press.
4. A *knockout,* a mechanism operating on the upstroke of a press, which ejects workpieces or blanks from a press tool.
5. A *cushion* is a press accessory located beneath or within a bolster for producing an upward motion and force; it is actuated by air, oil, rubber, or springs or a combination thereof.

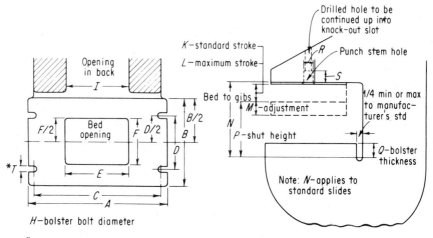

Tonnage	A	B	C	D	E	F	H	I	K	L	M	N	P	Q	R	S
22	20	12	18	7½	8	5	¾	9	2½	4	2	11¼	8½	2½	1⅝	2¼
32	24	15	22	9	11	8	¾	11	3	5	2¼	12¾	9½	2½	1⅝	2¼
45	28	18	25½	10½	14	8	1	13	3	6	2½	14¼	11	3	2⅛	3
60	32	21	29½	12	16	11	1	15	3½	7	2¾	16¾	13	3	2⅝	3
75	36	24	33	18	18	14	1¼	18	4	8	3	19¼	15	3½	2⅝	3
110	42	27	39	18	21	15	1¼	21	5	10	3½	23¼	18	4	3⅛	3
150	50	30	47	18	21	17	1¼	24	6	12	4	28¼	22	4½	3⅛	3
200	58	34	55	18	27	21	1¼	27	8	12	4½	32¼	24	5	3⅛	3

Fig. 3-2. JIC standard dimensions for OBI presses.

Press Types

Space does not permit discussion of all press types; manufacturers' catalogs and the *ASTME Die Design Handbook* contain data on hydraulic transfer, horn, and forging presses as well as four-slide, dieing, and eyelet machines.

A *straight slide press* of conventional design has columns (uprights) at the ends of the bed and is open at the front and back.

A *press brake* is essentially the same as a gap frame press except for its long bed—twenty feet or more. It accommodates a series of separate sets of press tools or a single long tool for operations on large metal sheets.

Open-back inclinable OBI presses have been previously discussed.

Chap. 3 Design of Pressworking Tools 169

CUTTING (SHEARING) OPERATIONS

In the following discussion, certain die nomenclature will be used frequently. Figure 3-3 presents the terms most commonly encountered.

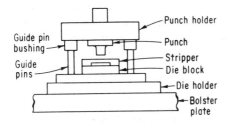

Fig. 3-3. Common components of a simple die.

Shear Action in Die Cutting Operations[1]*

The cutting of metal between die components is a shearing process in which the metal is stressed in shear between two cutting edges to the point of fracture, or beyond its ultimate strength.

The metal is subjected to both tensile and compressive stresses (Fig. 3-4); stretching beyond the elastic limit occurs; then plastic deformation,

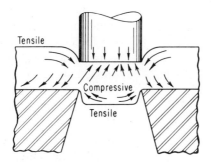

Fig. 3-4. Stresses in die cutting.

reduction in area, and, finally, fracturing starts through cleavage planes in the reduced area and becomes complete.

The fundamental steps in shearing or cutting are shown in Fig. 3-5. The pressure applied by the punch on the metal tends to deform it into the die opening. When the elastic limit is exceeded by further loading, a portion of the metal will be forced into the die opening in the form of an embossed pad on the lower face of the material and a corresponding depression on the upper face, as indicated at A. As the load is further increased, the punch will penetrate the metal to a certain depth and force an equal portion

---

* Superior numbers refer to specific references at the end of this chapter.

of metal thickness into the die, as indicated at *B*. This penetration occurs before fracturing starts and reduces the cross-sectional area of metal through which the cut is being made. Fractures will start in the reduced area at both upper and lower cutting edges, as indicated at *C*. If the clearance is suitable for the material being cut, these fractures will spread toward each other and eventually meet, causing complete separation. Further travel of the punch will carry the cut portion through the stock and into the die opening.

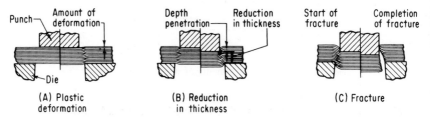

(A) Plastic deformation

(B) Reduction in thickness

(C) Fracture

Fig. 3-5. Steps in shearing metal.

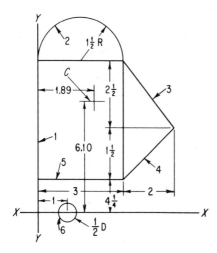

Fig. 3-6. Example of center-of-pressure calculation.

Center of Pressure

If the contour to be blanked is of irregular shape, the summation of shearing forces on one side of the center of the ram may greatly exceed the forces on the other side. This results in a bending moment in the press ram, and undesirable deflections and misalignment. It is therefore necessary to find a point about which the summation of shearing forces will be symmetrical. This point is called the center of pressure, and is the center of gravity of the *line* that is the perimeter of the blank. It is *not* the center of gravity of the area.

The press tool will be so designed that the center of pressure will be on the axis of the press ram when the tool is mounted in the press.

Mathematical Calculation of Center of Pressure. The center of pressure may be precisely determined by use of the following procedure:

1. Draw an outline of the actual cutting edges, as indicated in Fig. 3-6.
2. Draw axes *X-X* and *Y-Y* at right angles in a convenient position. If the figure is symmetrical about a line, let this line be one of the

Chap. 3 Design of Pressworking Tools

axes. The center of pressure will in this case be somewhere on the latter axis.

3. Divide the cutting edges into line elements, straight lines, arcs, etc., numbering each, 1, 2, 3, etc.
4. Find the lengths l_1, l_2, l_3, etc., of these elements.
5. Find the center of gravity of these elements. Do not confuse the center of gravity of the lines with the center of gravity of the area enclosed by the lines.
6. Find the distance x_1 of the center of gravity of the first element from the axis Y-Y, x_2 of the second, etc.
7. Find the distance y_1 of the center of gravity of the first element from the axis X-X, y_2 of the second, etc.
8. Calculate the distance X of the center of pressure C from the axis Y-Y by the formula:

$$X = \frac{l_1 x_1 + l_2 x_2 + l_3 x_3 + l_4 x_4 + l_5 x_5 + l_6 x_6}{l_1 + l_2 + l_3 + l_4 + l_5 + l_6}$$

9. Calculate the distance Y of the center of pressure from the axis X-X by the formula:

$$Y = \frac{l_1 y_1 + l_2 y_2 + l_3 y_3 + l_4 y_4 + l_5 y_5 + l_6 y_6}{l_1 + l_2 + l_3 + l_4 + l_5 + l_6}$$

In the accompanying illustration the elements are shown, numbered 1, 2, 3, etc. The length of l is obtained directly from the dimensions, and is seen to have the value 4. The center of gravity is evidently at the geometrical center of the line. Therefore

$$x_1 = 0 \quad \text{and} \quad y_1 = 4\tfrac{1}{4} + \tfrac{4}{2}$$

For the second element, x_2 is 1.5. The value for y_2 is found from the equation $CG = 2r/\pi$, where r is the radius of the element. To find the requirements for line 3, it is necessary to solve the right triangle of which it is the hypotenuse.

The requirements for the other elements are found in a similar manner, all values being entered in a table, as shown below. The products may be obtained with sufficient accuracy with a slide rule.

Element	l	x	y	lx	ly
1	4.00	0.00	6.25	0.00	25.00
2	4.71	1.50	9.20	7.05	43.33
3	3.20	4.00	7.00	12.80	22.40
4	2.50	4.00	5.00	10.00	12.50
5	3.00	1.50	4.25	4.50	12.75
6	1.57	1.00	0.00	1.57	00.00
	18.93			35.92	115.98

These values are then substituted in the preceding formulas, from which

$$X = \frac{35.92}{18.98} = 1.89 \text{ in.}; \qquad Y = \frac{115.98}{18.98} = 6.10 \text{ in.}$$

The center of pressure C is therefore located as indicated in Fig. 3-6.

Wire Method of Locating the Center of Pressure. The center of pressure of a blank contour may be located mathematically as shown above, but it is a tedious computation. Location of the center of pressure within one-half inch of true mathematical location is normally sufficient. A simple procedure accurate within such limits is to bend a soft wire to the blank contour. By balancing this frame across a pencil, in two coordinates, the intersection of the two axes of balance will locate the desired point.

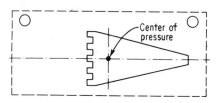

Fig. 3-7. Effect of center-of-pressure location on tool design.

As an example of the marked influence this factor may have on tool design, a rather unusual blank is shown in Fig. 3-7. Here, the center of pressure is near one end of the blank, and will require the indicated imbalance in the press tool design.

Clearances

Clearance is the space between the mating members of a die set. Proper clearances between cutting edges enables the fractures to meet, and the fractured portion of the sheared edge has a clean appearance. For optimum finish of a cut edge, proper clearance is necessary and is a function of the kind, thickness, and temper of the work material. Clearance, penetration, and fracture are shown schematically in Fig. 3-8. In Fig. 3-9, characteristics of the cut edge on stock and blank, with normal clearance, are schematically shown. The upper corner of the cut edge of the stock (indicated by A) and the lower corner of the blank (indicated by A^1-1) will

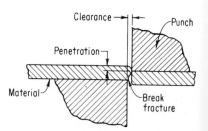

Fig. 3-8. Punch-and-die clearance; punch penetration into and fracture of die-cut metal.

have a radius where the punch and die edges, respectively, make contact with the material. This is due to the plastic deformation taking place. This edge radius will be more pronounced when cutting soft metals. Excessive

Chap. 3 Design of Pressworking Tools   173

clearance will also cause a large radius at these corners, as well as a burr on opposite corners.

In ideal cutting operations, the punch penetrates the material to a depth equal to about one-third of its thickness before fracture occurs, and forces an equal portion of the material into the die opening. That portion of the thickness so penetrated will be highly burnished, appearing on the cut edge as a bright band around the entire contour of the cut adjacent to the edge radius—indicated at B^1 and B^1-1 in Fig. 3-9. When the cutting clearance is

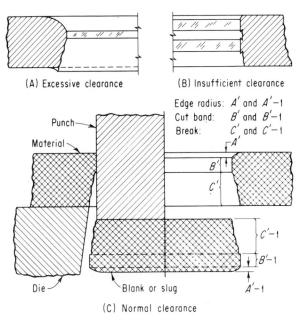

Fig. 3-9. Cut-edge characteristics of die-cut metal; effect of excessive and insufficient clearances.

not sufficient, additional bands of metal must be cut before complete separation is accomplished, as shown at B in Fig. 3-9. When correct cutting clearance is used, the material below the cut will be rough on both the stock and the slug. With correct clearance, the angle of fracture will permit a clean break below the cut band because the upper and lower fractures extend toward one another. Excessive clearance will result in a tapered cut edge since, for any cutting operation, the opposite side of the material which the punch enters will, after cutting, be the same size as the die opening.

The width of the cut band is an indication of the hardness of the material, provided that the die clearance and material thickness are constant; the wider the cut band, the softer the material. The harder metals require larger clearances and permit less penetration by the punch than the ductile metals; dull tools create the effect of too small a clearance as well as a burr on the die side of the stock. The effects of various amounts of clearance are shown in Figs. 3-6 and 3-8. Defective or nonhomogeneous material cut with the proper amount of clearance will produce nonuniform edges.

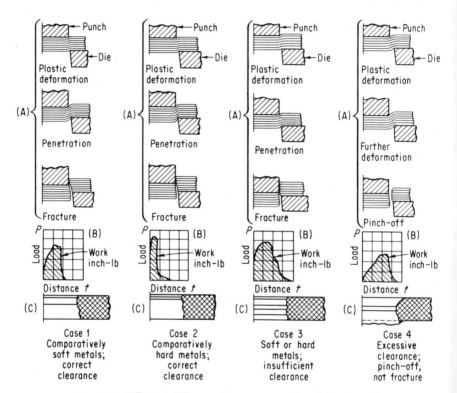

Fig. 3-10. Effect of different clearances on soft and hard metals.

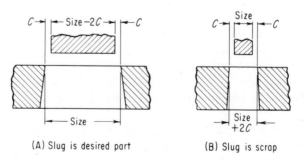

(A) Slug is desired part   (B) Slug is scrap

Fig. 3-11. Control of hole and blank sizes by clearance location.

Chap. 3 Design of Pressworking Tools 175

The edge conditions C and the hypothetical load curves B (Fig. 3-10, cases 1, 2, 3, and 4) are shown, as well as the amount of deformation and extent of punch penetration. Location of the proper clearance (Fig. 3-11) determines either hole or blank size; punch size controls hole size; die size controls blank size.

At A (Fig. 3-11), which shows clearance C for blanks of a given size, make die to size and punch smaller by total clearance 2C. At B, which shows clearance for holes of a given size, make punch to size and die larger by the amount of the total clearance 2C.

The application of clearances for holes of irregular shape is diagrammed in Fig. 3-12; at B the hole will be of punch size, while at A the blank will be of the same dimensions as the die.

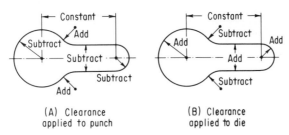

(A) Clearance applied to punch

(B) Clearance applied to die

Fig. 3-12. How to apply clearances.

One manufacturer charts clearances per side for groups of materials up to and including thicknesses of 0.125 in. (Fig. 3-13).

The die-clearance chart (Fig. 3-13) may be used to find the recommended die clearance to be allowed, and to be provided for, in designing a die for service as determined by the materials groups listed below, and for the pre-established percentage of material thickness of the original part which the die is designed to produce.

Group 1. 1100S and 5052S aluminum alloys, all tempers. An average clearance of $4\frac{1}{2}$ per cent of material thickness is recommended for normal piercing and blanking.

Group 2. 2024ST and 6061ST aluminum alloys; brass, all tempers; cold rolled steel, dead soft; stainless steel soft. An average clearance of 6 per cent of material thickness is recommended for normal piercing and blanking.

Group 3. Cold rolled steel, half hard; stainless steel, half hard and full hard. An average clearance of $7\frac{1}{2}$ per cent is recommended for normal piercing and blanking.

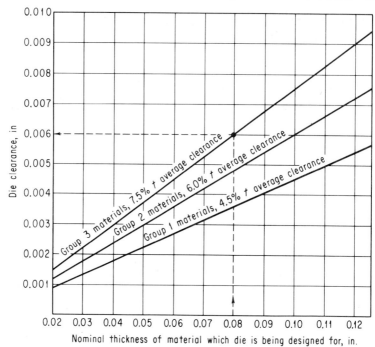

Fig. 3-13. Die clearances for different groups of metals.

Example. In Fig. 3-13 it is seen that, for a nominal stock thickness of 0.080 in., the die clearance for any group 1 materials would be 0.0036 in.; for any group 2 materials, 0.0048 in.; for any group 3 materials, 0.006 in.

Angular Clearance. Angular clearance is defined as that clearance below the straight portion of a die surface introduced for the purpose of enabling the blank or the slug (piercing operation) to clear the die (Fig. 3-11). Angular clearance is usually ground from ¼ to 1½° per side but occasionally as high as 2°, depending mainly on stock thickness and the frequency of sharpening. High-production dies usually have a ¼° clearance angle.

Cutting Forces

The pressure P required to cut (shear) work material is,
$$P = \pi DSt \text{ (for round holes)}$$
$$P = SLt \text{ (for other contours)}$$
where S = shear strength of material, psi
D = hole diameter, in.
L = shear length, in.
t = material thickness, in.

Chap. 3 Design of Pressworking Tools   177

The pressure needed to punch a 2-in. diam. hole in ⅛ in. thick SAE 1020 steel having a shear strength of 60,000 psi (see Table 3-1) is 47,124 lb.

TABLE 3-1. SHEAR STRENGTH, PSI, OF VARIOUS MATERIALS [1]

Ferrous Materials

0.10 carbon steel annealed ...	35,000	Nickel steel (drawn to 800°F and water-quenched):	
0.20 " " " ...	42,000		
0.30 " " " ...	52,000	SAE 2320	98,000
0.50 " " " ...	80,000	SAE 2330	110,000
1.00 " " " ...	110,000	SAE 2340	125,000
Chromium-molybdenum steel; SAE 4130:		Nickel-chromium steel (drawn to 800°F):	
90,000 u.t.s.*	55,000	SAE 3120	95,000
100,000 u.t.s.	65,000	SAE 3130	110,000
125,000 u.t.s.	75,000	SAE 3140	130,000
150,000 u.t.s.	90,000	SAE 3280	135,000
180,000 u.t.s.	105,000	SAE 3240	150,000
		SAE 3250	165,000

Nonferrous Materials

Aluminum and alloys	4000–41,000	Nickel: 68,000 u.t.s.	52,300
Copper and alloys	22,000–48,000	120,500 u.t.s.	75,300
Magnesium alloys	4000–21,000	Inconel (nickel-chromium-iron): 80,000 u.t.s.	59,000
Monel metal:		90,000 u.t.s.	63,000
69,000 u.t.s.	42,900	100,000 u.t.s.	66,000
108,000 u.t.s.	65,200	115,000 u.t.s.	71,000
K monel:		140,000 u.t.s.	78,000
97,500 u.t.s.	65,300	160,000 u.t.s.	84,000
155,600 u.t.s.	98,700	175,000 u.t.s.	87,000

Nonmetallic Materials

Asbestos board	5,000	Leather, rawhide	13,000
Cellulose acetate	10,000	Mica	10,000
Cloth	8,000	Paper †	6,400
Fiber, hard	18,000	Bristol board	4,800
Hard rubber	20,000	Pressboard	3,500
Leather, tanned	7,000	Phenol Fiber ‡	26,000

* u.t.s. = ultimate tensile strength.
† For hollow die, used one-half value shown for shearing strength.
‡ Blank and perforate hot.

Methods of Reducing Cutting Forces. Since cutting operations are characterized by very high forces exerted over very short periods of time, it is sometimes desirable to reduce the force and spread it over a longer portion of the ram stroke. Punch contours of large perimeter or a great many smaller punches will frequently result in tonnage requirements beyond the capacity of an available press. Also, whenever abnormally high tonnage requirements are concentrated in a small area, design difficulties are increased.

Two methods are generally used to reduce cutting forces and to smooth out the shock impact of heavy loads.

1. Step the punch lengths (see view C, Fig. 3-16). The load may thus be reduced approximately 50 per cent. The shorter punches should be about one stock thickness shorter for best results.
2. Grind the face of the punch or die at a small shear angle with the horizontal. This has the effect of reducing the area in shear at any one time, and may reduce cutting loads as much as 50 per cent. The angle chosen should provide a change in punch length of about $1\frac{1}{2}$ times stock thickness.

    It is usually preferable to use a double angle to maintain symmetry and prevent setup of lateral force components. Some distortion of the material in contact with the angular ground member will occur, and the shear angle should be applied to either the punch or the die, so that the distortion will be in the scrap of the material. Thus in a blanking operation the shear angle will be on the die member, while in a piercing operation the shear angle will be applied to the punch member.

In piercing, the direction of the shear angles must be such that the cut proceeds from the outer extremities of the contour toward the center. This avoids stretching the material before it is cut free.

Die Block General Design

Overall dimensions of the die block will be determined by the minimum die wall thickness required for strength and by the space needed for mounting screws and dowels and for mounting the stripper plate.

Wall thickness requirements for strength will depend on the thickness of the stock to be cut. Sharp corners in the contour may lead to cracking in heat treatment, and so require greater wall thickness at such points.

Two, and only two, dowels should be provided in each block or element that requires accurate and permanent positioning. They should be located

Chap. 3 Design of Pressworking Tools

as far apart as possible for maximum locating effect, usually near diagonally opposite corners. Two or more screws will be used, depending on the size of the element mounted. Screws and dowels are preferably located about 1½ times their diameters from the outer edges or the blanking contour.

Die block thickness (see Table 3-2) is governed by the strength necessary to resist the cutting forces, and will depend on the type and thickness of the material being cut. On very thin materials ½ in. thickness should be sufficient but, except for temporary tools, finished thickness is seldom less than ⅞ inch, which allows for blind screw holes and also builds up the tool to a narrower range of shut height for press room convenience.

Die Block Calculations

Method 1 ("Rule of Thumb"). Assuming a die block of tool steel, its thickness should be ¾ in. minimum for a blanking perimeter of 3 in. or less; 1 in. thick for perimeters between 3 and 10 in., and 1¼ in. thick for larger perimeters. There should be a minimum of 1¼ in. margin around the opening in the die block.

The die opening should be straight for a maximum of ⅛ in.; the opening should then angle out at ¼° to 1½° to the side (draft). The straight sides provide for sharpenings of the die; the tapered portion enables the blanks to drop through without jamming.

To secure the die to the die plate or die shoe, the following rules provide sound construction:

1. On die blocks up to 7 in. square, use two ⅜ in. cap screws and two ⅜ in. dowels.
2. On sections up to 10 in. square, use three cap screws and two dowels.
3. For blanking heavy stock, use cap screws and dowels of ½ in. diameter. Counterbore the cap screws ⅛ in. deeper than usual, to compensate for die sharpening.

Method 2. This method of calculating the proper size of the die was derived from a series of tests,[2] whereby die plates were made increasingly thinner until breakage became excessive. From these data the calculation of die thickness was divided into four steps:

1. Die thickness is provisionally selected from Table 3-2. This table takes into account the thickness of the stock and its ultimate shear strength (see Table 3-1).
2. The following corrections are then made:
    (a) The die must never be thinner than 0.3 to 0.4 in.

(b) Data in Table 3-2 apply to small dies, i.e. those with a cutting perimeter of 2 in. or less. For larger dies, the thicknesses listed in Table 3-2 must be multiplied by the factors in Table 3-3.

(c) Data in Tables 3-2 and 3-3 are for die members of tool-steel, properly machined and heat treated. If a special alloy of steel is selected, die thickness can be decreased.

TABLE 3-2. DIE THICKNESS PER TON OF PRESSURE

Stock thickness, in.	Die thickness, in.*	Stock thickness, in.	Die thickness, in.*
		0.06	0.15
0.01	0.03	0.07	0.165
0.02	0.06	0.08	0.18
0.03	0.085	0.09	0.19
0.04	0.11		0.20
0.05	0.13	0.10	

* For each ton per sq in. of shear strength.

TABLE 3-3. FACTORS FOR CUTTING EDGES EXCEEDING 2 INCHES

Cutting perimeter, in.	Expansion factor
2 to 3	1.25
3 to 6	1.5
6 to 12	1.75
12 to 20	2.0

(d) Dies must be adequately supported on a flat die plate or die shoes. Thickness data above do not apply if the die is placed over a large opening or is not adequately supported. However, if the die is placed into a shoe, the thickness of the member can be decreased up to 50 per cent.

(e) A grinding allowance up to 0.1 to 0.2 in. must be added to the calculated die thickness.

3. The critical distance A, Fig. 3-14, between the cutting edge and the die border must be determined. In small dies, A equals 1.5 to 2 times the die thickness; in larger dies it is 2 to 3 times the die thickness.

4. Finally, the die thickness must be checked against the empirical rule that the cross-sectional area $A \times T$ (Fig. 3-15) must bear a certain minimum relationship to the impact pressure for a die put on a flat

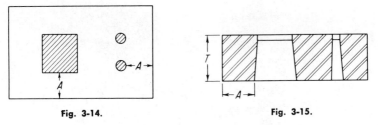

Fig. 3-14. Critical distance A must not be less than 1.5 to 2 times die thickness.

Fig. 3-15. The critical area between the die hole and the die border must be checked against minimum values in Table 3-4, and die thickness T corrected if necessary.

base. In Table 3-4, impact pressure equals stock thickness times the perimeter of the cut times ultimate shearing strength. If the die height, as calculated by steps 1 and 2 above, does not give sufficient area for the critical distance A (Figs. 3-11 and 3-12), the die thickness must be increased accordingly.

TABLE 3-4. MINIMUM CRITICAL AREA VS. IMPACT PRESSURE

Impact pressure, tons	Area between die opening border, sq in.
20	0.5
50	1.0
75	1.5
100	2.0

With the die block size determined, the exact size of the die opening can now be determined.

Assuming a clearance of approximately 10 per cent of the metal thickness, and by the rule-of-thumb method,

$$\text{metal thickness} = 0.064 \times 10\% = 0.006 \text{ in.}$$

If the finished die opening is 1.000 in. diam., then add 0.006, giving 1.006 ± 0.001 in.

If the blank were made according to size, the clearance would be applied to the punch.

Punch Dimensioning

The determination of punch dimensions has been generally based on practical experience.

When the diameter of a pierced round hole equals stock thickness, the unit compressive stress on the punch is four times the unit shear stress on the cut area of the stock, from the formula

$$\frac{4S_s}{S_c}\frac{t}{d} = 1$$

where S_c = unit compressive stress on the punch, psi
S_s = unit shear stress on the stock, psi
t = stock thickness, in.
d = diameter of punched hole, in.

The diameters of most holes are greater than stock thickness; a value for the ratio d/t of 1.1 is recommended.[8]

The maximum allowable length of a punch can be calculated from the formula

$$L = \frac{\pi d}{8}\left(\frac{E}{S_s}\frac{d}{t}\right)^{1/2}$$

where $d/t = 1.1$ or higher
E = modulus of elasticity

This is not to say that holes having diameters less than stock thickness cannot be successfully punched. The punching of such holes can be facilitated by:

1. Punch steels of high compressive strengths
2. Greater than average clearances
3. Optimum punch alignment, finish, and rigidity
4. Shear on punches or dies or both
5. Prevention of stock slippage
6. Optimum stripper design

Methods of Punch Support

Figure 3-16 presents a number of methods to support punches to meet various production requirements:[4]

View A. When cutting punch *A* is sharpened, the same amount is ground off spacer *B* to maintain the relative distance *C*.

Chap. 3 Design of Pressworking Tools 183

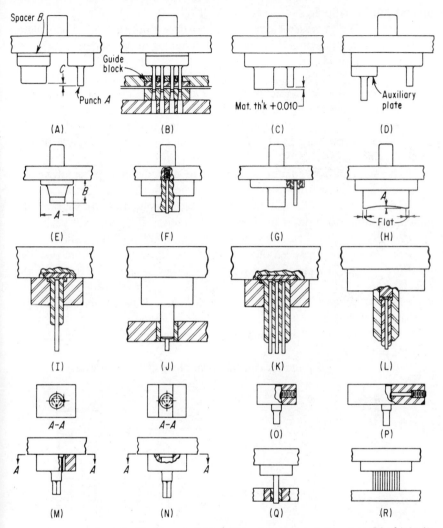

Fig. 3-16. Various methods of punch support. (*American Machinist.*[4])

View B. If delicate punches must be grouped closely together, a hardened guide block with the required number of holes should be used.

View C. A slender piercing punch (at left) should be made shorter than an adjacent large punch.

View D. If punches must protrude more than 4 in. beyond the punch holder, an auxiliary plate may be used to maintain stiffness.

View E. Flange width of the punch should be greater than height *B* to provide stability for unguided punches.

View F. In a large punch, push-off pins can prevent slugs from pulling up and causing trouble.

View G. To avoid cracking a large hardened punch or a punch plate, do not press a small punch directly into either of these members. Instead, use a soft plug or insert.

View H. Long slotting punches should be hollow ground so that dimension *A* equals the metal thickness, so as to put shear on the punch. The ends should be flat for $\frac{1}{8}$ in. to avoid bending the stock.

View I. A quill is useful for supporting pin punches.

View J. A bushing in the stripper plate can guide the quill for increased punch support.

View K. Quills need not be limited to a single punch. If prevented from turning, they can be used for pin punches on close centers.

View L. Two quills are used for a bit punch—one to support the punch, the other to support the inner quill, when a stripper is not used.

View M. A dowel can be used to prevent rotation of the punches.

View N. For high-speed dies, a flat on the punch head is more positive.

View O. In low-production dies, a setscrew is adequate to hold the punch.

View P. When a deep hole must be drilled, a drill-rod pin can be used to span the distance.

View Q. Light drill-rod punches are guided in the stripper plate to prevent buckling.

View R. Several punches can be set at close center distances.

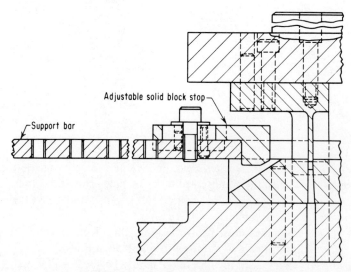

Fig. 3-17. Adjustable block stop for a parting die.

Chap. 3 Design of Pressworking Tools

Stock Stops

In its simplest form, a stock stop may be a pin or small block, against which an edge of the previously blanked opening is pushed after each stroke of the press. With sufficient clearance in the stock channel, the stock is momentarily lifted by its clinging to the punch, and is thus released from the stop. Figure 3-17 shows an adjustable type of solid block stop which can be moved along a support bar in increments up to 1 in. to allow various stock lengths to be cut off.

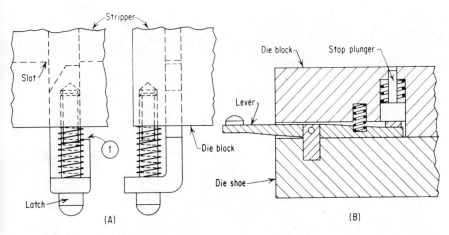

Fig. 3-18. Starting stops.

A starting stop, used to position stock as it is initially fed to a die, is shown in Fig. 3-18, view *A*. Mounted on the stripper plate, it incorporates a latch which is pushed inward by the operator until its shoulder (1) contacts the stripper plate. The latch is held in to engage the edge of the incoming stock; the first die operation is completed, and the latch is released.

The starting stop shown at view *B*, mounted between the die shoe and die block, upwardly actuates a stop plunger to initially position the incoming stock. Compression springs return the manually operated lever after the first die operation is completed.

Trigger stops incorporate pivoted latches (1, Fig. 3-19, views *A* and *B*). At the ram's descent, these latches are moved out of the blanked-out stock area by actuating pins, 2. On the ascent of the ram, springs, 3, control the lateral movement of the latch (equal to the side relief) which rides on the surface of the advancing stock, and drops into the blanked area to rest against the cut edge of the cut-out area.

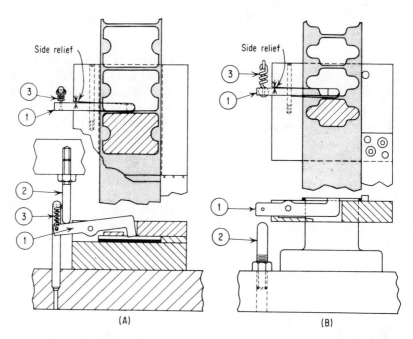

Fig. 3-19. Trigger stops: (*A*) top stock engagement; (*B*) bottom stock engagement.

Automatic Stops

When feeding the stock strip from one stage to another, some method must be used to correctly locate and stop the strip.

Automatic stops (trigger stops) register the strip at the final die station. They differ from finger stops in that they stop the strip automatically, the operator having only to keep the strip pushed against the stop in its travel through the die.

Figure 3-20 shows automatic stop designs ranging from simple pin to escapement type mechanisms.

View A illustrates a pin stop suitable for low- to medium-production dies. When the ram ascends, the strip clings to the punch, is stripped, and then is fed until the pin hits the edge of the hole.

View B shows the method of locating the pin stop so that it bears against the blank opening upon an angular edge, so that the strip is crowded against the back stop and accurate piloting is obtained.

View C. If no scrap is left between blanks, as in a double-action blank-and-draw die, a bent pin stop is suitable. The sharpened point of the stop

Chap. 3 *Design of Pressworking Tools* 187

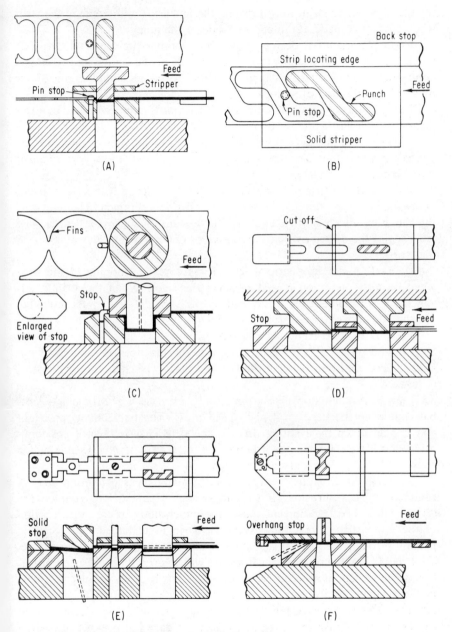

Fig. 3-20. Designs of automatic stops.

faces the incoming strip, thus thrusting the fins aside and stopping the strip when contact is made at the opposite side of the hole.

View D presents a design in which a combination stop and backup block locates the strip and prevents deflection of the cutoff punch of this two-station die. The part, a flat spring, drops to the punchholder and slides by gravity to the rear of an inclined press.

View E shows another solid stop. The part, after cutoff, can drop through a hole in the die set to a box directly beneath.

View F. Overhanging stops are useful when the press cannot be inclined, or when the size of the die or press will not allow part removal through the bolster plate.

View G. If the strip must be cut to accurate width, a trimming stop can be used against the shoulder formed by the trimming punch. The length of the trimming punch is made equal to the feed distance for part length. The left edge of the punch overlaps the previous cut to prevent leaving a fin on the strip.

View H. A notching strip is ideal for automatic stopping of the strip when the first operation partially blanks the sides. When the strip is advanced, it is automatically positioned by the shoulder left by the blanking punch. The only extra cost is that for the stop.

View I. Double trimming stops reduce the extent of the waste end. Stop A and punch B are used during starting and normal running of the strip. Stop C and punch D cut the last two parts when the end of the strip is reached.

View J. A latch stop, like a pin stop, is ideal for low-production jobs, but should not be used on a high-speed press. The latch pivots on a pin and is held down by a spring. In use, the strip is moved until the scrap bridge has lifted the latch and it drops into the hole. Then the strip is pulled back until the bridge is against the latch.

View K. Punch stops are applicable to many types of cutting dies. A round eccentric protrusion on a round or square punch body contacts the scrap bridge. The ram descent causes the punch body to cut out a portion of the scrap strip. Upon ram ascent the strip is advanced, the eccentric passing through the gap until contact is made with the next scrap bridge.

View L. Commercial trigger stops are widely used for high-speed operations. The design shown consists of gage pin A which fits loosely in a hole in the stripper, and spring B that normally holds up lever C. The strip advance crowds the gage pin to the slanted position shown. Ram descent causes pin D to push the lever down, and the gage pin is lifted above the strip. Spring E now pushes the gage pin on top of the scrap bridge. Upon forward movement of the strip, the gage pin falls inside the hole just blanked.

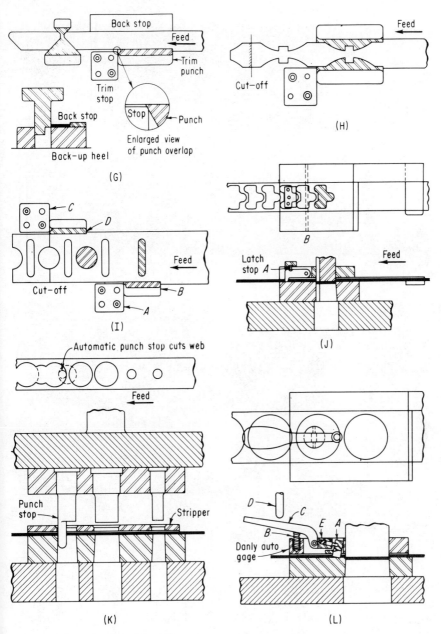

Fig. 3-20 (Contd.)

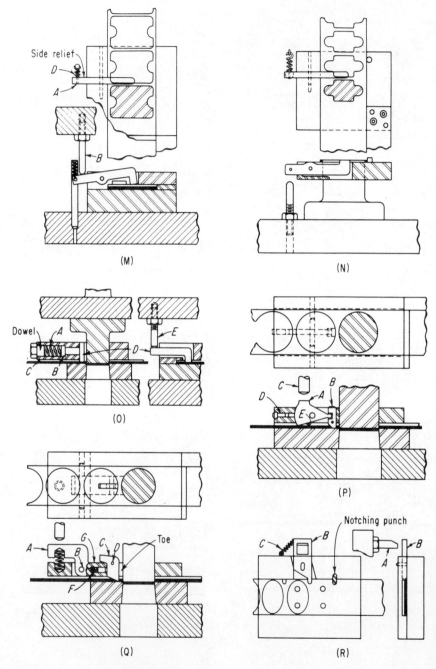

Fig. 3-20 (Contd.)

Chap. 3 Design of Pressworking Tools   191

Views M and N. Web trigger stops can be applied to simple blanking dies or to compound dies. At ram descent, stop *A* is lifted out of the strip by actuating pin *B*. Spring *D* rocks sideways so that it will fall onto the scrap at ram ascent. This permits the strip to be advanced until the cut edge resets the stop.

View O. Spring torsion and compression combine to operate a stop. Ends of spring *A* are entered in drilled holes in stop plunger *B* and cap plug *C*. At assembly the plug is rotated to apply sufficient torque so that stop *D* is kept firmly in contact with the die block. Strip advance slightly compresses the spring. At ram descent, pin *E* trips the stop and spring *A* pushes it forward to drop on the scrap bridge.

View P. An escapement stop eliminates misfeeds if the stock is fed forward to the stop every time. Rocker arm *A* lifts the square toe stop *B* above the strip, when struck by actuator *C* attached to the punch holder. Now spring-operated detent *D* holds the stop up, until the advancing scrap bridge strikes the cam surface *H* on the rocker arm. Thereupon, the toe is dropped in front of the cut edge, just as the next ram descent commences.

View Q. Long, high-speed runs on heavy stock often require an automatic toe stop. Lever *A* pivots on pin *B* to rock toe *C*, on pin *D*, above the strip surface, when the ram descends. Spring *F*, acting through plunger *G*, pushes the toe until the heel on top rests on the lever. When the ram goes up, the toe falls on the scrap bridge. It eventually contacts the cut edge, when the strip is advanced, thereby resetting the stop.

View R. A notch in the strip edge serves for strip location in a two-stage piercing and blanking die. At the first station a small notch is cut in the strip edge. At the second station, actuator *A* enters the cam hole in stop *B* to retract the toe from the notch. Spring *C* swings the toe counterclockwise so that it comes to rest on the strip edge. Upon strip advance, the toe springs into the notch just cut and stops the material in the correct position.

Pilots

Since pilot breakage can result in the production of inaccurate parts and the jamming or breaking of die elements, pilots should be made of good tool-steel, heat-treated for maximum toughness and to hardness of Rockwell C57 to 60.

Press-fit Pilots. Press-fit pilots (Figs. 3-21 and 3-22, view *C*), which may drop out of the punch holder, are not recommended for high-speed dies but are often used in low-speed dies. Recommended dimensions for press-fit pilots are listed in Table 3-5 for acorn types and flattened-point types. An alternate method of establishing the dimensions of pilots is to make the radius B equal to the pilot diameter A. The spherical nose radius C of the acorn type may be $0.25A$, approximately. Length C of a flattened-point-type pilot may be about $0.5A$.

Pilots may be retained by methods shown in Fig. 3-22. A threaded shank, shown at view A, is recommended for high-speed dies; thread

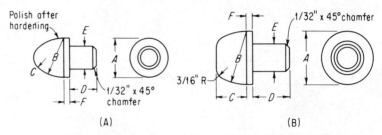

Fig. 3-21. Pressfit pilots: (A) acorn type; (B) flattened-point type.

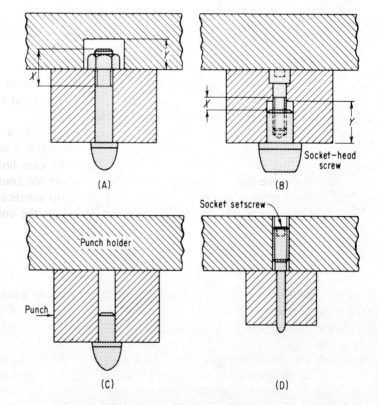

Fig. 3-22. Methods of retaining pilots: (A) threaded shank; (B) screw-retained; (C) pressfit; (D) socket setscrew.

length X and counterbore Y must be sufficient to allow for punch sharpening. For holes $3/4$ in. in diameter or larger, the pilot may be held by a socket-head screw, shown at B; recommended dimensions X and Y given for threaded-shank pilots also apply. A typical press-fit type is shown at C. Pilots of less than $1/4$ in. diameter may be headed and secured by a socket setscrew, as shown at D.

TABLE 3-5. DESIGN DIMENSIONS FOR DRILL-ROD PRESSFIT PILOTS
(for dimension diagram, see Fig. 3-21, views A and B)

Nominal A	B	C	D	E
ACORN TYPE				
$1/8$	$1/8$	$1/32$	$1/4$	$3/32$
$3/16$	$3/16$	$3/64$	$1/4$	$1/8$
$1/4$	$1/4$	$1/16$	$7/16$	$3/16$
$5/16$	$5/16$	$5/64$	$7/16$	$7/32$
$3/8$	$3/8$	$3/32$	$1/2$	$1/4$
$7/16$	$7/16$	$1/8$	$1/2$	$9/32$
$1/2$	$1/2$	$5/32$	$1/2$	$5/16$
$5/8$	$5/8$	$11/64$	$5/8$	$11/32$
$11/16$	$11/16$	$3/16$	$5/8$	$3/8$
FLATTENED-POINT TYPE				
$3/4$	$3/4$	$7/16$	$5/8$	$3/8$
$7/8$	$7/8$	$17/32$	$3/4$	$7/16$
1	1	$5/8$	$3/4$	$1/2$
$1 1/4$	$1 1/4$	$3/4$	1	$5/8$
$1 3/8$	$1 3/8$	$7/8$	1	$11/16$
$1 1/2$	$1 1/2$	$15/16$	$1 1/4$	$3/4$

Dimension $F = 1/16$ in., or stock thickness, whichever is greater.
All dimensions are in inches.

Indirect Pilots. Designs of pilots that enter holes in the scrap skeleton are shown in Fig. 3-23. A headed design, at A, is satisfactory for piloting in holes from $3/16$ in. to $3/8$ in. in diameter. A quilled design, at B, is suitable for pilots up to $3/16$ in. in diameter or less.

Spring-loaded pilots should be used for stock exceeding No. 16 gage. A bushed shouldered design is shown at C of Fig. 3-23. A slender pilot of drill rod shown at D is locked in a bushed quill which is countersunk to fit the peened head of the pilot.

Tapered slug-clearance holes through the die and lower shoe should be provided, since indirect pilots generally pierce the strip during a misfeed.

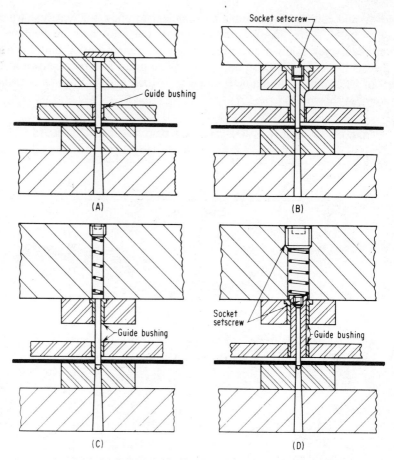

Fig. 3-23. Indirect pilots: (*A*) headed; (*B*) quilled; (*C*) spring-backed; (*D*) spring-loaded quilled.

Strippers

Strippers are of two types, fixed or spring-operated. The primary function of either type is to strip the workpiece from a cutting or noncutting punch or die. A stripper that forces a part out of a die may also be called a *knockout,* an *inside stripper,* or an *ejector.* Besides its primary function, a stripper may also hold down or clamp, position, or guide the sheet, strip, or workpiece.

The stripper is usually of the same width and length as the die block. In the simpler dies, the stripper may be fastened with the same screws and dowels that fasten the die block, and the screwheads will be counterbored

Chap. 3 Design of Pressworking Tools   195

into the stripper. In more complex tools and with sectional die blocks the die block screws will usually be inverted, and the stripper fastener will be independent.

The stripper thickness must be sufficient to withstand the force required to strip the stock from the punch, plus whatever is required for the stock strip channel. Except for very heavy tools or large blank areas, the thickness required for screwhead counterbores, in the range of $3/8$ to $5/8$ in., will be sufficient.

The height of the stock strip channel should be at least $1\frac{1}{2}$ times the stock thickness. This height should be increased if the stock is to be lifted over a fixed pin stop. The channel width should be the width of the stock strip, plus adequate clearance to allow for variations in the width of the strip as cut, as follows:

Stock thickness, in.	Add to strip width
up to 0.040	$5/64$ in.
0.040 to 0.080	$3/32$ in.
0.081 to 0.120	$7/64$ in.
over 0.120	$1/8$ in.

If the stripper length has been extended on the feed end for better stock guidance, a sheet metal plate should be fastened to the underside of the projecting stripper to support the stock. This plate should extend slightly in for convenience in inserting the strip. The entry edges of the channel should be beveled for the same reason.

Where spring-operated strippers are used, the force required for stripping is 3500 times cut perimeter times the stock thickness. It may be as high as 20 per cent of the blanking force, which will determine the number and type of springs required. The highest of these values should be used.

Die springs are designed to resist fatigue failure under severe service conditions. They are available in medium, medium-heavy, and heavy-duty grades, with corresponding permissible deflections ranging from 50 to 30 per cent of free length. The number of springs for which space is available and the total required force will determine which grade is required. The required travel plus the preload deflection will be the total deflection, and will determine the length of spring required to stay within allowable percentage of deflection limits. As the punch is resharpened, deflections will increase, and should also be allowed for.

To retain the stripper against the necessary preload of the springs, and to guide the stripper in its travel, a special type of shoulder screw known as a *stripper bolt* is used.

Choice of the method of applying springs to stripper plates depends on

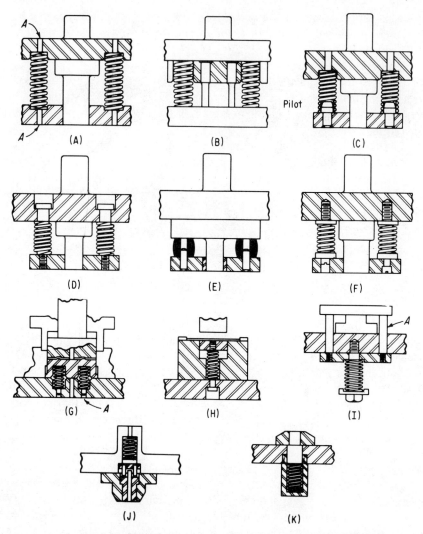

Fig. 3-24. Use of die springs with strippers. (*Am. Machinist.*[3])

the required pressure, space limitations, shape of the die and the nature of the work, and production requirements. Figure 3-24 [3] presents a number of such methods:

View A. The stripper plate has two counterbored holes, usually ¼ in. deep. The punch holder counterboring is deeper in order to apply the correct initial spring deflection. Hole edges are chamfered $\frac{1}{16} \times 45°$. In machining the counterbored holes, the stripper plate and the punch holder are clamped together, and the holes *A* are drilled. These holes are then used as pilots when counterboring.

View B. In this piercing die, the upper portions of the springs are retained in holes bored through the punch plate. The spring ends bear against the ground underside of the punch holder.

View C. The pilots shown here provide an excellent means of locating the springs.

View D. Here, the springs are placed around stripper bolts. Although this method is widely used, it is not recommended for high-grade dies.

View E. Standard rubber springs can provide sufficient stripping pressure, are low in cost, and are easily applied. However, they are adversely affected by pressworking compounds, and are indicated for short runs only.

View F. Standard "Strippit" spring units allow removal of the stripper plate without disturbing the springs. The left-hand strippit is so installed as to prevent sidewise movement of the lower portion of the unit. The right-hand unit is used to hold the locating plug of the unit to the stripper plate.

View G. This shows application of springs to a blank-and-draw die, where the pressure pad bottoms in such a way that the springs are entirely confined. In this case, the counterbored holes are chamfered at least $\frac{1}{8}$ in. $\times$ 45°. Safety pins A are pressed into the dieholder. If a spring should break, these pins hold the fractured coil and prevent it from getting between the pressure pad and the face of the die holder.

View H. This shows a bending die used to form a flat part to channel shape. Shoulder bolts limit stripper plate travel, and the springs are placed around them. The stripper bolts will hold the coil if spring breakage occurs.

View I. Here, the stripper plate is actuated by four pins A through one large die spring. A springholder stud threads into the dieholder unit, and is adjustable.

View J. An inverted compound die employs a spring-actuated knockout in the upper die. The spring is confined in a counterbored hole in the punch shank, and applies pressure to the knockout through pins.

View K. This shows a method of holding springs below the dieholder, if there is insufficient room for them within the dieset. A housing is screwed into a tapped hole, and the confined spring actuates the knockout.

Knockouts

Since the cut blank will be retained, by friction, in the die block, some means of ejecting on the ram upstroke must be provided. A knockout assembly consists of a plate, a push rod, and a retaining collar. The plate is a loose fit with the die opening contour, and moves upward as the blank is cut. Attached to the plate, usually by rivets, is a heavy pushrod which slides in a hole in the shank of the dieset. This rod projects above the shank, and a collar retains and limits the stroke of the assembly. Near the upper limit of the ram stroke, a knockout bar in the press will contact the pushrod and eject the blank.

It is essential that the means of retaining the knockout assembly be secure, since serious damage would otherwise occur.

In the ejection of parts positive knockouts offer the following advantages over spring strippers where the part shape and the die selections allow their use:

1. Automatic part disposal; the blank, ejected near the top of the ram stroke, can be blown to the back of the press, or the press may be inclined and the same result obtained.
2. Lower die cost; knockouts are generally of lower cost than spring strippers.
3. Positive action; knockouts do not stick as spring strippers occasionally do.
4. Lower pressure requirements, since there are no heavy springs to be compressed during the ram descent.

Figure 3-25 shows several good knockout designs.

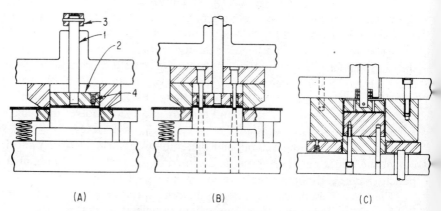

Fig. 3-25. Positive knockouts for dies.

View A. This one, applied to a plain inverted compound die, is of the simplest type. It consists of an actuating plunger A, knockout plate B, and a stop collar C doweled to the plunger A. Shedder D consists of a shouldered pin backed by a spring which is confined by a setscrew.

View B shows the knockout plate used as a means of guiding slender piercing punches through hardened bushings.

View C shows a design such that the flanged shell, upon completion, is carried upward in the upper die and ejected by a positive knockout.

TYPES OF DIE-CUTTING OPERATIONS

The operations of die cutting (shearing) of work materials are classified as follows:

Piercing (punching) (Fig. 3-26A) is the operation in which a round punch (or of other contour) cuts a hole in the work material which is supported by a die having an opening corresponding exactly to the contour of the punch. The material (slug) cut from the work material is often scrap.

Blanking (Fig. 3-26A) fundamentally differs from piercing only in that the part cut from the work material is usable, becoming a blank (workpiece) for subsequent pressworking or other processing.

Lancing combines bending and cutting along a line in the work material. It does not produce a detached slug and leaves a bent portion, or tab, attached to the work material (Fig. 3-26B).

A *cut-off* operation achieves complete separation of the work material by cutting it along straight or curved lines (Fig. 3-26C).

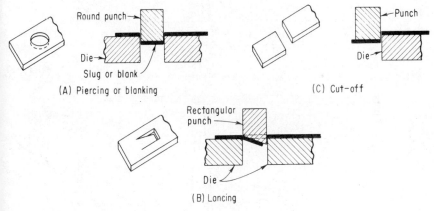

Fig. 3-26. Die-cutting (shearing) operations.

A *notching* operation cuts out various shapes from the edge of workpiece material (a blank or a part).

Shaving is a secondary shearing or cutting operation in which the surface of a previously cut edge of a workpiece is finished or smoothed. Punch and die clearance for a shaving die, considerably less than that for other cutting dies, allows a thin portion (or shaving) to be cleanly cut from such a surface of the workpiece.

PIERCING-DIE DESIGN

A complete press tool for cutting two holes in work material, at one stroke of the press, as classified and standardized by a large manufacturer as a single-station piercing die is shown in Fig. 3-27.

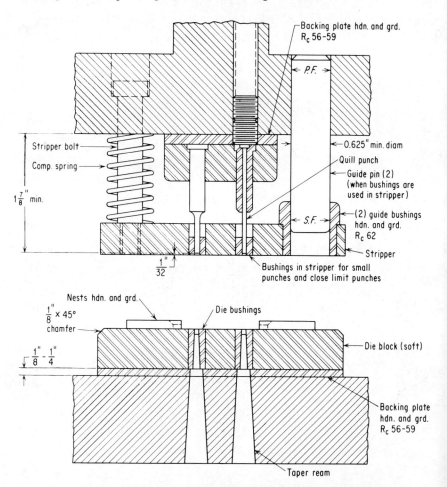

Fig. 3-27. Typical single-station die for piercing two holes.

Any complete press tool, consisting of a pair (or a combination of pairs) of mating members for producing pressworked (stamped) parts, including all supporting and actuating elements of the tool, is a die. Pressworking terminology commonly defines the female part of any complete press tool as a die.

The guide pins, or posts, are mounted in the lower shoe. The upper shoe contains bushings which slide on the guide pins. The assembly of the lower and upper shoes with guide pins and bushings is a *die set*. Die sets in many sizes and designs are commercially available. The guide pins shown in Figs. 3-28 and 3-29 guide the stripper in its vertical travel: for clarity the guide pins are not shown in Fig. 3-29.

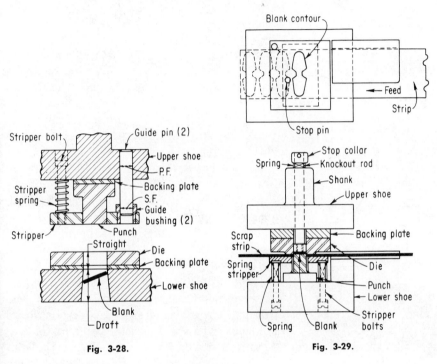

Fig. 3-28. A simple blanking die.

Fig. 3-29. An inverted blanking die.

A punch holder mounted to the upper shoe holds two round punches (male members of the die) which are guided by bushings inserted in the stripper (Fig. 3-27). A sleeve, or quill, encloses one punch to prevent its buckling under pressure from the ram of the press. After penetration of the work material, the two punches enter the die bushings for a slight distance.

The female member or die consists of two die bushings inserted in the die block. Since this press tool punches holes to the diameters required, the diameters of the die bushings are larger than those of the punches by the amount of clearance.

Since the work material stock or workpiece around a punch can cling to it on the upstroke, it may be necessary to strip the material from the punch. Spring-loaded strippers (Fig. 3-28 and 3-29) hold the work material against the die block until the punches are withdrawn from the punched holes. A workpiece to be pierced is commonly held and located in a *nest* (Fig. 3-27) composed of flat plates shaped to encircle the outside part contours. Stock is positioned in dies by pins, blocks, or other types of stops for locating before the downstroke of the ram.

BLANKING-DIE DESIGN

The design of a small blanking die shown in Fig. 3-28 is the same as that of the piercing die of Fig. 3-27 except that a die replaces the die bushings and the two piercing punches are replaced by one blanking punch, and a stock stop is incorporated instead of nest plates. The design is of the drop-through type since the finished blanks drop through the die, the lower shoe, and the press bolster.

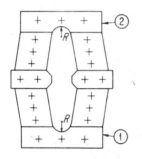

Fig. 3-30. Eight-section layout for die shown in Fig. 3-29.

Large blanks are commonly produced by an inverted blanking die (Fig. 3-29) in which the die is mounted to the upper shoe with the punch secured to the bottom shoe. The passing of a large blank through the bolster is often impractical but its size may necessitate sectional die design (Fig. 3-30).

Draft, or angular clearance, in an inverted die is unnecessary because the blank does not pass through it. The cutting edges of each section should not include points and intricate contours for ease of construction, regrinding, and for strength. Sections 1 and 2 of the die of Fig. 3-30 were laid out to include the entire semicircular contour, with straight contours included in the other six sections.

The spring-loaded stripper is mounted on the lower shoe; its travel is upward in stripping the stock from the punch fastened to the lower shoe. Stripper bolts hold and guide the stripper in its travel.

On the upstroke of the ram, the upper end of the knockout rods strikes an arm on the press frame, which forces the lower end of the rods downward and through the die and ejects the finished blank from the die cavity. A stop collar retains the rods and limits their travel.

Cutting-Rule Dies. Instead of a conventional female die, the sharp edges of steel rule form the cutting blades of the dies of Figs. 3-31 and 3-32. The rule, bent to the contour of the blank outline, is used principally for blanking cork, paper, and similar nonmetallic fibrous materials, although the economical blanking of aluminum stock up to 0.040 in. thick has been reported. The stripper can be of neoprene, cork, or similar resilient sheet material, to force the work material out of the shallow die cavity.

Chap. 3 Design of Pressworking Tools 203

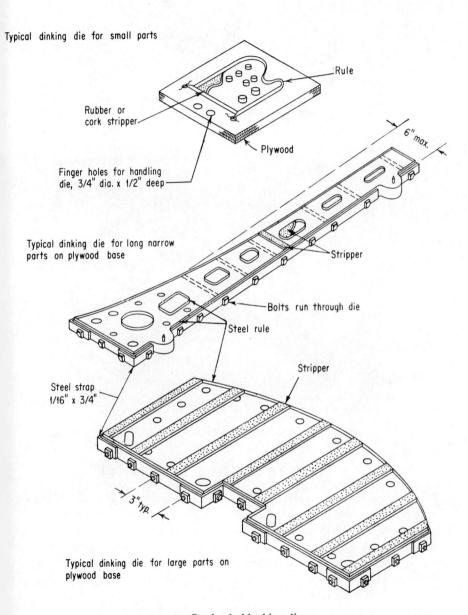

Fig. 3-31. Steel-rule blanking dies.

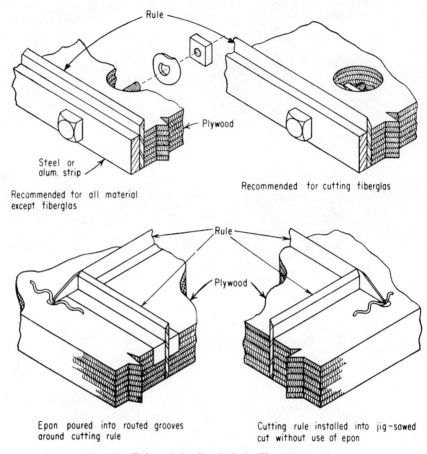

Fig. 3-32. Enlarged details of die in Fig. 3-31.

COMPOUND-DIE DESIGN

A compound die performs only cutting operations (usually blanking and piercing) which are completed during a single press stroke. A compound die can produce pierced blanks to close flatness and dimensional tolerances. A characteristic of compound dies is the inverted position of the blanking punch and blanking die. As shown in Fig. 3-33, the die is fastened to the upper shoe and the blanking punch is mounted on the lower shoe. The blanking punch also functions as the piercing die, having a tapered hole in it and in the lower shoe for slug disposal.

On the upstroke of the press slide, the knockout bar of the press strikes the knockout collar, forcing the knockout rods and shedder downward, thus pushing the finished workpiece out of the blanking die. The stock strip

Chap. 3 Design of Pressworking Tools

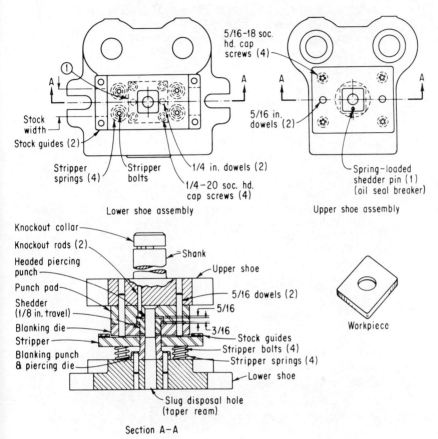

Fig. 3-33. A compound die.

is guided by stock guides screwed to the spring-loaded stripper. On the upstroke the stock is stripped from the blanking die by the upward travel of the stripper. Before the cutting cycle starts, the strip stock is held flat between the stripper and the bottom surface of the blanking die.

Four special shoulder screws (stripper bolts), commercially available, guide the stripper in its travel and retain it against the preload of its springs.

The blanking die as well as the punch pad is screwed and doweled to the upper shoe.

A spring-loaded shedder pin (oil-seal breaker) incorporated in the shedder is depressed when the shedder pushes the blanked part from the die. On this upstroke of the ram the shedder pin breaks the oil seal between the surfaces of the blanked part and shedder, allowing the part to fall out of the blanking (upper) die.

SCRAP-STRIP LAYOUT FOR BLANKING

In designing parts to be blanked from strip material, economical stock utilization is of high importance. The goal should be at least a 75 per cent utilization. A very simple scrap-strip layout is shown in Fig. 3-34.

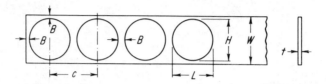

Fig. 3-34. Simple scrap-strip layout: t is the stock thickness; B is the space between part and edge of strip; C is the lead of the die, i.e., the distance from a point on one part to the corresponding point on the next part; L is the length of the part; H is the part width; W is the width of the scrap strip.

Scrap Allowance

A scrap-strip layout having insufficient stock between the blank and the strip edge, and between blanks, will result in a weakened strip, subject to breakage and thereby causing misfeeds. Such troubles will cause unnecessary die maintenance owing to partial cuts which deflect the punches, resulting in nicked edges. The following formulas are used in calculating scrap-strip dimensions for all strips over $\frac{1}{32}$ in. thick:

t = specified thickness of the material
$B = 1\frac{1}{4}t$ when C is less than $2\frac{1}{2}$ in.
$B = 1\frac{1}{2}t$ when C is $2\frac{1}{2}$ in. or longer
$C = L + B$, or lead of the die

Example. A rectangular part, to be blanked from #16 gage steel (Manufacturers Standard) is $\frac{3}{8} \times 1\frac{1}{16}$ in. If the scrap strip is developed as in Fig. 3-35, the solution is

$t = 0.0598$ in. for #16 gage steel
$B = 1.25 \times 0.0598 = 0.07475 = \frac{5}{64}$ in.
$C = \frac{3}{8} + \frac{5}{64} = \frac{29}{64} = 0.453$ in.
$W = H + 2B = 1\frac{1}{16} + \frac{5}{32} = 1\frac{7}{32}$ in.

Nearest commercial stock is $1\frac{1}{4}$ inch. Therefore, the distance B will equal 0.0897 in. This is acceptable since it exceeds minimum requirements.

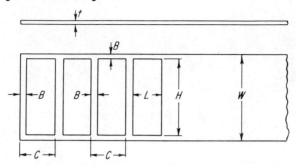

Fig. 3-35. Scrap strip development.

Minimum Scrap-Strip Allowance. If the material to be blanked is 0.025 in. thick or less, the formulas above should not be used. Instead, dimension B is to be as follows (all dimensions are in inches):

Strip width W	Dimension B
0–3	0.050
3–6	0.093
6–12	0.125
over 12	0.156

Other Scrap-Strip Allowance Applications. Figure 3-36 illustrates special allowances for *one-pass layouts:*

View A. For work with curved outlines, $B = 70$ per cent of strip thickness t.

View B. For straight-edge blanks: where C is less than $2\frac{1}{2}$ in., $B = 1t$; where C is $2\frac{1}{2}$ to 8 in., $B = 1\frac{1}{4}t$; where C is over 8 in., $B = 1\frac{1}{2}t$.

View C. For work with parallel curves, use the same formulas as for view B.

View D. For layouts with sharp corners of blanks adjacent, $B = 1\frac{1}{4}t$.

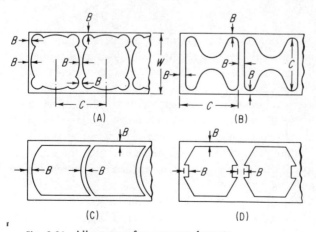

Fig. 3-36. Allowances for one-pass layouts.

Figure 3-37 illustrates special allowances for *two-pass layouts:*

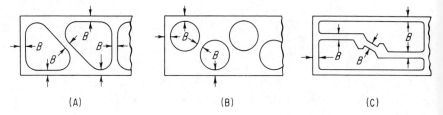

Fig. 3-37. Allowances for two-pass layouts.

View A. Single-row layout intended for two passes through the die: $B = 1\frac{1}{2}t$.
View B. Double-row layout of blanks with curved outlines: $B = 1\frac{1}{4}t$.
View C. Double-row layout of parts with straight and curved outlines: $B = 1\frac{1}{4}t$.

Percentage of Stock Used. If the area of the part is divided by the area of the scrap strip used, the result will be the percentage of stock used.

If A = total area of strip used to produce a single blanked part, then $A = CW$ (Fig. 3-35), and a = area of the part = LH.

If $C = 0.453$ in. and $W = 1.25$ in., then $A = 0.453 \times 1.25 = 0.56625$ sq. in.

If $L \times 0.375$ in. and $H = 1.0625$ in., then $a = 0.375 \times 1.0625 = 0.3984375$ sq. in.

Percentage of stock used:

$$\frac{a}{A} = \frac{0.3984375}{0.56625} = 70\% \text{ approx.}$$

COMMERCIAL DIE SETS

Two-post commercial die sets are available in many styles and sizes. Very commonly used die sets are the round-series diagonal type (Table 3-6), and the back-post type (Table 3-7). These are covered in American Standard ASA B5.25-1950, "Punch and Die Sets."

In addition to providing enlarged mounting bases for the punch and die elements, the die set is equipped with heavy guide posts which maintain alignment of the two members. Most presses not in new condition will have fairly large running clearances in the ram gibs. Such looseness, if free from horizontal forces due to press misalignment, can be minimized by the guide posts of the die set.

The die holder should be at least $\frac{1}{4}$ in. larger all around than the die block. If the bolster opening is excessively large, an oversize die holder may

Chap. 3 Design of Pressworking Tools

be needed to bridge the opening. Between the rear edge of the die block (or any portion which requires resharpening) and the guide posts there should be a ⅝-in. minimum clearance to allow for the grinding-wheel guard.

Bushings are available in various lengths, of different materials, and of plain or ball-bearing type.

The punch holder is usually equipped with an integral round shank which is gripped by a clamp in the press ram. The shank is located in the center of the die space area, as cataloged by the maker of the die set. The die set must be so laid out that the center of pressure of the blank is approximately central with this shank. The shank diameter is determined by the press in which the tool is to be used, and must be specified on the tool drawing.

The determined shut height will establish the length of the guide posts, which must be at least ½ in. shorter to allow for the reduced shut height due to resharpening.

TABLE 3-6. DIMENSIONS OF ROUND DIAGONAL-POST DIE SETS

Die Area			Thickness		Min
Rectangular Area		Diam	Die holder diam	Die holder	Punch holder	guide-post diam
A	B	D	C	J	K	P
1¾	3½	2¾	5	1¾	1¼	½
2¼	4½	3½	6	1¾	1¼	⅝
2¾	5½	4	7	2	1½	¾
3½	7	5¼	9	2	1½	1
4½	9	7	11	2	1½	1⅛

All dimensions are given in inches.

Material: This standard does not specify the material for the punch holder or die holder. They are generally made from cast iron, semisteel, or steel plate and should be free from dirt, slag, and detrimental blowholes and be sufficiently thick to permit machining of outer scale to provide uniform structure as to the finish of the surface on which the punches and dies are mounted. The material should be of sufficient hardness to stand up on service and have good machining qualities.

Note: A and B dimensions may be plus or minus ½ in. during a 5-year transition period so that suppliers may use present patterns.

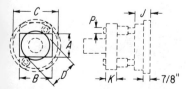

See Table 3-6.

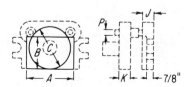

See Table 3-7.

TABLE 3-7. DIMENSIONS OF BACK-POST DIE SETS

Die Area			Thickness				Min guide-post diam P
Right to left A	Front to back B	Diam C	Die Holder J		Punch Holder K		
			From	To	From	To	
3	3	3	1	1¼	1	..	¾
4	4	4	1⅜	1¾	1¼	..	1
4	6	..	1½	2¾	1¼	2¼	1
5	4	..	1⅜	1¾	1¼	..	1
5	5	5	1½	2	1¼	1¾	1
5	8	..	1½	3	1¼	2¼	1
6	3	..	1½	2	1¼	1¾	1
6	4	5	1½	2¾	1¼	2¼	1
6	6	6½	1½	2½	1¼	2¼	1
6	9	..	1½	3¼	1¼	2¼	1¼
7	5	5¾	1½	3	1¼	2¼	1
7	7	7½	1½	2½	1¼	2¼	1
7	10	..	1⅝	3¼	1⅜	2¼	1¼
8	4	..	1½	2½	1¼	2¼	1
8	6	7	1½	3	1¼	2¼	1
8	8	8½	1½	2½	1¼	2¼	1
9	12	..	1¾	3½	1½	2¼	1½
10	5	..	1½	2½	1¼	1¾	1¼
10	7	..	1⅜	2¾	1⅜	2¼	1¼
10	10	10	1⅝	2¾	1⅜	2¼	1¼
10	14	..	1⅞	3¾	1⅝	2¾	1½
11	9	10	1¾	3½	1½	2¼	1¼
12	4	..	1¾	2¼	1½	2	1¼
12	6	..	1½	2½	1½	2	1¼
12	12	12½	1¾	3½	1¾	2¼	1½
12	16	..	2	3¾	1¾	2¾	1½
14	8	..	1¾	3¼	1⅝	2¾	1½
14	10	11¼	1¾	3¼	1⅝	2¾	1½
14	14	14	1¾	3¼	1⅝	2¼	1½
15	5	..	1½	2½	1½	2	1½
15	9	..	1½	2½	1½	2	1½
18	8	..	1½	2½	1½	2	1½
18	10	..	1½	2½	1½	2¼	1½
18	14	15	2	3	1¾	2¼	1½
18	16	17	2	3	1¾	2¼	1½
20	5	..	1¾	2½	1½	2	1½
22	6	..	1¾	2½	1½	2¼	1½
22	12	..	2	3	1½	2	1½
25	7	..	1¾	3	1½	2¼	1½
25	14	..	1¾	3	1½	2¼	1½

All dimensions are given in inches.
Material and tolerances: same as for Table 3-6.

Chap. 3 Design of Pressworking Tools   211

EVOLUTION OF A BLANKING DIE

In the planning of a die, the examination of the part print immediately determines the shape and size of both punch and die as well as the working area of the die set.

Die Set Selection

A commercially available standardized two-post die set (Tables 3-6 and 3-7) with 6-in. overall dimensions side-to-side and front-to-back allows the available 3-in. wide stock to be fed through it. It is large enough for mounting the blanking punch on the upper shoe (with the die mounted on the lower shoe) for producing the blank shown in Fig. 3-38, since the guide posts can be supplied in lengths of from 4 to 9 in.

Since the stock, in this case, was available only in a width of 3 in., the length of the blanked portions extended across the stock left a distance between the edges of the stock and the ends of the blank of 0.250 in., or twice the stock thickness; this allowance is satisfactory for the 0.125-in. stock.

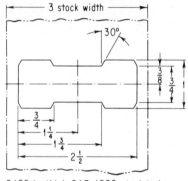

0.125 in. thick SAE 1020 steel to be blanked from 3 in. stock

Fig. 3-38. Part to be blanked.

Die Block Design

By the usual "rule-of-thumb" method previously described, die block thickness (of tool steel) should be a minimum of 0.75 in. for a blanking perimeter up to 3 in., and 1 in. for a perimeter between 3 and 4 in. For longer perimeters, die block thickness should be 1.25 in. Die blocks are seldom thinner than ⅞-in. finished thickness to allow for grinding and for blind screw holes. Since the perimeter of the blank is approximately 7 in., a die block thickness of 1.5 in. was specified, including a ¼-in. grinding allowance.

There should be a margin of 1.25 in. around the opening in the die block; its specified size of 6 in. by 6 in. allows a margin of 1.75 in. in which four ⅜-in. cap screws and ⅜-in. dowels are located at the corners 0.75 in. from the edges of the block.

The wall of the die opening is straight for a distance of 0.125 in. (stock thickness); below this portion or the straight, an angular clearance of $1\frac{1}{2}°$ allows the blank to drop through the die block without jamming.

The dimensions of the die opening are the same as that of the blank; those of the punch are smaller by the clearance (6 per cent of stock thickness, or 0.075 in.) which result in the production of blanks to print (and die) size.

The top of the die was ground off a distance equal to stock thickness (Fig. 3-39) with the result that shearing of the stock starts at the ends of the die and progresses towards the center of the die, and less blanking pressure is required than if the top of the die were flat.

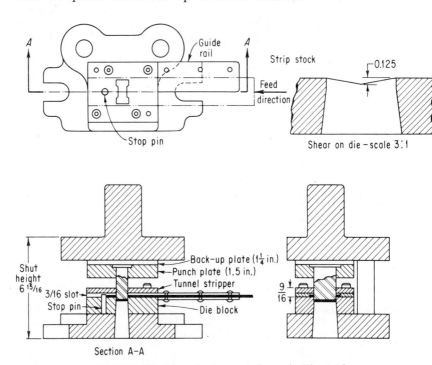

Fig. 3-39. Blanking die for part shown in Fig. 3-38.

Punch Design

The shouldered punch (2.25 in. long) is held against a ¼-in. thick hardened steel backup plate by a punch plate (¾ in. thick) which is screwed and doweled to the upper shoe. The shut height of the die can be accommodated by a 32-ton (JIC Standard) open-back inclinable press, leaving a shut height of 9.5 in. For the conditions of this case study, shear strength $S = 60,000$ lb, blanked perimeter length $L = 7$ in. approx., and thickness $T = 0.125$ in. From the equation $P = SLT$, the pressure $P = 60,000$ lb ×

Chap. 3 Design of Pressworking Tools

7 in. × 0.125 = 26.25 tons. This value is well below the 32-ton capacity of the selected press.

The shut height (Fig. 3-39) is 7 in. less the $\frac{1}{16}$-in. travel of the punch into the die cavity.

Stripper Design

The stripper that was designed is of the fixed type with a channel or slot having a height equal to $1\frac{1}{2}$ times stock thickness and a width of 3.125 in. to allow for variations in the stock width of 3 inches. The same screws that hold the die block to the lower shoe fasten the stripper to the top of the die block.

If, instead of 0.125-in. stock, thin (0.031-in.) stock were to be blanked, a spring-loaded stripper such as shown in Figs. 3-27 and 3-28 would firmly hold the stock down on top of the die block and could, to some extent, flatten out wrinkles and waves in it. A spring-loaded stripper should clamp the stock until the punch is withdrawn from the stock. The pressure which strips the stock from the punch on the upstroke is difficult to evaluate exactly. A formula frequently used is

$$P_s = 3500Lt \text{ lb}$$

where P_s = stripping pressure, lb
 L = perimeter of cut, in.
 t = stock thickness, in.

Spring design is beyond the scope of this book; die spring data are available in the catalogues of spring manufacturers.

Stock Stops

The pin stop pressed in the die block is the simplest method for stopping the hand-fed strip. The right-hand edge of the blanked opening is pushed against the pin before descent of the ram and the blanking of the next blank. The $\frac{3}{16}$-in. depth of the stripper slot allows the edge of the blanked opening to ride over the pin and to engage the right-hand edge of every successive opening.

The design of various types of stops adapted for manual and automatic feeding is covered in a preceding discussion.

EVOLUTION OF A PROGRESSIVE BLANKING DIE

Figure 3-40 gives the blanked dimensions of a linkage case cover of cold rolled steel, stock size $\frac{1}{8} \times 2\frac{3}{8} \times 2\frac{3}{8}$ in. Production is stated to be 200 parts made at one setup, with the possibility of three or four runs per year.

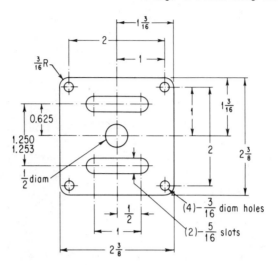

Fig. 3-40. Linkage case cover.

Step 1, Part Specification

1. The production is of medium class; therefore a second-class die will be used.
2. Tolerances required: Except for location of the slots, all dimensions are in fractions. The slot locations, though specified in decimals, are not very close. Thus a compound die is not necessary; a two-or three-station progressive die will be adequate.
3. Type of press to be used: Available for this production are presses of 5-ton, 8-ton, or 10-ton capacity, with a shut height of 7 or 7¾ in.
4. Thickness of material: Specified as ⅛-in. standard cold rolled steel.

Step 2, Scrap-Strip Development

From the production requirements, a single-row strip will suffice. After several trials, the scrap strip shown in Fig. 3-41 was decided upon. Owing to the closeness of the holes it was decided to make a four-station die.

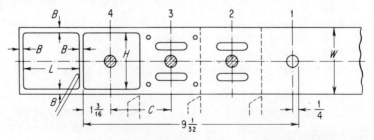

Fig. 3-41. Scrap-strip development for part shown in Fig. 3-40.

Chap. 3 Design of Pressworking Tools   215

From discussion related to Fig. 3-34:

$$B = 1\tfrac{1}{4}t = 1\tfrac{1}{4} \times \tfrac{1}{8} = \tfrac{5}{32} \text{ in.}$$
$$W = H + 2B = 2\tfrac{3}{8} + \tfrac{1}{4} = 2\tfrac{5}{8} \text{ in.}$$
$$C = L + B = 2\tfrac{3}{8} + \tfrac{5}{32} = 2\tfrac{17}{32} \text{ in.}$$

The scrap strip would be fed into the first finger stop, and the center hole would be pierced. The strip would then be moved in to the second finger stop, and the two holes would then be pierced. At the third stage and third finger stop, a pilot would locate the strip and the four corner holes would then be pierced. At the fourth and final stage, a piloted blanking punch would blank out the finished part.

Step 3, Press Tonnage

It is now in order to determine the amount of pressure needed. Only the actual blanking in the fourth stage need be calculated, since the work in the first three stages will be done by stepped punches.

From Table 3-1, the shear strength S of cold rolled steel is 58,000 psi. The length L of the blanked perimeter equals $2\tfrac{3}{8} \times 4 = 9.500$ in. The depth of cut (stock thickness t) equals 0.125 in. From the equation $P = SLt$:

$$P = 58{,}000 \text{ psi} \times 9.5 \text{ in.} \times 0.125 = 34.4375 \text{ or } 34\tfrac{1}{2} \text{ tons}$$

This tonnage is greater than can be handled by the available presses. To lower the pressure, shear is ground on the blanking punch to reduce the needed pressure by one-third. Thus, $\tfrac{1}{3} \times 34.5 = 11.5$; $34.5 - 11.5 = 23$ tons. A punch press of 25-ton capacity would do, but there is reported available only a 30-ton press with a $7\tfrac{1}{2}$-in. shut height and a 2-in. stroke. This press is selected. The bolster plate is found to be 12 in. deep, $5\tfrac{1}{2}$ in. from centerline of ram to back edge of bolster, and 24-in. wide. Shank diameter is $2\tfrac{1}{2}$ in.

Step 4, Calculation of the Die

(a) *The die*. The perimeter of the cut equals $9\tfrac{1}{2}$ in., and therefore the thickness of the die must be 1 in.

The width of our scrap-strip opening is $2\tfrac{3}{8}$ in. With $1\tfrac{1}{4}$-in. extra material on each side of the opening, it will be $2\tfrac{3}{8}$ in. $+ 2\tfrac{1}{2}$ in. $= 4\tfrac{7}{8}$ in., or 5-in. width.

The distance from the left side of the opening in stage 4 to the edge of the opening in stage 1 equals $3C + 1\tfrac{3}{16} + \tfrac{1}{4} = 7.59375 + 1.1875 +$

0.25 = 9.03125 and plus 2½ in. = 11.53125 in., or 11⅝ in. long. Therefore the die should be 1 by 5 by 11⅝ in. long.

(b) *The die plate.* As a means of filling in between the die and the die shoe, a die plate of machinery steel is used. To secure the die plate to the die shoe, ½-in. cap screws and dowels are used. A minimum of twice the size of the cap screw for the distance from the edge of the die to the edge of the die plate is needed, which will equal 1 inch. Twice this distance = 2 in., and 2 in. added to the size of the die will result in a die plate of 1 × 7 × 13⅝ inches.

Figure 3-42 shows the die and die plate fitted together, and with the holes which show the sharpening portion and the relief portion.

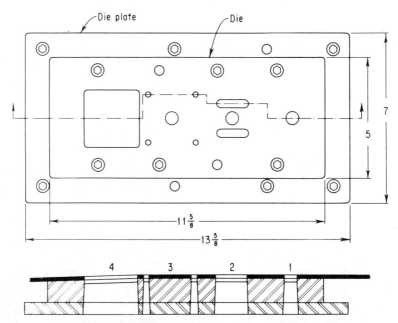

Fig. 3-42. The fitting of the die and die plate. Note the shear on Stage 4, also the straight edge and the relief at die opening.

Step 5, Calculation of Punches

Good practice requires 10 per cent of the metal thickness to be removed from the basic dimension of the blanking punch. This same value is used on the die opening, since holes are to be pierced in the blank. The clearance rule will be applied to the die opening in Stages 1, 2, and 3, and to the punch in Stage 4 (see Fig. 3-43).

For Stage 4: Blank to be 2.375 square; stock thickness = 0.125 in.; 10% = 0.0125. Punch = 2.375 − 0.0125 = 2.3625 in.

Chap. 3 Design of Pressworking Tools   217

Therefore the die opening will equal 2.3755 to 2.3750 in., and the punch will equal 2.3625 to 2.3620 in.

For Stage 2: Slot to be 0.3125 in. wide by 1.3125 in. long. Die = 0.3125 + 0.0125 = 0.3250 in. wide = 0.3250 to 0.3255 in. Die = 1.3125 + 0.0125 = 1.3250 in. long = 1.3250 to 1.3255 in. Punch will equal 0.3130 to 0.3125 in. wide, and 1.3150 to 1.3125 in. long.

The punch and the die opening will have straight sides for at least $\frac{1}{8}$ in. for sharpening, and then will have a taper relief of about $1\frac{1}{2}$ deg to the side. Figure 3-43 also shows a $\frac{1}{8}$-in. shear for the die at Stage 4 and a $\frac{1}{8}$-in. shear for the punches of Stage 2, and also the stepped arrangement of the punches for all stages.

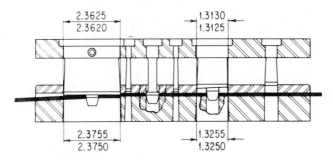

Fig. 3-43. Illustrates calculation of clearance. shear on punches and die, and stepped arrangement of punches to reduce cutting pressure.

Step 6, Springs

A solid stripper plate can be used for this job.

Step 7, Piloting

Figures 3-41 and 3-43 illustrate the arrangement for piloting. In this case it is direct piloting. However, if the part did not have a center hole, and the slots and other holes were too small, indirect piloting would have to be provided.

Step 8, Automatic Stops

Finger stops, illustrated in Fig. 3-46, will act as stops when a new scrap strip is being inserted but, after that, an automatic spring drop stop must be used to halt the scrap strip.

Figures 3-44 through 3-47 illustrate details of the completed drawing of the die.

218 Design of Pressworking Tools Chap. 3

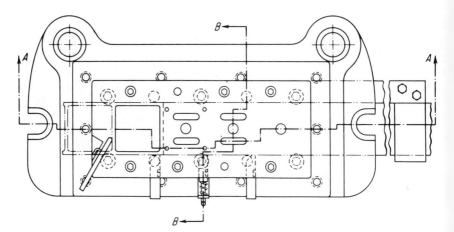

Fig. 3-44. Top view of die with the punch holder removed.

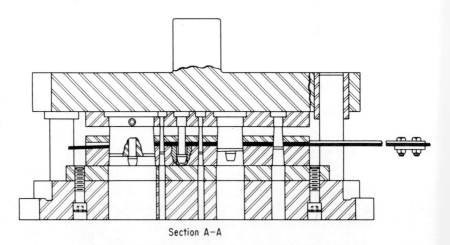

Section A–A

Fig. 3-45. Front sectional view of completed die.

Chap. 3 Design of Pressworking Tools

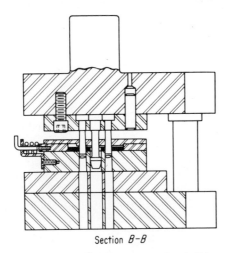

Fig. 3-46. Side sectional view of completed die.

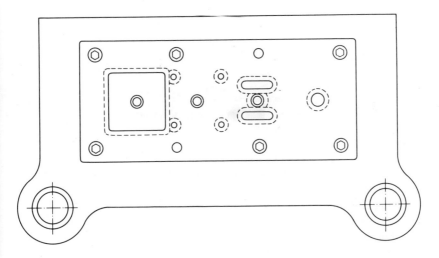

Fig. 3-47. Bottom view of punch assembly.

PROBLEMS

1. Calculate the press tonnage for blanking the parts of Fig. 3-48.
2. Prepare three different layouts for each of the three parts of Fig. 3-48 and estimate the per cent of scrap stock (waste).

3. What punch-and-die clearances should be used for producing the parts of Fig. 3-48?

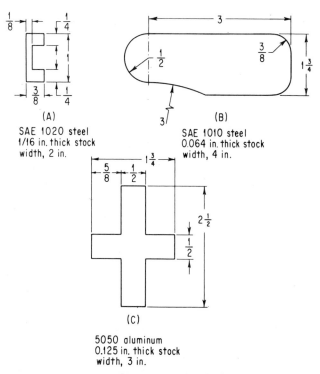

Fig. 3-48. Parts to be blanked.

4. What is the function of a stripper? When should a spring-loaded stripper be used?
5. When should clearances be subtracted from the punch dimensions?
6. What is the purpose of the shear angle found on a punch or die?
7. Define the die cutting of metal.
8. What is the effect of excessive clearance upon die-cut metals?
9. What are proper scrap allowances?
10. Define a compound die. It is often of what design?
11. Why are steel-rule dies used?
12. What are the general rules for die block thicknesses and its margins?

REFERENCES

1. American Society of Tool and Manufacturing Engineers, *Die Design Handbook,* McGraw-Hill Book Company, Inc., New York, 1955.

2. Strasser, F., "Should Die Thickness Be Calculated?", *Am. Machinist,* February 19, 1951.

4

BENDING, FORMING, AND DRAWING DIES

BENDING DIES

Bending is the uniform straining of material, usually flat sheet or strip metal, around a straight axis which lies in the neutral plane and normal to the lengthwise direction of the sheet or strip. Metal flow takes place within the plastic range of the metal, so that the bend retains a permanent set after removal of the applied stress. The inner surface of a bend is in compression; the outer surface is in tension. A pure bending action does not reproduce the exact shape of the punch and die in the metal; such a reproduction is one of forming.

Terms used in bending are defined and illustrated in Fig. 4-1. The neutral axis is the plane area in bent metal where all strains are zero.

Bend Radii. Minimum bend radii vary for the various metals; generally most annealed metals can be bent to a radius equal to the thickness of the metal without cracking or weakening.

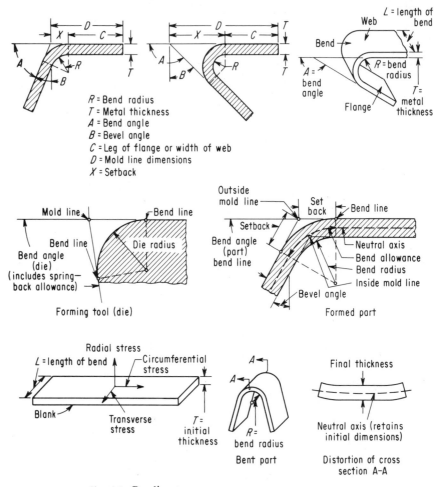

Fig. 4-1. Bending terms.

Bend Allowances. Since bent metal is longer after bending, its increased length, generally of concern to the product designer, may also have to be considered by the die designer if the length tolerance of the bent part is critical. The length of bent metal may be calculated from the equation.

$$B = \frac{A}{360} 2\pi(IR + Kt) \qquad (4\text{-}1)$$

where B = bend allowance, in. (along neutral axis)
A = bend angle, deg
IR = inside radius of bend, in.
t = metal thickness, in.
K = 0.33 when IR is less than $2t$ and is 0.50 when IR is more than $2t$

Bending Methods. Two bending methods are commonly made use of in press tools. Metal sheet or strip, supported by a V block (Fig. 4-2A), is forced by a wedge-shaped punch into the block. This method, termed *V bending,* produces a bend having an included angle which may be acute, obtuse, or of 90°. Friction between a spring-loaded knurled pin in the vee of a die and the part will prevent or reduce side creep of the part during its bending.

Edge bending (Fig. 4-2B) is cantilever loading of a beam. The bending punch, 1, forces the metal against the supporting die, 2. The bend axis is parallel to the edge of the die. The workpiece is clamped to the die block by a spring-loaded pad, 3, before the punch contacts the workpiece to prevent its movement during downward travel of the punch.

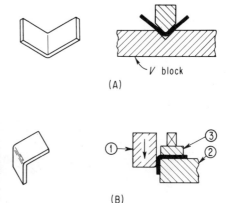

Fig. 4-2. Bending methods: (*A*) V bending; (*B*) edge bending.

Bending Pressures. The pressure required for V bending is

$$P = \frac{KLSt^2}{W} \quad (4\text{-}2)$$

where P = bending force, tons
K = die opening factor: 1.20 for a die opening of 16 times metal thickness, 1.33 for an opening of 8 times metal thickness
L = length of part, in.
S = ultimate tensile strength, tons per sq in.
W = width of V or U die, in.

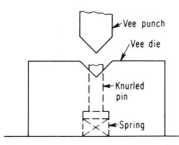

Fig. 4-3. Clamping a part in a V die.

For U bending (channel bending) pressures will be approximately twice those required for V bending; edge bending requires about one-half those needed for V bending.

Springback. After bending pressure on metal is released, the elastic stresses are also released, which causes metal movement resulting in a decrease in the bend angle (as well as an increase in the included angle between the bent portions). Such a metal movement, termed spring-back, varies in steel from ½ to 5°, depending upon its hardness; phosphor bronze may spring back from 10 to 15°.

224 Bending, Forming, and Drawing Dies Chap. 4

V-bending dies customarily compensate for spring-back with V blocks and wedge-shaped punches having included angles somewhat less than that required in the part. The part is bent through a greater angle than that required but it springs back to the desired angle.

Parts produced in other types of bending dies are also overbent through an angle equal to the spring-back angle with an undercut or relieved punch.

Evolution of A Bending Die

The production of a workpiece of Fig. 4-4 in the die of Fig. 4-5 required blank development before die design began.

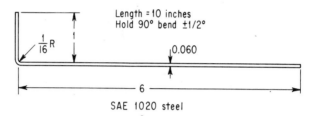

Fig. 4-4. Part bent in the die of Fig. 4-5.

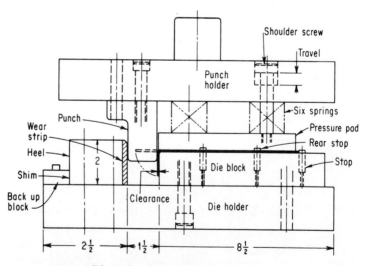

Fig. 4-5. Die design for part of Fig. 4-4.

The straight length of the vertical leg is $1.000 - 0.060$ or 0.940 in.; the straight length of the horizontal leg is $6.000 - 0.060$ or 5.940 in.

Chap. 4 Bending, Forming, and Drawing Dies

The bend length (since *IR* is less than twice metal thickness) is, from Eq. (4-1),

$$B = \frac{90}{360} 2\pi \left(0.625 + \frac{0.060}{3}\right)$$
$$= 0.1295 \text{ in.}$$

The developed length is

$$0.940 + 5.940 + 0.1295 = 7.0095 \text{ in.}$$

To hold the tolerance of $\pm\frac{1}{2}$ deg allowed for the 90-deg bend, the designer decided that an edge-bending die, with a slight ironing action on the stock, be used.

Based upon Eq. (4-2), the bending pressure needed without ironing is

$$F = \frac{1.33 \times 10 \times 30 \times (0.060)^2}{2 \times 1.414} = 0.51 \text{ ton}$$

The total spring pressure required of six springs in the pressure pad is 1020 lb.; each spring will supply a pressure of 170 lb. Commercial 1-in. diam die springs, 2 in. long, will easily supply this pressure. Almost any small OBI press will supply these pressures with an ample allowance for slight ironing of the blank, and has a bed area large enough to accommodate a die set. There are no formulae for determining ironing pressure; it can be approximated by multiplying the yield strength of the metal by the thickness of the metal after reduction times its length.

Since the size of the blank to be sent is 10×7 in., the area of a die set 14 in. (right-to-left) by 10 in. (front-to-back) allows for mounting of the punch and pressure pad on the upper shoe and the die block and heel to the lower shoe.

The blank is located on the die block against an end-stop pin and two rear-stop pins. On the downstroke of the press, the pressure pad clamps the blank in this location.

The descent of the punch forces the end of the blank against the end of the die block. Its wiping action results in some ironing of the blank, the amount of which is determined by the clearance between the heel block and the punch. To establish optimum clearance and to allow for wear on punch and heel block, shims can be inserted between the backup and heel blocks.

The surface of the heel block against which the punch rubs can be hardened or can have a bronze wear strip as shown.

FORMING DIES

Forming dies, often considered in the same class with bending dies, are classified as tools that form or bend the blank along a curved axis instead of straight axis. There is very little stretching or compressing of the material. The internal movement or the plastic flow of the material is localized and has little or no effect on the total area or thickness of the material. The operations classified as forming are embossing, curling, beading, twisting, and hole flanging.

A large percentage of stampings used in the manufacturing of products require some forming operations. Some are simple forms that require tools of low cost and conventional design. Others may have complicated forms, which require dies that produce multiple forms in one stroke of the press. Some stampings may be of such nature that several dies must be used to produce the shapes and forms required.

A first consideration in analyzing a stamping is to select the class of die to perform the work. Next to be considered is the number of stampings required, and this will govern the amount of money that should be spent in the design and building of the tools. Stampings of simple channels in limited production can be made on a die classed as a *solid form die*. It would be classified under *channel forming dies*. Others—the block and pad type—are also channel forming dies. Such operations as curling, flanging, and embossing as well as channeling employ pressure pads.

A forming die may be designed in many ways and produce the same results; at this point the cost of the tool, safety of operation, and also the repairing and reworking must be considered. The tool that is cheapest and of the simplest design may not always be best because it may not produce the stamping to the drawing specifications. Where limited production is required, and a liberal tolerance is allowed in a stamping, a solid form die can be used.

Solid Form Dies

The solid form die is usually of simple construction and design. Stampings produced in these dies are usually of the flanged clamp type, such as pipe straps, etc. They are made of metal which is of a soft grade, mostly strip stock, and the grain runs parallel to the form. Some distortion is encountered; this can be compensated for in the designing of the templates. A male and a female template are usually made. The male template is made to the contour that will be shaped on the punch of the die—the same as the inside contour of the stamping. The female template is made to the outside contour of the stamping and will be shaped on the die halves. By so doing, allowance has been made for the thickness of the metal, when both die halves are set in place. Figure 4-6 illustrates both templates in place.

A forming die of this kind need not be mounted on a die set. A die set should be considered because of the amount of time that can be saved in setting the die for production; also it eliminates the chances for misalignment in setting the die. A die that is not properly set could cause some pinching of the metal, thereby causing the stamping to break. Considering that a die shoe and punch holder are required in each case, the cost of adding the leader pins to complete the die set is nominal, and should be saved in a short time. A great deal of side pressure is exerted on the die blocks, and must be considered in the designing. The die block should be made of more than one piece of tool-steel. This is necessary to eliminate the possibility of cracking the die block if the operator should feed a double blank or if metal of a thicker gage is used. The die blocks should be tied together by means of cross pieces. The die blocks should be sunk into the die shoe, to obtain some support from the edge of the sinking. The blocks should be constructed wider than they are high, a proportion of at least $1:1\frac{1}{2}$, with large sturdy dowel pins. The form edge of the blocks must be of proper radius to prevent digging on the side of the stamping. The radius should not be less than twice material thickness, and for best results the edge should have a high polish. A smaller radius could cause some fatigue in the material when formed.

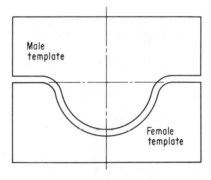

Fig. 4-6. Templates for forming.

The punch is made of tool-steel, hardened according to the severeness of the operation. It should be designed long enough to allow complete forming of the part without interference with the punch holder. The width and shape are governed by the part to be formed, and at all times it must be twice material thickness smaller than the width or span of the die blocks.

The screws and dowel pins should be spaced properly and should be located so that they will not mar the stamping. Consideration must be given to stripping the formed part from the punch. This can be accomplished by means of a knockout, stripper hooks, or stripper-pin (spring) construction. It is important to consider and plan for the removing of the formed part. Figure 4-7 shows these details and also illustrates the gages necessary for locating the stamping.

The shoe (1) should be thick enough to withstand the pressure required for the forming operation, and in selecting its thickness the size and shape of the hole in the bolster plate of the press must also be considered. The A and B dimensions of the shoe when placed over the hole of the bolster plate should also be long and wide enough to allow ample space for clamping it to the bolster plate. The depth of the sunk-in d section should be considered when selecting shoe thickness. The die shoes can be had in cast iron, semisteel, or steel. It is important that the proper material be selected for the die shoe, because the shoe must withstand a good share of the force applied during the forming operation. A die shoe made of steel will give good service, and costs only a little more than a cast or semicast shoe. The diameter of the leader pins (3) should be selected according to the working area of the shoe, or by consulting a die set manufacturers catalog. The length of the pins should be at least $\frac{1}{4}$ in. shorter than the shut height of the die, as listed on the drawing. The guide bushings (4) should be of the shoulder type, and for a die of this kind can be the regular-length type. (Guide bushings are made in three lengths—regular, long, and extra long.)

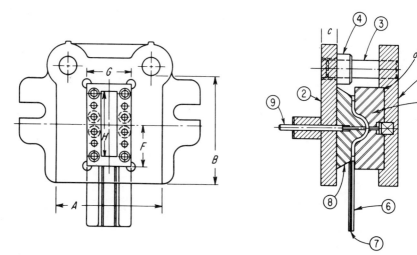

Fig. 4-7. A solid form die.

Die sets are made in two types—precision and commercial. Precision die sets are considered for stamping work that requires great accuracy in alignment, such as between punch and die parts of cutting dies. For secondary operations, such as bending, forming, or other noncutting operations, commercial die sets should be specified.

The punch holder (2) is the same as the shoe in the a and b dimensions. The thickness c should be $1\frac{1}{4}$ in. and have a 2-in. diam shank. The shank is always placed on the centerline of the punch holder of a regular stock die set. If it is necessary to locate it off center, the set becomes a special, therefore costing more. The shanks of die sets come in various diameters—$1\frac{1}{2}$, $1\frac{9}{16}$, 2, $2\frac{1}{2}$, and 3 in.—also can be had in special diameters when specified, at extra cost. The shank is used to align the centerlines of the die with the centerlines of the press. It also is clamped securely in the press ram, and must help lift the punch and stamping from the die. Punch shank diameters are selected according to the hole in the press ram. The 2-in.

diam shank is the most popular one, because it provides a strong shank and eliminates the use of collars. The shanks 1½ and 1 9/16 in. in diameter are 2⅛ in. long; the others are 2⅞ in. long.

Die blocks (5) are of a two-piece construction. They should be made of tool-steel that will withstand excessive wear and galling. Toughness and shock resistance should be considered in selecting the kind of tool-steel. It should be an oil- or air-hardening steel, and in a good many applications a double draw will produce some added toughness and abrasion (galling) resistance in the steel. The die blocks take most of the wear, so care must be taken to design them properly. The length F of the die blocks must always be longer than the height G, at least one and one-half times as long. The width should be governed by the width of the stamping to be formed, taking into account also the space required to fasten gages, to locate the stamping. The screws and dowels should be spaced to help withstand some of the forces that will be encountered in the forming operation. The screw and dowel holes should be located so that they will not mar the stamping. The dowel pins should be large enough in diameter to help withstand some of the spreading force that will be encountered. The screws should be large enough to compensate for these forces and should be located to help control them. The screw and dowel hole locations should also be considered for locating the gage plate (6). Designing the gage plate shape when designing the die blocks, and then placing the screw and dowel holes accordingly, accomplishes a dual purpose. It helps eliminate some of the holes required in the die blocks, as well as the work of drilling and tapping the holes and the cost of the extra screws.

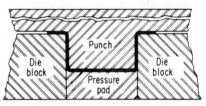

Fig. 4-8. A pad-type form die.

The shape of the stamping to be formed should be studied, and if small radius corners are required it should be designed for a pad type rather than a solid form die (Fig. 4-8). The radius in the corners should be at least five times metal thickness in the solid form die design.

The gaging of the stamping should be studied, and whenever possible the blank should be fed the long way. The die should be located on the die set so as to permit the blank to feed the long way. When blanks are fed the short way, they have a chance to twist, and the operator loses control over them. This causes a loss of time in locating the blank in the die, and can cause mislocation of the blank.

The gage plate should be made so that the blank can slide into the proper location for forming. A slide should be provided with sides (7) whose height is at least one and one-half times metal thickness. The pocket for locating the blank in the length should stop it on both ends and the clearance should be governed by the stamping drawing tolerances. The width of the slide pocket should be sufficient to include the mill tolerance of the strip. When the blanks are wide and flat it is necessary to provide some ribs or wires to help slide the blank into the die and reduce some of the friction caused by the oil on the blank surfaces. The slide should be long enough to prevent the operator from sliding his fingers under the punch. The safety factor most commonly used is to have it long enough for the operator to have the web of his thumb rest at the front edge of the gage, and with his hand spread forward, have his center finger clear the punch or die opening (approximately 6 to 6½ in.).

The punch (8) should be of tool-steel; it is usually made of the same kind as the die blocks. The width of the punch is usually governed by the width of the stamping being formed. When narrow-width stock, ½ in. or less, is used, care must be taken to add flanges to the sides to make it wide and strong enough at the base.

The screw and dowel holes should be placed to eliminate the possibility of marring the stamping. The screws and dowels should be large enough to take care of the forces encountered by the punch in the forming operation. The screws and dowels should not be located to cut into the punch shank. The only hole we should consider putting into the punch shank at any time is for a knockout rod, and this is most always in the center of the shank diameter. When possible, a knockout arrangement using a knockout rod (9) should be employed for releasing the finished stamping from the punch. This usually is simple to design, and making it is most economical. Also, if the press bar for pushing the knockout rod is properly adjusted, the removing of the stamping is foolproof and should increase production.

All the comments made in this section (on the solid form die) relating to die sets and their selection, screws and dowels, die blocks, punches, etc., apply to the design of other forming dies. They are necessary details to be considered for most die construction, and will not be referred to again.

Forming Dies With Pressure Pads

When the forming of stampings requires accuracy, dies employing pressure pads are often designed. The pressure pad helps to hold the stock securely during the forming and eliminates shifting of the blank. The pressure can be applied to the pad by springs or by the use of an air cushion (Fig. 4-9 A and B). When springs are used, they can be located directly under the pad and confined in the die shoe (Fig. 4-9A). They may also be located in or under the press bolster plate; and by the use of pressure pins, which are located under the pad, and through the die shoe, pressure is applied to the pressure plate.

Chap. 4 Bending, Forming, and Drawing Dies   231

Pressure pins are also used with an air cushion. The construction of the pressure plate and pressure pins would be the same as shown in Fig. 4-9B except that an air cushion is substituted for the springs.

When springs are used to apply pressure to a pressure pad, spring pressure increases (in pounds) with the pad travel. Each fraction of an inch of travel increases the pressure on the pad. This could cause some trouble in stampings of light-gage material, because too much pressure may cause the metal to stretch. When springs are used, a certain amount of pressure is lost owing to the springs' setting (losing height after being worked). When an air cushion is used, the proper amount of pressure on the pressure pad is assured as long as air supply is set properly. It is important to have a set amount of pressure on the pressure pad to control the quality of the stampings.

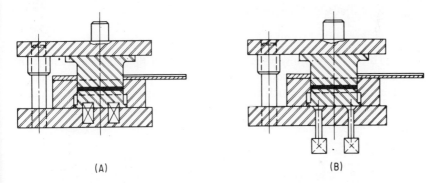

Fig. 4-9. Pressure-pad-type form dies.

The pressure pad, the moving member of the die, must always be controlled in its travel between the die blocks. This can be done by means of retaining shoulders, or by shoulder screws. When using the retainer shoulder construction, a recess is machined into the form blocks, and a corresponding shoulder is machined on the pad. The retainer shoulder should always be made strong enough to withstand the pressure applied by either springs or air cushion. The size of the shoulder to be used varies according to the size and metal thickness of the stamping. A good rule to employ is to have the height of the shoulder one and one-half times the width (Fig. 4-10).

Always design the shoulders of the pad with a radius in the corner. When the pad is made of hardened tool-steel, heat treatment should specify a double draw of the shoulder section.

When using shoulder screws to control the travel of the pad, the **die** shoe must be thick enough to permit sufficient travel.

The pressure pad should always travel so that it extends slightly above the die blocks. This will insure uniform parts, because there will be pressure to lock the part between the punch and pad faces, before the actual forming takes place.

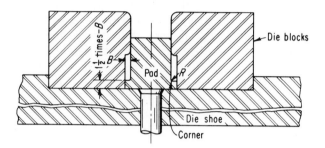

Fig. 4-10. Pressure pad design.

The amount of travel the pad should have depends upon the height of the form die. It is not always necessary to travel the full height; in many cases half the die's form height is sufficient. When a blank is distorted, or has a tendency to curl, which may cause the completed blank to be out of square, it may be necessary for the pad to travel the full length. It is necessary for the pad to bottom on the die shoe, to allow the punch to give

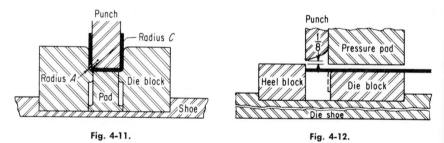

Fig. 4-11. **Fig. 4-12.**

Fig. 4-11. Radii considerations for form die design.

Fig. 4-12. Die with a heel block and relieved punch.

the part a definite set at the bottom of the stroke. When a stamping must have sides that are square with the bottom, after forming, the corner radius should be set. This is done by designing the die blocks with the correct radius A, Fig. 4-11. The pressure pad is made to match the height of the die blocks' radius edge. The punch radius C is made slightly smaller, approximately 10 per cent less than the die block radius.

It may be necessary to machine a slight angle on the side of the punch to allow a slight overbending of the side being formed. This ensures that the sides of the formed part will be square with the base after forming.

Single and multiple pressure pads are used in die construction. The single pressure pad is used when the forming is done in one direction. It is most commonly used for forming short flanges, tabs, lugs, or ears at right angles to the base of the part. The pressure pad is used to support the base of the part accurately, either by pilot pins or other gages, and by its securing the part properly, the part is formed with great accuracy. The side or tab to be formed may be bent downward as well as upward. When the side or tab is being bent downward, the length may vary slightly, because the metal is stretched or drawn more. Some of this may be overcome by having an angle and radius on the punch as shown in Fig. 4-12. The greater the angle and radius, the less bending pressure is required. When a side is bent down, a heel block is required to help support the punch before it starts to do any forming. It should be at least two metal thicknesses higher than the die block. The pressure pad must travel at least $\frac{1}{8}$ in. beyond the edge of the form punch. This is done to assure holding pressure before any forming work is done. The punch should travel far enough beyond the corner radius to smooth out the formed side.

Multiple pressure pads are used when a series of forms are necessary; they are used mostly in progressive dies, when several bends are required on small precision parts. A combination of stationary form blocks, supplied by pressure pads, helps lift the strip so it can be advanced from one station to the next.

Embossing Operations

In an embossing operation a shallow surface detail is formed by displacement of the metal between two opposing mated tool surfaces. In one surface we have the depressed detail, on the other the relief detail. The metal is stretched into the detail rather than being compressed. Embossing is used for various purposes, the most common being the stiffening of the bottom of a pan or container; the embossing is designed to follow the outside profile of the part. A round can may have an embossed circle or raised grooves of various widths or panels. When the can or box is square or rectangular, such embossments follow their contours. Embossings often are ribs or crosses stamped in the metal to help make a section of a blank stronger by stiffening. An embossing die can be a male and female set of lettering dies, or a profile of one of various shapes.

The method of constructing the die blocks for an embossing operation depends on the size and shape of the form, also the accuracy and flatness required. When embossing simple shapes such as stiffening ribs, it is not necessary to fit up the die to strike the bottom. The metal stretches over the punch and across the two radius edges of the die hole (Fig. 4-13).

The die opening has the same width of the rib or embossing *a,* and a slight radius *b* is added to the edges of the opening to allow the metal to

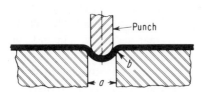

Fig. 4-13. An embossing die.

flow freely. The punch is made slightly smaller than the required metal thickness per side, so that it does not strike along this area. By constructing it in this manner the pressure required to stamp the embossing is reduced.

When embossings are of the lettering type, such as depressed or raised letters for name plates, care must be taken to see that both dies are properly located and doweled in place (Fig. 4-14).

The male die is located in a pocket or recess, and keyed in place. By lining up the female die profile to correspond with the male die profile, and keying it in place, a good stamping or embossing can be made. Stamping operations of this kind require precision work by the toolmaker; the dies are easily damaged by misalignment.

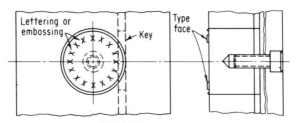

Fig. 4-14. Lettering (embossing) die detail.

A small embossing is often used as a weld projection nib. These nibs are used to weld piece-parts together. There are two kinds: a button type, which we use for light-gage metal, $3/32$ in. or less thick, and a cone type for heavier-gage metal (Fig. 4-15). Care must be taken in their design, because if the projection is too weak the nib will collapse before a good weld can be made. A nib that is too thick or heavy through the section will require too much pressure to produce good welds. The piece-parts are heated electrically, and this causes the projections to melt, so if the projections are too heavy or too light, the heat and pressure required can cause trouble. Table 4-1 lists the correct design for both kinds of projections.

One or more projections can be used on a part for welding purposes. The design of the part and its use govern the number and size of the projections.

Chap. 4 Bending, Forming, and Drawing Dies

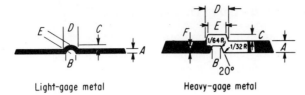

Fig. 4-15. Die for embossing weld-projection nibs.

When embossing ribs or shapes in blanks, it is best to have the die blocks large enough to cover the whole blank. A die block that is too small holds the metal between the die and pressure pad. The portion of the blank not held will distort and cause the metal to twist, wrinkle, and pucker.

When designing a die with embossings, regardless of whether they are formed up or down, a way must be provided to lift the form out of the die pocket. This is usually done by pressure pads or ejector pins.

TABLE 4-1. DESIGN GUIDES FOR PROJECTIONS (24- TO 5-GAUGE STEEL) [1] *

(for projections in 1010 SAE steel)

Type	U.S.S. Ga.	A	B	C	D	E	F	G
Button-type projection	24	0.025	0.050	0.025	0.109	0.025	..	..
	23	0.0281	0.050	0.025	0.109	0.025	..	..
	22	0.0312	0.050	0.030	0.125	0.030	..	..
	21	0.0344	0.050	0.030	0.125	0.030	..	..
	20	0.0375	0.050	0.035	0.125	0.035	..	..
	19	0.0437	0.050	0.040	0.125	0.035	..	..
	18	0.050	0.050	0.040	0.156	0.040	..	..
	17	0.0562	0.055	0.040	0.156	0.045	..	..
	16	0.0625	0.060	0.045	0.172	0.050	..	..
	15	0.0703	0.075	0.045	0.172	0.055	..	..
	14	0.0781	0.075	0.050	0.180	0.065	..	..
	13	0.0937	0.075	0.050	0.180	0.065	..	..
Cone-type projection	12	0.1093	0.080	0.055	0.172	0.080	0.090	0.131
	11	0.125	0.080	0.055	0.172	0.080	0.100	0.138
	10	0.1406	0.080	0.060	0.172	0.080	0.110	0.145
	9	0.1562	0.080	0.060	0.172	0.080	0.122	0.155
	8	0.1718	0.080	0.060	0.190	0.080	0.138	0.166
	7	0.1875	0.080	0.070	0.203	0.094	0.166	0.185
	6	0.2031	0.080	0.070	0.203	0.094	0.182	0.206
	5	0.218	0.080	0.075	0.210	0.100	0.200	0.220

* Superior numbers indicate references listed at the end of the chapter.
All dimensions are given in inches.

Beading and Curling Dies

In beading and curling operations, the edges of the metal are formed into a roll or curl. This is done to strengthen the part or to produce a better-looking product with a protective edge. Curls are used in the manufacturing of hinges, pots, pans, and other items. The size of the curl should be governed by the thickness of the metal; it should not have a radius less than twice metal thickness. To make good curls and beads, the material must be ductile, otherwise it will not roll and will cause flaws in the metal. If the metal is too hard the curls will become flat instead of round. If possible, the burr edge of the blank should be the inside edge of the curl. This location facilitates metal flow and also helps keep the die radius from wearing or galling. In making curls and beads a starting radius is always helpful and should be provided if possible (Fig. 4-16).

The curling radius of the die must always be smoothly polished and free of tool marks. Any groove or roughness will tend to back up the metal while it is rolling and cause defective curls. The inside surface of the blank must be held positively in line with the inside curling radius of the punch (Fig. 4-17).

When curling or beading pots, pans, cans, or pails, wires are often rolled inside the curls to make them stronger. The wire is made to the contour of the pan and placed on a spring pad. When the curling die descends, the edge of the pan is forced to curl around the wire as shown in Fig. 4-18.

Fig. 4-16. Starting curl radius.

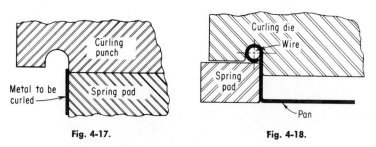

Fig. 4-17.   Fig. 4-18.

Fig. 4-17. Curling punch design.

Fig. 4-18. Curling die design.

Bulging Operations

In a bulging operation a die forms or stretches the metal into the desired contour. By using rubber, heavy grease, water, or oil the metal is forced, under pressure, to take the shape of the die. To facilitate the removal of the part from the die after bulging, the die is of a split construction. Dies that are split in halves or sections must have strong hinges and latches. The sections when operated must work freely and rapidly and, when under pressure, be tight enough so that the sections will not mark the stamping. The pressures required for these operations are usually found by experimenting during the first setup of the die. This is necessary because other variables besides metal thickness and the annealed condition of the stamping must be considered. Most stampings that are to be bulged have been work-hardened by previous drawing operations and therefore need annealing. Most metal that is not soft enough will rupture in the bulging operation.

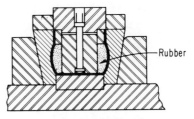

Fig. 4-19. A bulging die.

Rubber is the cleanest and easiest material to use for bulging operations (Fig. 4-19). Once it is designed and tried out, the shape and hardness can be duplicated for replacement. Neoprene rubber of medium hardness is the most suitable.

The rubber should be confined between two punch sections. The upper section slides on the arbor on which the rubber pad is assembled. The lower section is fastened to the arbor. It acts as a stop to locate the rubber in the correct place for the bulging operation. As the punch decends, the rubber forces the metal into the exact shape of the die. Once the press is set to the required height, all the stampings should be of equal quality.

The use of grease or tallow results in a handling problem. First of all, the same amount must be used for each stamping. Secondly, these substances are messy to handle both before and after the bulging operation. They also require extra work in the cleaning of the finished stamping because all traces of grease or oil must be removed. The die and punch must be designed to confine the grease securely in the desired area. A loss of grease owing to escaping or squirting out of the die area results in inferior stampings.

Water or oil requires a pump to produce the required pressure. This also causes other undesirable conditions such as operators' clothing getting wet, rusting of the tools and press, and also the necessity for pump maintenance. The die for bulging using water pressure must be designed with care. Its open end or the ends of the stamping should be sealed securely.

For tube forming, enough extra length must be allowed on each end for the end-plugs (Fig. 4-20).

One end-plug must be designed so that water and pressure can be fed into the tube. The other end-plug can be designed without any feed lines, but for each end-plug a seal-off gasket must be designed. The die is usually split along the centerline of the contour to be formed, so that it can be lifted out of the die without any trouble. The press or press brake that is used to hold the die halves down during the forming operations should have a bed or bolster area that is longer than the die for proper die support.

The press must have enough tonnage to keep the die halves securely locked during the pumping cycle of the bulging operation; otherwise a large flashing line will be produced on the stamping along the parting line of the die. This is not desirable because it requires extra labor to remove the line; also it tends to break down the edges of the die along the profile of the form. Bulging dies are costly, and extra care should be taken to keep them in good shape.

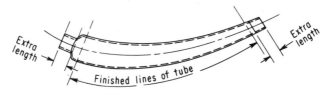

Fig. 4-20. Tube design.

When using a die to bulge a tube with water pressure the first step is to see that the tube is properly plugged on one end with a solid plug (no feed line). Next the tube is filled with water, leaving only enough room for the end-plug with the feed line or hose attached. After fastening the end-plug in place, the tube is laid in the die. The next step is to lower the press ram and close the die. The pump is started, but before any pressure is applied to the water, the tube must be filled with water to capacity and there must be no air pocket. The tube is now ready for water pressure application and expansion.

The water pressure necessary to expand a stamping varies according to the thickness and kind of metal to be formed. The pump must be large to do a variety of work.

In most work the pumping cycle is of short duration, from 15 to 30 seconds to complete the form. It takes a much longer time to get the tube ready for the operation; therefore, care must be taken to have everything in order at all times. For uniform stampings in bulging operations using water, the metal must be controlled for thickness of the wall and hardness, and the press and water pressures must be the same at all times.

Chap. 4 Bending, Forming, and Drawing Dies   239

Twisting Operations

A twisting operation usually is done on flat strip blanks. The demand for this operation is limited; therefore, the cost of the die and the amount of production required must be considered. On simple strip blanks of light-gage stock that require a twist on one end, a hand fixture may be most economical. A good operator can produce almost as many stampings with a hand fixture as with a press die. Most stampings, after being twisted, are difficult to remove from the tools. The twisted stamping usually lies in the die, and, if the blank is short, it generally falls between the form blocks. A twist on flat strip blanks is usually 90°, but the angle can vary to suit the stamping design. The metal is usually hot rolled SAE 1010 steel, but must be $\frac{1}{4}$ hard or softer in temper. Too hard a strip will show fractures

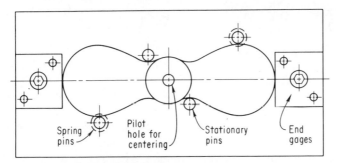

Fig. 4-21. Twisting die design.

along the edges after twisting. It is best to use rolled strip stock, with the mill edge, because a sheared edge could produce fractures in the twisting of the metal.

Twisted strips or brackets make sturdy braces or brackets at low cost. Some items, such as fan blades, also require a twisting operation with dies that are made with a great deal of care. Each blade must be bent the same amount, and the spring-back of the metal must be taken into consideration. In designing a die of this kind the feeding of the blank into the die, its location in the gages, and the removing of the finished stamping must be considered. Uniform location is the first requirement, and next is the ease of removing the blade without distorting it. Blanks for fan blades usually have a hole in the center, which is used for assembling the finished blade to the drive shaft. This hole should always be used to locate the blank centrally, and other gages should be provided along the outside profile as shown in Fig. 4-21.

A pilot can be used for the center hole locator. Spring pins which disappear when the punch comes down can also be used to locate the outside contour of the blade. For stationary pins, clearance must be provided in the punch. Gages always hold the blank in the same position. In twisting blades, the blank must be prevented from rolling while being formed. The dies for twisting blades must be made with care, so that each blade is made with a uniform and equal amount of form. A slight difference could cause the blade to vibrate. This would mean a costly balancing operation, which should be eliminated if possible.

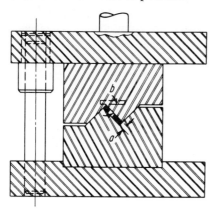

Fig. 4-22. Twisting die block considerations.

The die blocks for twisting flat strip blanks are simple in design. In fact the upper and lower dies can be made or shaped in one piece and then cut to the desired length. When locating them on the die shoe and punch shoe, care must be taken to provide the proper amount of opening for the metal thickness (Fig. 4-22a). The angle on the die block faces (b) can be made to the desired angle of the part, but an allowance must be made for spring-back of the metal.

A die of this kind can be made for a blank with a single twist, or it can be made to make any number of twists in a part. The designer must be sure that the blank area that is to remain flat is held securely in the die before the twisting operation is performed. This is usually accomplished by having a stationary block on the bottom or die shoe, and a block with spring pressure on the upper die shoe. The spring pressure should be strong enough to keep the part from tipping or moving while being twisted.

For flat blanks that require a twist, and where accuracy is not a requirement, a fork lever can be made. The part can be placed and gaged in a bench vise for length location, and the operator slips the forked (slot in the lever) lever over the area to be twisted. The amount of twist is governed by the operator and can be controlled by setting a gage to show him where to stop. The slot width, A, Fig. 4-23, in the fork lever is made to suit the metal thickness. The depth B of the slot should always be made $\frac{1}{4}$ in. or more than the width of the part. This will give the operator control over the full face width of the part being twisted. The fork ends must be made strong enough to withstand the pressure that is required to twist the blank.

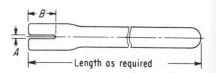

Fig. 4-23. A fork lever for twisting.

Coining Dies

Coining dies are used to emboss on the part the detail engraved on the faces of the punch and die. By compressing the metal between the punch and die, metal flows into the detail or embossing. Coining dies produce coins, medals, medallions, jewelry parts, ornamental hardware, plates, and escutcheons.

Hydraulic or knuckle-joint presses are usually used to perform coining operations. To force the metal to flow into the detail requires very high tonnage. One of the reasons for the high tonnage is that when metal is compressed it hardens and toughens. This is called *work-hardening* of the metal, and the more complex a part, the greater its resistance and the tonnage required.

During the coining operation the part must be properly confined. The thickness of the part is controlled by the die block surfaces. The closed height of the die is checked by the surfaces of the blocks. The die block surfaces, on which the detail is inscribed, must be highly polished and free of scratches or mars. That is

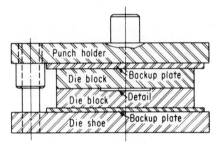

Fig. 4-24. A coining die.

necessary because the slightest scratch or mark will be embossed into the part. The sides of the die that control the outside contour of the part must be slightly tapered to allow removal of the part from the dies.

When designing dies for a coining operation, extra caution must be taken to build the dies strong enough to withstand severe pressures. The pressures of the metal being coined, regardless of the kind, are severe, and heavy die sections are necessary. Polish and the finish of the die surfaces must be the best. The slightest mark could cause the die to split. The matching sections of the dies must be made to very close tolerances, especially if the die has moving parts. The slightest opening will allow the metal to squirt, or be forced into the opening, and could cause the die to burst. The die shoe and punch holder are designed thicker, or heavier, than for most dies because of the heavy tonnage required. Tool-steel backup plates hardened to a good tough hardness of R/C 50-52 are used to back up the dies on the shoe and punch holder to prevent the dies from sagging, or bending, under pressure (Fig. 4-24).

The plates should be wider and longer than the die blocks, so that they will support the full face of the die blocks. The bolster plate of the press should be solid, or have as small a hole as possible, to help support the die.

The screws and dowels must be located away from the coining detail, so that they will not mar the part, nor weaken the dies. The screws and dowels should be large enough to hold the dies firmly and securely. Dies for coining receive rough wear, and they wash out owing to the pressure of the metal as it flows into the detail. This makes for short die life, adding to the cost of the operation.

Swaging Dies

A swaging operation is similar to a coining operation, except that the part contours are not as precise. The metal is forced to flow into depressions in the tool faces, but the remaining metal is unconfined, and it flows generally at an angle to the direction of the applied force. The flow of the metal is restricted somewhat by the tool faces, but an overflow flash is usually encountered, and must be removed in a subseqeunt operation. The upsetting of heads of bolts, rivets, pins, and many cold- and hot-forging operations are classified as swaging operations. The sizing of faces or areas on castings, which is often referred to as planishing, is also classified as swaging. The planishing of faces, especially around hubs or bosses of casting or forging, is considered desirable, because it increases the wear resistance of the area as much as 80 per cent, as compared to a similar machined surface. The faces of connecting rods and piston rings are cold sized, in order to increase the hardness of the wearing surfaces and to make them smooth. A surface of this kind cannot be accomplished by milling or machining; moreover, the squeezing operation can be done with a die about ten times as fast as by milling. Copper electrical terminals are also made by swaging dies.

The presses used for swaging operations must be selected according to the size of the work and the interval of time necessary to complete the operation. Knuckle-joint presses and hydraulic presses, of extra-heavy tonnage capacities, are usually used. Hydraulic presses have a decided advantage over knuckle-joint presses because of the extra dwell at the bottom of the stroke, which puts a definite "set" in the work. Knuckle-joint presses have short powerful strokes and can compress the metal but, lacking the extra-dwell feature at the bottom of the stroke, slight variations of the swaged parts result.

Pressures up to 100 tons per square inch are applied to metals in swaging operations. Presses ranging from 25 to 2500 tons are used.

The most practical way of determining the size of the press for swaging a part is to squeeze the first parts with the dies placed in a hydraulic press provided with tonnage gages which will eliminate any guesswork. If a mechanical or knuckle-joint press is used, one should be selected with a safety factor of three or more times the maximum work pressure. Swaged and cold-sized parts are highly compressed; the metal becomes harder (work-hardened) and more dense. The more metal to be moved, the greater

Chap. 4 Bending, Forming, and Drawing Dies   243

its resistance. All these factors should be considered when selecting a press for swaging to avoid the possibility of broken machinery. When computing the pressures required for swaging or cold sizing, the correct ultimate compression strength of the metal must be used. The formula used to compute the pressure is

$$P = \frac{A \times S}{2000} \qquad (4\text{-}3)$$

where P = pressure required, tons
A = area to be sized, sq in.
S = ultimate compressive strength of the metal, lb. per sq in.

Tool-steel used for swaging and sizing operations must be of high strength. Chrome-tungsten oil-hardening steel, which combines high hardness with maximum toughness, is used for swaging dies. High-carbon high-chrome steels are also used, but they are more difficult to machine, and their resistance to shock is not as high as that of chrome-tungsten oil-hardening steels.

The die body sizes must be extra heavy. The shut height of the die is checked by the surfaces of the blocks. The area of the stop and the top and bottom surfaces of the block should be large enough to allow the block to withstand three times the yield strength of the workpiece metal. The contour or profile of the part to be sized is usually machined into the

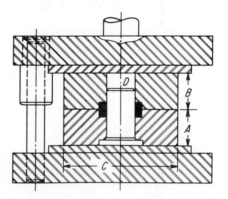

Fig. 4-25. A swaging die.

surface of the die blocks. These surfaces are in a plane coinciding with the longitudinal center plane of the workpiece. To avoid using excessive pressures, the dies are relieved where no pressing is done. A draft is provided along all edges of the work that are not squeezed. There is also a 45-deg draft around the bosses to be sized to facilitate removal of the finished work. In some cases the parts are machined about $\frac{1}{32}$ in. oversize before being placed in the sizing dies, which eliminates the chances of overtaxing the dies and also improves the dimensional quality of the parts.

Die shoes and punch holders are similar to the ones used for coining dies. They must be heavy, with wide and long base surfaces. Figure 4-25 illustrates general principles for designing a swaging or sizing die. The die blocks have identical thicknesses, A and B, and lengths, C. The center of the part is located in exactly the same position, D, in each block. The screw and dowel holes must not be located near or in the area of the part to avoid marking it.

Hole Flanging or Extruding Dies

The forming or stretching of a flange around a hole in sheet metal is termed *hole flanging* or *extruding*. The shape of the flange can vary according to the part requirements. Flanges are made as countersunk, burred, or dimpled holes.

When countersunk shaped extruded holes are made in steel, it is necessary to coin the metal around the upper face and beveled sides to set the material. The holes are also made about 0.005 in. deeper than the required height of the rivet or screw head, which allows bunching that occurs when squeezing the rivet in place. A section of a die for this purpose is shown in Fig. 4-26.

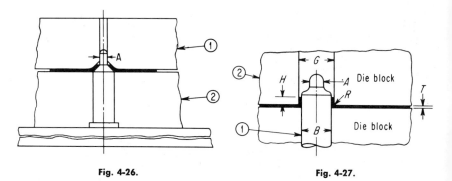

Fig. 4-26.   Fig. 4-27.

Fig. 4-26. A forming die for countersunk holes.

Fig. 4-27. A die for punching and countersinking a hole.

The hole can be pierced before it is placed in the countersinking die, or it can be formed and pierced in a single stroke of the press.

As shown in Fig. 4-26, the sheet is placed over the pilot diam A which locates it centrally in the die. The die body, 1, descends and forces the metal down around the flange surface of the punch. Spring pressure strips the part from the punch and releases the formed part from the die.

Figure 4-27 shows a two-step punch, 1, which first punches the hole in the part and then forces the metal around to the countersunk shape of the die block, 2. The hole punched by this method is always somewhat smaller than the size of the hole in the finished part. Spring pressure is used to strip the finished part from the punch. A shedder pin should be provided in the piercing point of the punch to remove the slug.

The size of the pierced hole for a 90° hole flange can be calculated, but should never be used until it has been proved correct by using the same tools that will be employed in the die. To calculate the hole size the same principles are employed when finding a 90° bend.

Dimensional details of Fig. 4-27 are identified as follows:

T = thickness of metal to be flanged
A = diam of calculated hole
B = diam of hole inside of flange (punch body size)
G = diam of hole in die (outside of flange)
R = radius on edge of die (usually one-third metal thickness)
H = height of flanged hub

R can be specified as from $\frac{T}{3}$ to $\frac{T}{4}$

When the flanges are stretched more than 2½ times metal thickness in height, the wall can split. This can be prevented to some extent by burring the edge around the hole before the extruding operation.

90° Hole Flanging. Forming a flange around a previously pierced hole at a bend angle of 90° (the most common operation) is nothing more than the formation of a stretch flange at that angle.

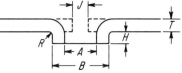

Fig. 4-28. Hole flange design.

One manufacturer has standardized flange widths (Fig. 4-28, dimension H) for holes to be tapped in low-carbon-steel stamping stock, as follows:

$$B = A + \frac{5T}{4} \quad \text{when } T \text{ is less than 0.045 in.} \quad (4\text{-}4)$$

$$B = A + T \quad \text{when } T \text{ is more than 0.045 in.} \quad (4\text{-}5)$$

$$H = T \quad \text{when } T \text{ is less than 0.035 in.} \quad (4\text{-}6)$$

$$H = \frac{4T}{5} \quad \text{when } T \text{ is 0.035 to 0.050 in.} \quad (4\text{-}7)$$

$$H = \frac{3T}{5} \quad \text{when } T \text{ is more than 0.050 in.} \quad (4\text{-}8)$$

$$R = \frac{T}{4} \quad \text{when } T \text{ is less than 0.045 in.} \quad (4\text{-}9)$$

$$R = \frac{T}{3} \quad \text{when } T \text{ is more than 0.045 in.} \quad (4\text{-}10)$$

$$J = \sqrt{\frac{TB^2 + 4TA^2 + 4HA^2 - 4HB^2}{9T}} \quad (4\text{-}11)$$

The radius P on the nose of the punch should be blended into the body diameter, eliminating any sharpness which could cause the metal to score as it passes over it. The radius on the body B or hole-sizing portion of the punch must be as large as possible, and smooth. The portion between the *A* and *B* diameters of the punch should have a radius *C* which should be as large as possible. (See Fig. 4-29.)

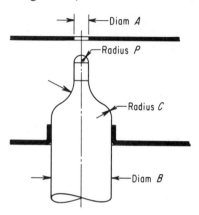

Fig. 4-29. Flanging punch design.

DRAWING DIES

Drawing is a process of changing a flat, precut metal blank into a hollow vessel without excessive wrinkling, thinning, or fracturing. The various forms produced may be cylindrical or box-shaped with straight or tapered sides or a combination of straight, tapered, or curved sides. The size of the parts may vary from ¼ in. diameter or smaller, to aircraft or automotive parts large enough to require the use of mechanical handling equipment.

Metal Flow

When a metal blank is drawn into a die, a change in its shape is brought about by forcing the metal to flow on a plane parallel to the die face, with the result that its thickness and surface area remain about the same as the blank. Figure 4-30 shows schematically the flow of metal in circular shells. The units within one pair of radial boundaries have been numbered and each unit moved progressively toward the center in three steps. If the shell were drawn in this manner, and a certain unit area examined after each depth shown, it would show (1) a size change only as the metal moves toward the die radius; (2) a shape change only as the metal moves over the die radius. Observe that no change takes place in area 1, and the maximum change is noted in area 5.

The relative amount of movement in one unit or in groups of units is shown in Fig. 4-31 *A* and *B*, in which two methods of marking the blanks

Chap. 4 Bending, Forming, and Drawing Dies   247

are used to illustrate size, shape, and position of the units of area, before and after drawing. The blank in view *A* is marked with radial lines and concentric circles, and in view *B* with squares. If, after these blanks are marked and drawn, sections are cut out of the shell, flattened, and compared with the original triangular portions, a change in shape of the triangular pieces will be found. The illustration shows that the inner portion of the triangle, which becomes the base of the shell, remains unchanged throughout the operation. The portion which becomes the side wall of the

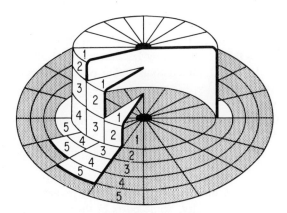

Fig. 4-30. A step-by-step flow of metal.

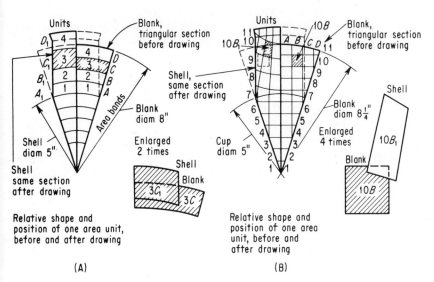

Fig. 4-31. Two methods of marking blanks to illustrate size, shape, and position of the units of area, before and after drawing.

shell is changed from an angular figure to a longer parallel-sided one as it is drawn over the die radius, from which point no further change takes place. The particular areas observed have been enlarged and superimposed upon each other, respectively, to show more clearly their size, shape, and position before and after drawing.

The general change in circular draws, due to flow, may be summarized as follows:

1. Little or no change in the bottom area because no cold work was done in this area.
2. All radial boundaries of the units of area remain radial in the bottom area. The units in the top flange area remain radial until they move over the die radius; then they become parallel and assume dimensions equal to their dimensions at the point where they move over the die radius.
3. There is a slight decrease in surface area and increase of thickness in the units involving maximum flow. The increase in thickness is limited to the space between the punch and die.
4. The flow lines on a circular shell indicate that the metal movement is uniform.

Flow in Rectangular Shells. The drawing of a rectangular shell involves varying degrees of flow severity. Some parts of the shell may require severe cold working and others, simple bending. In contrast to circular shells, in which pressure is uniform on all diameters, some areas of rectangular and irregular shells may require more pressure than others. True drawing occurs at the corners only; at the sides and ends metal movement is more closely allied to bending. The stresses at the corner of the shell are compressive on the metal moving toward the die radius and are tensile on the metal that has already moved over the radius. The metal between the corners is in tension only on both the side wall and flange areas.

The variation in flow in different parts of the rectangular shell divides the blank into two areas. The corners are the *drawing area,* which includes all the metal in the corners of the blank necessary to make a full corner on the drawn shell. The sides and ends are the *forming area,* which includes all the metal necessary to make the sides and ends full depth. To illustrate the flow of metal in a rectangular draw, the developed blank in Fig. 4-32B has been divided into unit areas by two different methods. In Fig. 4-32A the corners of the shell drawn from the blank in view B are shown. The upper view is the corner area which was marked with squares, and the lower view is the corner area which was marked with radial lines and concentric circles. The severe flow in the corner areas is clearly shown in the lower view by the radial lines of the blank being moved parallel and close together, and

Chap. 4 Bending, Forming, and Drawing Dies   249

the lines of the concentric circles becoming farther apart the nearer they are to the center of the corner and the edge of the blank. The relatively parallel lines of the sides and ends show that little or no flow occurred in these areas. The upward bending of these lines indicates the flow from the corner area to the sides and ends to equalize the height where these areas on the blank were blended to eliminate sharp corners.

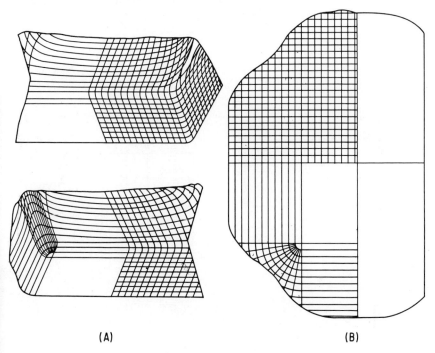

Fig. 4-32. Metal flow in rectangular draws: (*A*) blank marked before drawing; (*B*) corner areas after drawing.

Single-action Dies

The simplest type of draw die is one with only a punch and die. Each component may be designed in one piece without a shoe by incorporating features for attaching them to the ram and bolster plate of the press. Figure 4-33*A* shows a simple type of draw die in which the precut blank is placed in the recess on top of the die, and the punch descends, pushing the cup through the die. As the punch ascends, the cup is stripped from the punch by the counterbore in the bottom of the die. The top edge of the shell expands slightly to make this possible. The punch has an air vent to eliminate suction which would hold the cup on the punch and damage the cup when it is stripped from the punch.

The method by which the blank is held in position is important, because successful drawing is somewhat dependent upon the proper control of blankholder pressure. A simple form of drawing die with a rigid flat blankholder for use with 13-gage and heavier stock is shown in Fig. 4-33B. When the punch comes in contact with the stock, it will be drawn into the die without allowing wrinkles to form.

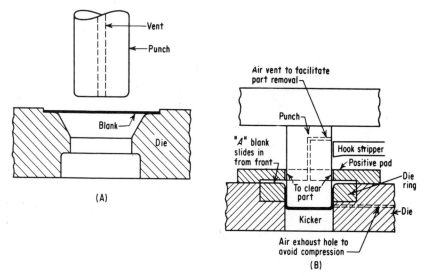

Fig. 4-33. Draw die types: (A) simple type; (B) simple draw die for heavy stock.

Another type of drawing die for use in a single-action press is shown in Fig. 4-34. This die is a plain single-action type where the punch pushes the metal blank into the die, using a spring-loaded pressure pad to control the metal flow. The cup either drops through the die or is stripped off the punch by the pressure pad. The sketch shows the pressure pad extending over the nest, which acts as a spacer and is ground to such a thickness that an even and proper pressure is exerted on the blank at all times. If the spring pressure pad is used without the spacer, the more the springs are depressed the greater the pressure exerted on the blank, thereby limiting the depth of draw. Because of limited pressures obtainable, this type of die should be used with light-gage stock and shallow depths.

A single-action die for drawing flanged parts, having a spring-loaded pressure pad and stripper, is shown in Fig. 4-35. The stripper may also be used to form slight indentations or re-entrant curves in the bottom of a cup, with or without a flange. Draw tools in which the pressure pad is attached

Chap. 4 Bending, Forming, and Drawing Dies   251

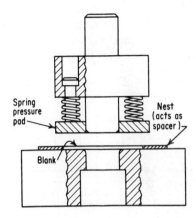

Fig. 4-34. Draw die with spring pressure pad.

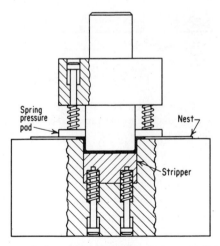

Fig. 4-35. A draw die with spring pressure pad and stripper.

to the punch are suitable only for shallow draws. The pressure cannot be easily adjusted, and the short springs tend to build up pressure too quickly for deep draws. This type of die is often constructed in an inverted position with the punch fastened to the lower portion of the die.

Deep draws may be made on single-action dies, where the pressure on the blankholder is more evenly controlled by a die cushion or pad attached to the bed of the press. The typical construction of such a die is shown in Fig. 4-36. This is an inverted die with the punch on the die's lower portion.

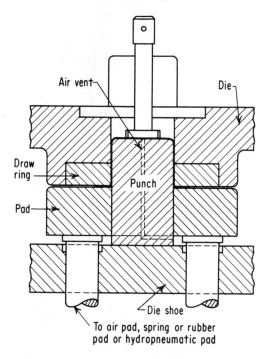

Fig. 4-36. Cross section of an inverted draw die for a single-action press; die is attached to the ram; punch and pressure pad are on the lower shoe.

Double-action Dies

In dies designed for use in a double-action press, the blankholder is fastened to the outer ram which descends first and grips the blank; then the punch, which is fastened to the inner ram, descends, forming the part. These dies may be a push-through type, or the parts may be ejected from the die with a knockout attached to the die cushion or by means of a delayed action kicker. Figure 4-37 shows a cross section of a typical double-action draw die.

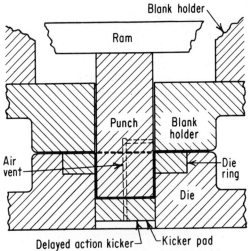

Fig. 4-37. A typical double-action cylindrical draw die.

Development of Blanks

The development of the approximate blank size should be done first (1) to determine the size of a blank to produce the shell to the required depth and (2) to determine how many draws will be necessary to produce the shell. This is determined by the ratio of the blank size to the shell size. Various methods have been developed to determine the size of blanks for drawn shells. These methods are based on (1) mathematics alone; (2) the use of graphical layouts; (3) a combination of graphical layouts and mathematics. The majority of these methods are for use on symmetrical shells.

It is rarely possible to compute any blank size to close accuracy or to maintain perfectly uniform height of shells in production, because the thickening and thinning of the wall vary with the completeness of annealing. The height of ironed shells varies with commercial variations in sheet thickness, and the top edge varies from square to irregular, usually with four more or less pronounced high spots resulting from the effect of the direction of the crystalline structure of the metal. Thorough annealing should largely remove the directional effect. For all these reasons it is ordinarily necessary to figure the blank sufficiently large to permit a trimming operation. The drawing tools should be made first; then the blank size should be determined by trial before the blanking die is made. There are times, however, when the metal required to produce the product is not immediately available from stock and must be ordered at the same time as the tools are ordered. This situation makes it necessary to estimate the blank size as closely as possible by formula or graphically in order to know what sizes to order.

Blank Diameters

The following equations may be used to calculate the blank size for cylindrical shells of relatively thin metal. The ratio of the shell diameter to the corner radius (d/r) can affect the blank diameter and should be taken into consideration. When d/r is 20 or more,

$$D = \sqrt{d^2 + 4dh} \qquad (4\text{-}12)$$

When d/r is between 15 and 20,

$$D = \sqrt{d^2 + 4dh} - 0.5r \qquad (4\text{-}13)$$

When d/r is between 10 and 15,

$$D = \sqrt{d^2 + 4dh} - r \qquad (4\text{-}14)$$

When d/r is below 10,

$$D = \sqrt{(d - 2r)^2 + 4d(h - r) + 2\pi r(d - 0.7r)} \qquad (4\text{-}15)$$

where D = blank diameter
d = shell diameter
h = shell height
r = corner radius

The above equations are based on the assumption that the surface area of the blank is equal to the surface area of the finished shell.

In cases where the shell wall is to be ironed thinner than the shell bottom, the volume of metal in the blank must equal the volume of the metal in the finished shell. Where the wall-thickness reduction is considerable, as in brass shell cases, the final blank size is developed by trial. A tentative blank size for an ironed shell can be obtained from the equation

$$D = \sqrt{d^2 + 4dh\,\frac{t}{T}} \qquad (4\text{-}16)$$

where t = wall thickness
T = bottom thickness

Reduction Factors

After the approximate blank size has been determined, the next step is to estimate the number of draws that will be required to produce the shell and the best reduction rate per draw. As regards diameter reduction, the area of metal held between the blankholding faces must be reasonably proportional to the area on which the punch is pressing, since there is a limit to the amount of metal which can be made to flow in one operation. The greater the difference between blank and shell diameters, the greater the area that must be made to flow, and therefore the higher the stress required to make it flow. General practice has established that, for the first draw, the area of the blank should not be more than three and one-half to four times the cross-sectional area of the punch.

One of the important factors in the success or failure of a drawing operation is the thickness ratio, or the relation of the metal thickness to the blank or previous shell diameter; this ratio is expressed as t/D. As this ratio decreases, the tendency to wrinkling increases, requiring more blankholding pressure to control the flow properly and prevent wrinkles from starting. The top limit of about 48 per cent seems to be substantiated by practice for single-action first draws. The 30 per cent limit for double-action redraws is dictated by practice and is modified by corner radii, friction, and the angle of the blankholding faces with respect to the shell wall. Because of strain-hardening stresses set up in the metal, third and subsequent draws should not exceed 20 per cent reduction without an annealing operation.

Pressures

Drawing Pressure. The pressure applied to the punch, necessary to draw a shell, is equal to the product of the cross-sectional area and the yield strength s of the metal. Taking into consideration the relation between the blank and shell diameters and a constant C of 0.6 to 0.7 to cover friction and bending, the pressure P for a cylindrical shell may be expressed by the empirical equation

$$P = \pi dts \left(\frac{D}{d} - C\right) \quad (4\text{-}17)$$

Blankholder Pressure. The amount of blankholder pressure required to prevent wrinkles and puckers is largely determined by trial and error. The pressure required to hold a blank flat for a cylindrical draw varies from very little to a maximum of about one-third or more of the drawing pressure. On cylindrical draws, the pressure is uniform and balanced at all points around the periphery because the amount of flow at all points is the same. On rectangular and irregularly shaped shells, the amount of flow around the periphery is not uniform; hence the pressure required varies also.

EVOLUTION OF A DRAW DIE

The first step in planning a die (Fig. 4-39) for the part of Fig. 4-38 is not the development of the die but that of the blank for this drawn shell or cup.

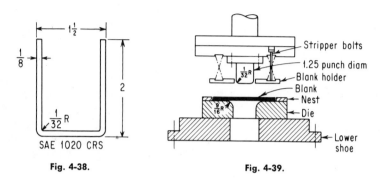

Fig. 4-38.   **Fig. 4-39.**

Fig. 4-38. A drawn shell, the basis for blank development and draw die design (Fig. 4-39).

Fig. 4-39. Draw die for producing the shell of Fig. 4-38.

Since the ratio d/r is more than 20, the diameter of the blank D is, from Eq. (4-12):

$$D = \sqrt{(1.5)^2 + 4 \times 1.5 \times 2}$$
$$= 3.77 \text{ in.}$$

The area of this blank is slightly less than three times the cross-sectional area of the punch (approximately 4.9 sq in.). This blank can be drawn into a shell of the dimensions specified in one draw because the ratio of its area to that of the punch nose is less than 4:1.

The length of the draw radius (radius of the toroidal zone at the entrance to the die) must be determined before any other die design details are considered. The radius of the draw die should be kept as large as possible to aid in the flow, but if it is too large, the metal will be released by the blankholder before the draw is completed and wrinkling will result. When the radius is too small, the material will rupture as it goes over the radius, or against the face of the punch. Table 4-2 gives the practical drawing radii for certain stock thicknesses. The values in this table are based on a radius of approximately four times the stock thickness. In some cases the radius may vary from four to six times the stock thickness. The length of the draw radius from Table 4-2 is $9/16$ in.

TABLE 4-2. PRACTICAL DRAWING RADII FOR CERTAIN THICKNESSES OF STOCK

Thickness of stock, in.	Drawing radius, in.
1/64	1/16
1/32	1/8
3/64	3/16
1/16	1/4
5/64	3/8
3/32	7/16
1/8	9/16

The required drawing pressure is determined; Eq. (4-17) is used:

$$P = 17 \text{ tons}$$

Blankholder pressure is determined largely by trial and error; an allowable range is from a few pounds up to one-third of the drawing pressure; approximately 6 tons is more than adequate. An ordinary 25-ton OBI press has enough capacity for the total possible maximum pressure of 23 tons required for the operation.

The blank is nested in a semicircular plate. Its thickness is established by the optimum blankholder pressure on the blank during drawing of the shell to specifications. The unit pressure on the blank becomes greater as it is pushed into the die.

Shimming the nest plate will vary the pressure on both nest and blank and results in a critical pressure on the latter without the production of defective shells.

The blankholder is a 5/8-in. thick circular plate suspended from the punchholder by six stripper bolts equally spaced around its circumference.

A like number of die springs fit around the bolts and are retained by pockets counterbored in the punchholder and blankholder.

In addition to its function of holding the blank, the blankholder strips the shell from the punch.

The draw clearance (space between the punch and die) from Table 4-3 should be from 0.1375 to 0.1500 in. The ID of the die is 1.6375 in. for tryout of the die. If there is too much ironing of the shell this ID can be increased up to approximately 1.650 in.

The OD of the draw punch is the same as the ID of the drawn shell.

The methods of determining die dimensions (and holddown size) discussed in Chap. 3 apply to the same elements in a drawing die but may generally be increased approximately in proportion to the press forces applied to and the stresses in die blocks and blankholders.

TABLE 4-3. DRAW CLEARANCE

Blank thickness, in.	First draws	Redraws	Sizing-draw *
Up to 0.015	1.07t to 1.09t	1.08t to 1.1t	1.04t to 1.05t
0.016 to 0.050	1.08t to 1.1t	1.09t to 1.12t	1.05t to 1.06t
0.051 to 0.125	1.1t to 1.12t	1.12t to 1.14t	1.07t to 1.09t
0.136 and up	1.12t to 1.14t	1.15t to 1.2t	1.08t to 1.1t

* Used for straight-sided shells where diameter or wall thickness is important, or where it is necessary to improve the surface finish in order to reduce finishing costs.
$t =$ thickness of the original blank.

Draw Rings. The draw ring of many draw dies is the die itself, made of tool-steel with the edge of the cavity forming a spherical zone, over which the metal is drawn. In many large drawing dies the die is not made in one piece but has an inserted draw ring for tool-steel economy and lower replacement cost.

Draw Die Materials. The selection of material for a draw punch and a draw die is determined the same as for any other tool. It is determined largely by the number of parts to be drawn and their ultimate unit cost. For low production, material cost is a major factor. If less than 100 parts are to be drawn, a plastic or zinc alloy material is satisfactory.

For production of around 1000 parts, a plain cast iron ring is suitable. Cast iron punches and dies can be chrome plated for medium-production runs. A plating thickness of 0.003 in. is satisfactory.

The best tool-steels should be used for runs of 10,000 or more parts. The choice of the type of tool-steel is based upon factors of wear, strength, etc., considered in the selection for any other tool.

Carbide punches and dies or carbide inserts for these tools have proven to be the most economical material for extremely large runs of around 1,000,000 parts.

Material selection for blankholders, knockouts, and other parts of a draw die is based upon these same factors.

Lubricants. The functional requirements of lubricants for drawing operations are much more severe than for shafts and bearings.

Possible corrosive action of some compounds on some metals and the ease of their removal from drawn metal parts must be considered.

Handbooks and manufacturers of various metals should be consulted for lubrication recommended for shallow and deep drawing operations.

The following lubrication data for drawing mild steel are included as a guide:

Mild Operations:

1. Mineral oil of medium-heavy to heavy viscosity
2. Soap solutions (0.03 to 2.0 per cent, high-titer soap)
3. Fat, fatty-oil, or fatty- and mineral-oil emulsions in soap-base emulsions
4. Lard-oil or other fatty-oil blends (10 to 30 per cent fatty oil)

Medium Operations:

1. Fat or oil in soap-base emulsions containing finely divided fillers such as whiting or lithopone
2. Fat or oil in soap-base emulsions containing sulfurized oils
3. Fat or oil in soap-base emulsions with fillers and sulfurized oils
4. Dissimilar metals deposited on steel plus emulsion lubricant or soap solution
5. Rust or phosphate deposits plus emulsion lubricants or soap solution
6. Dried soap film

Severe Operations:

1. Dried soap or wax film, with light rust, phosphate, or dissimilar metal coatings
2. Sulfide or phosphate coatings plus emulsions with finely divided fillers and sometimes sulfurized oils
3. Emulsions or lubricants containing sulfur as combination filler and sulfide former
4. Oil-base sulfurized blends containing finely divided fillers

PROGRESSIVE DIES

A progressive die performs a series of fundamental sheet-metal operations at two or more stations during each press stroke in order to develop a workpiece as the strip stock moves through the die. This type of die is sometimes called *cut-and-carry, follow,* or *gang* die. Each working station performs one or more distinct die operations, but the strip must move from the first through each succeeding station to produce a complete part. One or more idle stations may be incorporated in the die, not to perform work on the metal but to locate the strip, to facilitate interstation strip travel, to provide maximum-size die sections, or to simplify their consruction.

The linear travel of the strip stock at each press stroke is called the *progression, advance,* or *pitch* and is equal to the interstation distance.

The unwanted parts of the strip are cut out as it advances through the die, and one or more ribbons or tabs are left connected to each partially completed part to carry it through the stations of the die. Sometimes parts are made from individual blanks, neither a part of, nor connected to, a strip; in such cases, mechanical fingers or other devices are employed for the station-to-station movement of the workpiece.

The operations performed in a progressive die could be done in individual dies as separate operations but would require individual feeding and positioning. In a progressive die, the part remains connected to the stock strip which is fed through the die with automatic feeds and positioned by pilots with speed and accuracy.

SELECTION OF PROGRESSIVE DIES

The selection of any multioperation tool, such as a progressive die, is justified by the principle that the number of operations achieved with one handling of the stock and the produced part is more economical than production by a series of single-operation dies and a number of handlings for each single die.

Where total production requirements are high, particularly if production rates are large, total handling costs (man-hours) saved by progressive fabrication compared with a series of single operations are frequently greater than the costs of the progressive die.

The fabrication of parts with a progressive die under the above-mentioned production conditions is further indicated when

1. Stock material is not so thin that it cannot be piloted or so thick that there are stock-straightening problems.
2. Overall size of die (functions of part size and strip length) is not too large for available presses.
3. Total press capacity required is available.

STRIP DEVELOPMENT FOR PROGRESSIVE DIES

Individual operations performed in a progressive die are often relatively simple, but when they are combined in several stations, the most practical and economical strip design for optimum operation of the die often becomes difficult to devise.

The sequence of operations on a strip and the details of each operation must be carefully developed to assist in the design of a die to produce good parts.

A tentative sequence of operations should be established and the following items considered as the final sequence of operations is developed:

1. Pierce piloting holes and piloting notches in the first station. Other holes may be pierced that will not be affected by subsequent noncutting operations.

Chap. 4 Bending, Forming, and Drawing Dies

2. Develop blank for drawing or forming operations for free movement of metal.
3. Distribute pierced areas over several stations if they are close together or are close to the edge of die opening.
4. Analyze the shape of blanked areas in the strip for division into simple shapes so that punches of simple contours may partially cut an area at one station and cut out remaining areas in later stations. This may suggest the use of commercially available punch shapes.
5. Use idle stations to strengthen die blocks, stripper plates, and punch retainers and to facilitate strip movement.
6. Determine whether strip grain direction will adversely affect or facilitate an operation.
7. Plan the forming or drawing operations either in an upward or a downward direction, whichever will assure the best die design and strip movement.
8. The shape of the finished part may dictate that the cutoff operation should precede the last noncutting operation.
9. Design adequate carrier strips or tabs.
10. Check strip layout for minimum scrap; use a multiple layout if feasible.
11. Locate cutting and forming areas to provide uniform loading of the press slide.
12. Design the strip so that scrap and part can be ejected without interference.

Figure 4-40 illustrates the use of a three-station die to avoid weak die blocks. At *A* the pierced hole is near the edge of the part where it is cut off, thereby weakening the die block at this point. If an idle station is added so that the piercing operation is moved ahead one station, the die block is stronger and there is less chance of cracking in operation or fabrication. At *B,* the pierced holes are centered on the strip but close together. In this case the holes should be pierced in two stations to avoid thin sections in the die block between the holes. The adding of stations also provides better support for the piercing punches.

Figure 4-41 shows the use of one die station instead of two stations to maintain a close-toleranced dimension. If two stations were used, the variation in the location of the stock guides and cutting punches could make it difficult to hold the ± 0.001-in. tolerance.

The strip development for shallow and deep drawing in progressive dies must allow for movement of the metal without affecting the positioning of the part in each successive station. Figure 4-42 shows various types of cutouts and typical distortions to the carrier strips as the cup-shaped parts are formed and then blanked out of the strip. Piercing and lancing of the

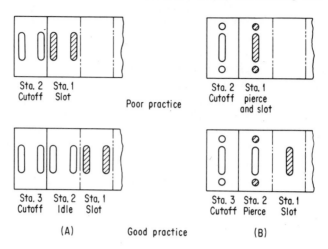

Fig. 4-40. Use of three-stage die to avoid weak die blocks: (*A*) pierced hole close to edge of part; (*B*) pierced holes close together.

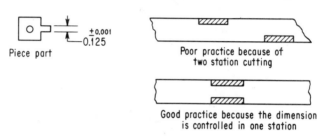

Fig. 4-41. Use of one station versus two to hold a close tolerance.

strip around the periphery of the part as shown at *A*, leaving one or two tabs connected to the carrier strip, is a commonly used method. The semicircular lancing as shown at *B* is used for shallow draws. The use of this type of relief for deeper draws places an extra strain on the metal in the tab and causes it to tear. The carrier strip is distorted to provide stock for the draw. A popular cutout for fairly deep draws is shown at *C*. This double-lanced relief suspends the blank on narrow ribbons, and no distortion takes place in the carrier strips. Two sets of single rounded lanced reliefs of slightly different diameters are placed diametrically opposite each other to produce the ribbon suspension. The hourglass cutout in *D* is an economical method of making the blank for shallow draws. The connection to the carrier strips is wide, and a deep draw would cause considerable distortion. An hourglass cutout for deep draws is shown in *E*, which provides a narrow tab connecting the carrier strip to the blank. The cupping operations narrow the width of the strip as the metal is drawn into the cup shape.

Chap. 4 Bending, Forming, and Drawing Dies   263

The hourglass cutout may be made in two stations by piercing two separated triangular-shaped cutouts in one station, and lancing or notching the material between them in a second station. The cutouts shown at F and G provide an expansion-type carrier ribbon that tends to straighten out when the draw is performed. These cutouts are made in two stations to allow for stronger die construction. Satisfactory multiple layouts may be designed using most of the reliefs by using a longitudinal lance or slitting station to divide the wide strip into narrower strips as the stock advances. The I-shaped relief cutout in H is a modified hourglass cutout used for relatively wide strips from which rectangular or oblong shapes are produced.

Straight slots or lances crosswise of the stock are sometimes used on very shallow draws or where the forming is in the central portion of the blank. On the deeper draws, this type of relief tends to tear out the carrier strips or cause excessive distortion in the blank and is not too satisfactory to use.

Stock Positioning

Of prime importance in the strip development is the positioning of the stock in each station. The stock must be positioned accurately in each station so that the operation can be done in the proper location. A commonly used method of stock positioning is the incorporation of pilots in the die.

There are two methods of piloting in dies: direct and indirect. *Direct piloting* consists of piloting in holes punched in the part at a previous station. *Indirect piloting* consists of piercing holes in the scrap-strip and locating these holes with pilots at later operations. Direct piloting is the ideal method for locating the part in subsequent die operations. Unfortunately, ideal conditions may not exist, and in such cases indirect piloting must be used to achieve the desired results of part accuracy and high production speeds.

The advantages of locating pilots in the scrap material area are:

1. Not readily affected by workpiece change.
2. Size and location not as limited.

Disadvantages of locating pilots in the scrap section of strip are:

1. Material width and lead may increase.
2. Scrap-strip carriers distort on certain types of operations and make subsequent station use impossible.

How to pilot is an arbitrary decision that the tool designer must make. It is impossible to give definite rules and formulas, because the material and

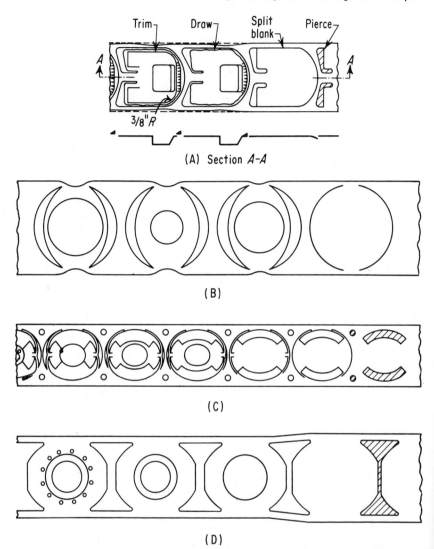

Fig. 4-42. Cutout reliefs for progressive draws: (*A*) lanced outline; (*B*) circular lance; (*C*) double lance suspension; (*D*) hourglass cutout.

the hardness of the stock influence the decision. However, in the indirect piloting method, it is possible to use pilots of greater diameter than if holes in the part are used for piloting such as in Fig. 4-43*A*. The greater the diameter of the pilot, the less chance there is of distortion of either the strip or the pilot. Also, small-diameter pilots introduce the possibility of broken pilots.

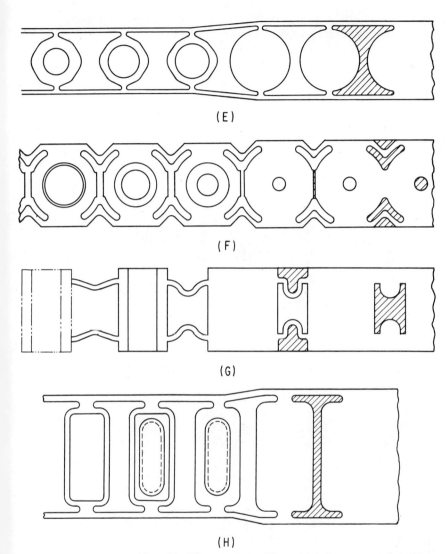

Fig. 4-42 (*Contd.*) (*E*) cutout providing expansion-type carrier ribbon for circular draws; (*G*) cutout providing expansion-type carrier ribbon for rectangular draws; (*H*) I-shaped relief for rectangular draws.

When holes in the part are held to close tolerances (Fig. 4-43*B*) it is possible for the pilots to affect the hole size in their effort to move the strip to proper location.

When holes in the part are too close to the edges (Fig. 4-43*C*) the weak outer portions of the part are likely to distort upon contact with the pilots, instead of the strip's moving to the correct location. This possibility is often overlooked in planning a progressive die, and gives to subsequent runs of scrap parts and expensive die alterations.

Just what constitutes a condition where the edge of the hole is too close to the edge of the part is, like many aspects of design, a matter of personal judgment. Many designers use the rule-of-thumb: The distance between the two must be at least twice the stock thickness.

A similar problem exists when the part holes are located in a weak portion of the inside area of the part (Fig. 4-43D). Here, there is a possibility of the part's buckling before the pilots can position the stock strip. In this case it is advisable to pilot in the scrap strip.

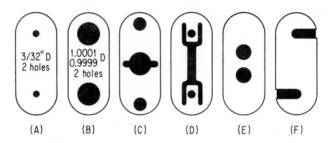

Fig. 4-43. Part conditions that require indirect pilots: (*A*) small holes; (*B*) close-tolerance holes; (*C*) holes too near edge; (*D*) holes in fragile areas of the part; (*E*) holes too close together; (*F*) slots in parts.

To achieve accurate part location, the pilots must be placed as far apart as possible. When the holes in the workpiece are close together, as in Fig. 4-43*E*, holes in the scrap strip should be used for piloting. A second method would be to place a pilot in one hole in one station and in the same hole in a succeeding station. The feasibility of the second method depends upon the availability of an additional die station.

When slots are punched in the blank parallel to the stock movement (Fig. 4-43*F*) the slots are not suitable piloting holes. Therefore, indirect pilots must be used.

Disposition of Scrap Strip

A strip development is illustrated in Fig. 4-44*B*, utilizing pierce, trim, form, and blank-through operations and carriers on both sides of the strip. The workpiece is dropped through the die, while the carrier bars continue to the scrap cutters to be cut into short lengths. The dropping of the workpiece through the die is the most desirable method of part ejection, but cannot always be obtained. Cutting the scrap into small sections simplifies the material handling problems and produces a greater dollar return when sold as scrap metal. Figure 4-44C shows an alternate strip development with one side carrier. The workpiece is pierced, trimmed, cut off, and formed on a pad with air or gravity ejection, and the carrier bar is cut into short pieces by the scrap cutter. It is well to remember that if a part is to be ejected as this one is, the double carrier bar design in Fig. 4-44B should be avoided, because the part may become trapped in these bars and cause die damage.

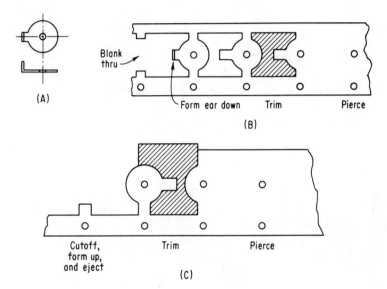

Fig. 4-44. Alternate strip developments for a workpiece.

The design of the part in Fig. 4-44A requires that the carrier be outside the part configuration. This necessitates the use of stock wider than the part width plus the normal trimming allowance. The part shown in Fig. 4-45A can be made of stock the same width as the part.

The strip development of Fig. 4-45D illustrates how the strip is pierced, trimmed, and the part cut off and formed. A slug-type cutoff punch is used and the flange is formed downward. The part is then ejected by an air jet or by gravity. This arrangement is often referred to as a *scrapless development* since no carrier strips remain after the part is cut from the strip.

Figure 4-45E shows a strip development for the same part using a shear-type cutoff. The flange is formed upward as the combination cutoff

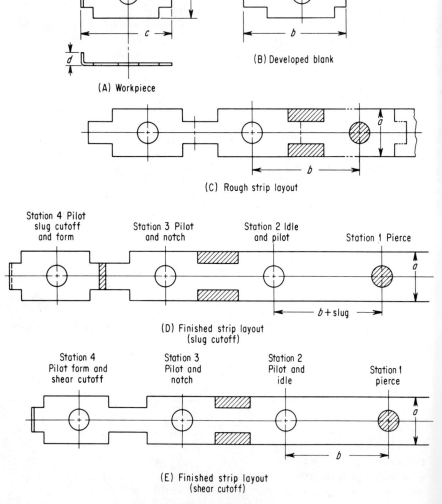

Fig. 4-45. Scrapless strip development.

and form punch descends. A spring-loaded pad supports the workpiece during forming and assists in ejecting the part from the die. The progression of this type of development is shortened by the width of the cutoff slug.

Figure 4-46 shows another development in which the stock is the same as the developed width of the workpiece. The strip is pierced in station 1; piloted and notched in station 2; piloted, pierced, and formed in station 3.

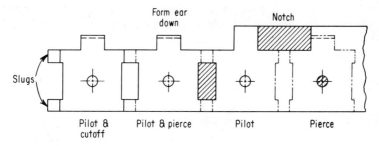

Fig. 4-46. Four-station strip development.

Progressive Die Elements

The die elements used in progressive dies such as punches, stops, pilots, strippers, die buttons, punch guide bushings, die sets guide posts, and guide post bushings are of similar design to those used in other types of dies. Refer to Chap. 3 and previous text in this section for their design.

General Die Design

A progressive die should be heavily constructed to withstand the repeated shock and continuous runs to which it is subjected. Precision or antifriction guide posts and bushings should be used to maintain accuracy. The stripper plates (if spring-loaded and movable), when also serving as guides for the punches, should engage guide pins before contacting the strip stock. Lifters should be provided in die cavities to lift up or eject the formed parts, and carrier rails or pins should be provided to support and guide the strip when it is being moved to the next station. A positive ejector should be provided at the last station. Where practical, punches should contain shedder or oil-seal-breaker pins to aid in the disposal of the slug. Adequate piloting should be provided to ensure proper location of the strip as it advances through the die. For more die-design considerations, see Chap. 3.

EVOLUTION OF A PROGRESSIVE DIE

Figure 4-45*A* shows a workpiece to be blanked and formed at high volume in a progressive die.

The first step in the design of the die is the development of the blank (Fig. 4-45*B*). The grain of coiled metal is normally parallel to the length of the strip. Forming operations are normally performed perpendicular to the grain to forestall any tendency to fracture. The workpiece is therefore sketched with the bend line perpendicular to the strip.

The *a* dimension appears to coincide with stock width. A comparison is made between the exact dimension with tolerance, and available purchased coil stock. If the *a* dimension is not within the slitting tolerance of the purchased stock, the part must be blanked from a wider strip and perhaps shaved to the finished dimension. In this case the *a* dimension is within the slitting tolerance, the slit edge will be satisfactory, and no work need be performed to obtain the dimension.

The *b* dimension is developed by adding the *c* dimension, the *d* dimension, and a bend allowance, minus twice the outside radius. Tolerances must be carefully considered in developing the *b* dimension, and will determine whether the dimension can be achieved with a single shear or whether shaving will be required. In this case the *b* dimension is not critical and is, for the moment, accepted as the station-to-station distance.

With the *a* and *b* dimensions established, a centerline is drawn, and the template of the blank is traced along the centerline (Fig. 4-45C). The areas from which stock must be removed are clearly indicated. The straight edges of the stock appear acceptable for guidance. The absence of scrap stock on either side of the workpiece determines that the strip must be piloted by registry of workpiece features. The hole *e* appears most suitable for piloting purposes. The hole dimensions are examined to determine whether the hole can be pierced conventionally (hole diameter not less than $1\frac{1}{2}$ times stock thickness), or whether a secondary operation may be needed. The hole dimensions on this case are not critical, so the hole can be pierced in the first station and can be used as a pilot hole in subsequent stations. The first station can now be drawn and will include a guidance method for stock direction and elevation, the punch, and the die. The punch location can be transposed to the other stations where it will indicate the presence of a bullet-nosed pilot.

As the strip enters the second die station, it will register on the pilot. The next operation required on the workpiece is the notching of the edges. The workpiece dimensions are such that the notching should be performed by one stroke in one station. A preliminary layout discloses, however, that a notching operation in the second die station would be close to the piercing operation of the first station. The second station is tentatively designated as an idle station.

Idle stations are of great value. They enable the designer to distribute the work load uniformly throughout the length of the die. Individual die details can consequently be less complex, and can be easier to build and maintain. Idle stations also permit later changes in workpiece design by providing space for added operations.

The third die station can now be drawn to include a pilot and the notching punches.

Chap. 4   Bending, Forming, and Drawing Dies   271

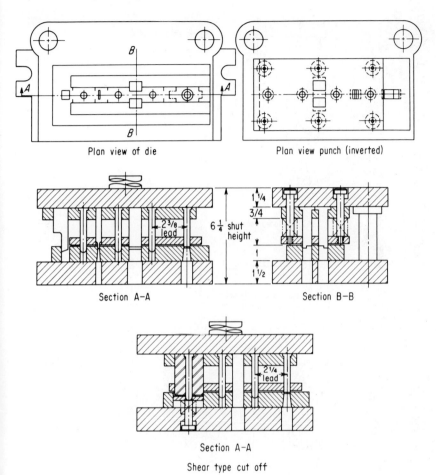

Fig. 4-47. Die for part of Fig. 4-45.

The stock strip will enter the fourth die station and register on the pilot. Two operations are required to finish the workpiece: the leading edge must be formed 90°, and the workpiece must be separated from the strip. Once the workpiece has been separated, no further operations are normally possible, but the workpiece must be separated for forming. The two operations can, however, be performed simultaneously in one station, just as the two sides were notched in one station. Two methods may be considered. If the stock is to be separated by removing a slug, the width of the slug must be added to the *b* dimension and the station-to-station distance of the entire layout must be correspondingly amended. This method, however, would permit a second punch to simply wipe down (form) and thereby complete the part without changing the elevation of the stock. If the separation is to

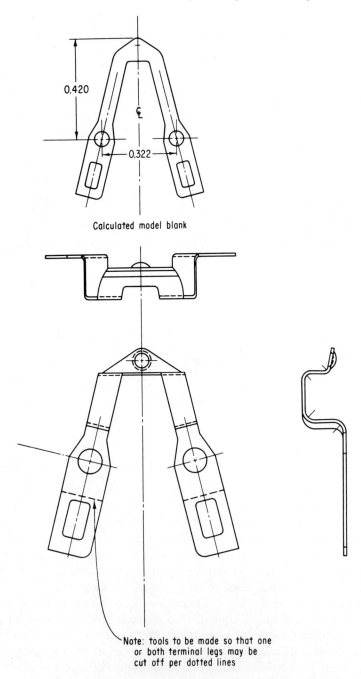

Fig. 4-48. Phosphor bronze clip blanked in a progressive die.

be performed by direct shear without a slug, the workpiece must change elevation. The die station would be spring-loaded to permit the shear action, and the forming action would take place at the depressed level.

Figure 4-47 shows the proposed die including stock elevation, stripper pads, and the ejection method. Figure 4-47A-A shows the die construction for a shear-type cutoff.

Force Determination. The force required for each station is computed by the methods detailed in Chap. 3. If the work load has been properly distributed throughout the length of the die, no rocking action should occur. If the required force is concentrated in any one station or area of the die, the planned operations should be shifted to achieve balance. When reasonable balance is assured, the forces may be added to determine the required press tonnage.

Press Selection. The press selected must be of adequate tonnage for the planned operation and must also be able to withstand the overloads that may accidentally be encountered. It must be equipped with a precise stock feed mechanism. The bolster plate must not deflect under the planned load even after it has been weakened by perforation for slug disposal. The sets are commercially available in a wide variety of sizes. An inexpensive die set would suffice. If precise shaving operations are necessitated by close workpiece tolerances, an expensive precise die set will be required.

EXAMPLES OF PROGRESSIVE DIES

The part to be made is a small clip, Fig. 4-48, made of 0.008-in. thick phosphor bronze, 8 numbers hard. Because of the delivery date for the parts and the anticipated total production, it was decided to pierce and blank the part in a progressive die, then form in a separate die.

The product designer indicated that the grain direction for optimum part performance is to be perpendicular to the bends. The grain direction favors the bends shown on the part and will decrease the tendency to crack at the bend lines.

The pilots were located in the scrap area of the strip because of possible design changes, thin walls on the blanking punch if pilots were inserted in the holes of the part, and the opportunity to use larger-diameter pilots since the holes are only 0.0655 ± 0.0015 in. diameter.

The perimeter of the blank and holes to be pierced is about 2.54 in. Multiplying the length of cut by the stock thickness of 0.008 in. and a shear stress of 80,000 psi gives a shear load of approximately 1600 lb. Therefore, the size of press will be more dependent on the die area than the capacity.

Figure 4-49 shows the strip development for the part. In the first station the two 0.067/0.064-in. diam holes are pierced, the tip of the blank dimpled, and a 0.094-in. diam pilot hole is pierced. When the full-length legs are required, station 2 is a piloting station only. When one or both of the legs are to be cut off, punches are inserted in this station.

In station 3, the strip is piloted and the two rectangular holes are pierced. In station 4, the strip is piloted while the part is blanked through the die.

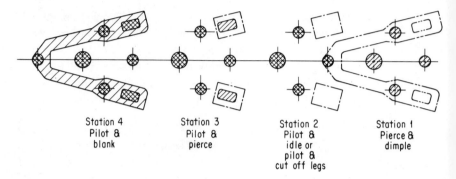

Station 4　　　　Station 3　　　　Station 2　　　　Station 1
Pilot &　　　　　Pilot &　　　　　Pilot &　　　　　Pierce &
blank　　　　　　pierce　　　　　idle or　　　　　dimple
　　　　　　　　　　　　　　　　pilot &
　　　　　　　　　　　　　　cut off legs

Fig. 4-49. Strip development for part of Fig. 4-48.

There are clearance spots on the die blocks for the dimple, thus permitting the strip to be flat on the die blocks. The die blocks are made in three sections for ease of machining as shown in Fig. 4-50. A channel-type stripper, 1, was selected for this die because of its more reasonable cost, and less design and build time. The second-operation forming die would remove some of the curvature caused by the blanking operation. Hardened guide bushings, 2, were used for the pilot hole piercing punch, 3, and the punches, 4, for the 0.067/0.064-in. diam holes. Guide bushings were used with these punches since they were small in diameter and were for the most important holes in the part.

The punches, 5, for cutting off the legs were backed up with a socket set screw, 6. When the punches were not in use, the screws were backed off a few turns and the tight fit between the punch and punch holder held them away from the stock.

A 5 × 5-in. two-post precision die set was selected for this die. The lower shoe was $1\frac{1}{2}$ in. thick and the upper shoe was $1\frac{1}{4}$ in. thick with a $1\frac{9}{16}$-in. diam by $2\frac{1}{8}$-in. long shank.

The stripper was extended beyond the die set close to the push-type roll feed. This was to support the stock and avoid buckling and subsequent damage to the die.

Figure 4-51 illustrates the die layout for a small, irregularly shaped stamping having eight small holes and one large hole. For accurate positioning of the strip, pilots are provided in each station. There are six stations in the die with a progression of 1.718 in. between stations.

Chap. 4   Bending, Forming, and Drawing Dies   275

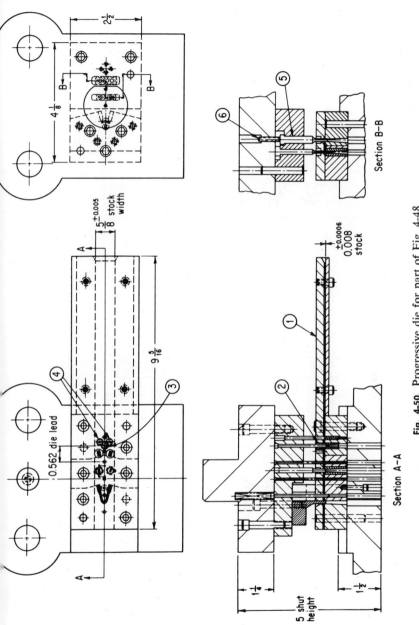

Fig. 4-50. Progressive die for part of Fig. 4-48.

276   Bending, Forming, and Drawing Dies   Chap. 4

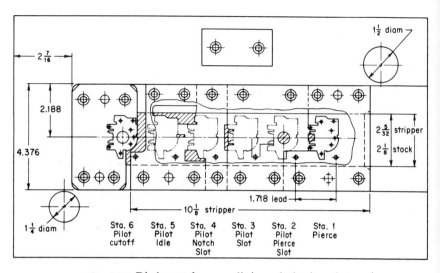

Fig. 4-51. Die layout for a small, irregularly-shaped stamping.

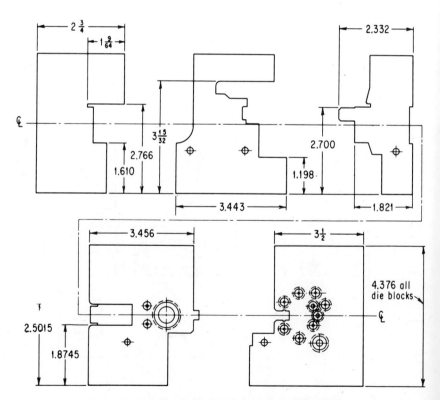

Fig. 4-52. Details of die blocks for die of Fig. 4-51.

In station 1, the eight holes, the 0.1875-in. diam pilot hole, and a 0.302 × 0.259-in. hole are pierced. In station 2, a 0.500-in. diam hole, two 0.062 × 0.068-in. holes, and a 0.095 × .379-in. slot are pierced.

In station 3, a slot is pierced symmetrically about the centerline. The strip is notched and slotted in the next station. Station 5 is an idle station because of the length of the notching punch in the previous station. The last station, 6, incorporates a slug-type cutoff punch. This punch trims the left-hand side of the blank piloted in station 5 and the right-hand side of the blank in station 6.

For ease of grinding, the die was made in five segments as shown in Fig. 4-52. The die set was specially made with diagonally placed guide posts. One post is $1\frac{1}{2}$ in. diam while the other is $1\frac{1}{4}$ in. diameter. Guide posts of two different diameters were used to prevent placing the upper shoe on the lower shoe incorrectly and damaging the punches.

Die buttons were used for each of the piercing punches; the die also incorporated guide bushings in the channel stripper for the piercing punches.

EXTRUSION DIES

Extrusion Principles

Impact extrusion, also known as cold extrusion or cold forging, is closely allied to coining, sizing, and forging operations. The operations are generally performed in hydraulic or mechanical presses. The press applies sufficient pressure to cause plastic flow of the workpiece material (metal) and to form the metal to a desired shape. A metal slug is placed in a stationary die cavity into which a punch is driven by the press action. The metal is extruded upward around the punch, downward through an orifice, or in any direction to fill the cavity between the punch and die. The shape of the finished part is determined by the shape of the punch and the die.

Product design is influenced by methods of manufacturing such as machining from the solid, drawing, spinning, stamping, or casting. Impact extrusion may combine more manufacturing processes into a single operation than any other metalworking method. Cold extrusion can result in excellent surface finish and accurate size and can improve the mechanical and physical properties of the workpiece material. Cold extruded parts may have smooth surfaces ranging between 30 and 100 micro-inches. Very close tolerances can be achieved by cold extrusion. Owing to severe cold working of the metal a fine, dense grain structure is developed parallel to the direction of metal flow. These continuous flow lines increase the fatigue resistance of the material.

Table 4-4 lists the pressures required to extrude common metals. The pressures required depend upon the alloy, its microstructure, the restriction to flow, the severity of work hardening, and the lubricant used. With high speed and restricted flow the press load may suddenly increase to three times the anticipated tonnage requirement. Pressures have been recorded up to 165 tons/in.² Owing to the high pressures required, presses must be carefully selected.

TABLE 4-4. EXTRUSION PRESSURES FOR COMMON METALS

Material	Pressure, tons/in.²
Pure aluminum "extrusion grade"	40.70
Brass (soft)	30.50
Copper (soft)	25.70
Steel C1010 "extrusion grade"	50.165
Steel C1020 (spheroidized)	60.200

The pressures required for extrusion also depend upon the percentage of reduction in area. When the percentage of reduction of area, for parts made of various aluminum alloys, is known, the required pressure may be taken from Fig. 4-53. Although the alloys mentioned are usually extruded at room temperature, a reduction in press pressure can be achieved by extruding at elevated temperatures. Closely related to, and a factor in establishing, reduction of area is part wall thickness. The increasing effect on punch pressure required as the wall becomes thinner is illustrated in Fig. 4-54.

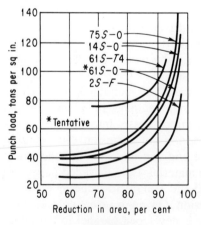

Fig. 4-53.

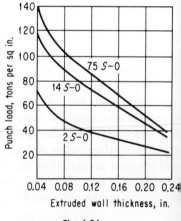

Fig. 4-54.

Fig. 4-53. The effect of reduction in area to the punch load in extruding.

Fig. 4-54. The effect of extruded wall thickness on the punch load.

Chap. 4 Bending, Forming, and Drawing Dies   279

The relationship between reduction of area and extrusion pressures for a series of plain carbon steels is shown in Fig. 4-55. The steels referred to in the different curves have carbon content in the range of 0.05 to 0.50 per cent, and less than 0.03 per cent each of sulfur and phosphorus. Steel 11 contained 0.58 per cent chromium, 0.11 per cent carbon, and 0.36 per cent manganese, and 0.03 per cent each sulfur and phosphorus.

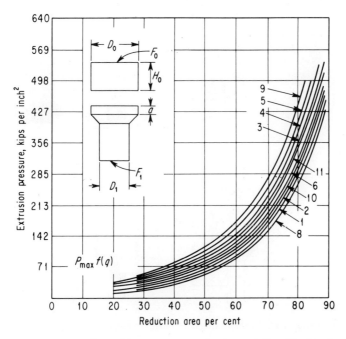

Fig. 4-55. Extrusion pressures: reduction relationship for the forward extrusion of a series of steels with carbon contents in the range 0.05 to 0.50 per cent.

Correct lubrication reduces considerably the pressures required and makes possible the cold extrusion of steel. Ordinary die lubricants break down because of the high pressures and the excessive surface heat generated by the plastic flow of material. A bonded steel-to-phosphate layer, and a bonded phosphate-to-lubricant layer is a satisfactory lubricant to prevent metal-to-metal welding or pickup. The workpiece is usually phosphatized before extrusion. If the workpiece is subjected to several extrusion operations, it is usually annealed and phosphatized between operations.

Because the slugs used for impact extrusion are cut from commercial rod or bar it is desirable to coin them to a desired size and shape for better die-fitting characteristics prior to extrusion. Using coiled wire stock, headers can be used to cut and coin or upset as a continuous high-speed operation.

Figure 4-56 illustrates slug coining and upsetting. Figure 4-57 illustrates how a profiled slug fits the die cavity and how the material flows when pressure is applied to cause plastic flow. Voids in the die cavity are filled as the slug collapses under initial pressure. Profiled slugs thus cushion impact of the punch, and allow higher ram velocity. Plastic flow then continues through small orifices between the diepot and the punch.

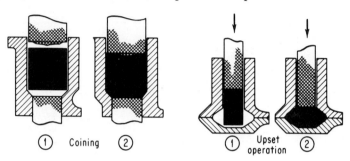

Fig. 4-56. Slug coining and upsetting. (*Courtesy American Machinist.*)

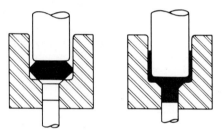

Fig. 4-57. Profiled slugs. (*Courtesy American Machinist.*)

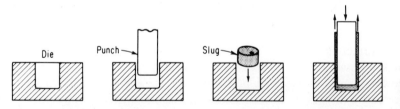

Fig. 4-58. Backward extrusion. (*Courtesy Design Engineering.*)

Basic methods of impact extrusion are backward, forward, and combination methods. In backward extrusion (Fig. 4-58) metal flows in a direction opposite to the direction of punch movement. The punch speed can be from 7 to 14 in. per second. As the punch strikes the slug, the heavy pressure causes the metal to flow through the orifice created by the punch and the die, and forms the side wall of the part by extrusion. In forward

Chap. 4 Bending, Forming, and Drawing Dies   281

extrusion (Fig. 4-59) plastic flow of metal takes place in the same direction as punch travel. The orifice in this case is formed between the extension on the punch and the opening through the die. To prevent reverse flow of metal, the body of the punch seals off the top of the die. The finished part is ejected backward after the punch is retracted.

The design of a punch and die with double orifices to permit the plastic flow of metal in forward and reverse extrusion simultaneously will produce parts by combination impact extrusion. Figure 4-60 illustrates examples of combination extrusion.

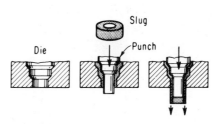

Fig. 4-59. Forward extrusion. (*Courtesy Design Engineering.*)

Impact extrusion dies are also classified as open or closed. In open dies the metal does not completely fill the die cavity. In closed dies the material fills out the details of the die. To compensate for variations in slug volume it is desirable to relieve closed dies and thus minimize excessive pressures.

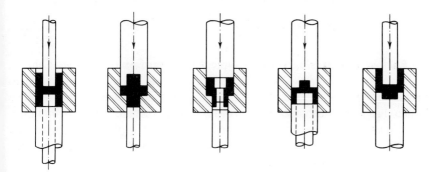

Fig. 4-60. Combination extrusion. (*Courtesy American Machinist.*)

Die Design Principles

Coining Dies. In backward extruding dies the punch is always smaller in diameter than the die cavity in order to give the clearance between punch and die equaling the desired wall thickness of the part to be produced. The punch is loaded as a column. To minimize punch failure it is desirable to coin the slugs to a close fit in diameter to assure concentricity. Figure 4-61 illustrates a coining die to prepare a slug for backward extrusion. Coining the slug to fit the diepot and coining the upper end to fit and guide the free end of the punch will minimize punch breakage of the extruding die.

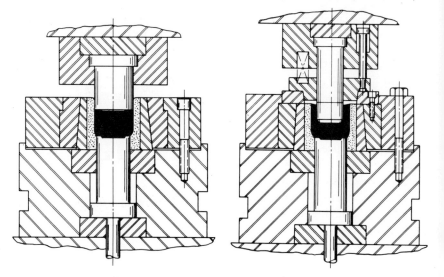

Fig. 4-61.   **Fig. 4-62.**

Fig. 4-61. Coining die for slug preparation.

Fig. 4-62. Backward extrusion die. Note that the centering ring for the punch and the carbide die cavity are preloaded in shrink rings and supported on toughened load-distributing steels. (*Courtesy E. W. Bliss Company.*)

Backward Extrusion Dies. A typical backward extrusion die is shown in Fig. 4-62. The use of a carbide die cavity will minimize wear due to excessive pressures. The carbide insert is shrunk into a tapered holder. The holder has a 1° side taper that prestresses the carbide insert to minimize expansion and fatigue failure. The inserts are well supported on hardened blocks. The extruding punch is guided by a spring-loaded guide plate which in turn is positioned by a tapered piloting ring on the lower die. Ejection of the finished part from the die is by cushion or pressure cylinder. Figure 4-63 illustrates a backward extrusion die with an unusual punch penetration ratio of 5 to 1 made possible with a modified flat-end punch profile.

Forward Extrusion Dies. Figure 4-64 is an example of a typical forward extrusion die in which the metal flows in the same direction as the punch, but at a greater rate owing to change in the cross-sectional area. The lower carbide guide ring is added to maintain straightness. The nest above the upper carbide guide ring serves as a guide for the punch during the operation. Figure 4-65 illustrates another forward extrusion die in which the punch creates the orifice through which the metal flows. The extruding pressure is applied through the punch guide sleeve.

Combination Extruding Dies. A typical combination forward and backward extrusion die is shown in Fig. 4-66. In this die the two-piece pressure

Chap. 4 Bending, Forming, and Drawing Dies

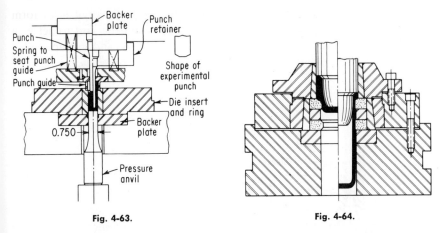

Fig. 4-63.   Fig. 4-64.

Fig. 4-63. Backward extrusion die. (*Courtesy American Machinist.*)

Fig. 4-64. Forward extrusion die.

anvil acts as a bottom extruding punch and a shedder. The upper extruding punch is guided by a spring-loaded guide plate into which the guide sleeve is mounted. To maintain concentricity between the punch and die, the punch guide sleeve is centered into the die insert.

Punch Design. The most important feature of punch design is end profile. A punch with a flat end face and a corner radius not over 0.020 in. can penetrate three times its diameter in steel, four to six times its diameter in aluminum. A punch with a bullet-shaped nose or with a steep angle will cut through the phosphate coat lubricant quicker than a flat-end punch. When the lubricant is displaced in extrusion, severe galling and wear of the punch will take place. The punch must be free of grinding marks and requires a 4 microinch finish, lapped in the direction of metal flow. The punch should be made of hardened tool-steel or carbide. In some backward extrusion dies a shoulder is provided on the punch to square up the metal as it meets this shoulder.

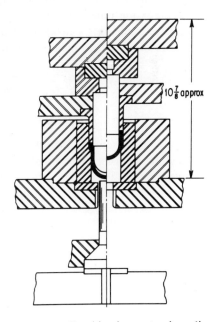

Fig. 4-65. Combination extrusion die. (*Courtesy American Machinist.*)

Pressure Anvil Design. The function of the pressure anvil is to form the base of the diepot, to act as a bottom extruding punch, and to act as a shedder unit to eject the finished part. Heat treatment and surface finish requirements are the same for pressure anvils and for punches.

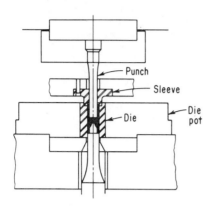

Fig. 4-66. Combination extrusion die with two-piece pressure anvil. (*Courtesy American Machinist.*)

Diepot Design. To resist diepot bursting pressure, the tool-steel or carbide die ring is shrunk into the shrink ring or die shoe. The die shoe is normally in compression. A shrink fit of 0.004 in. per in. of diameter of the insert is desirable. Material, heat treatment, and finish requirements of the diepot are the same as for the punch. The recommended material for shrink rings is a hot-worked alloy tool-steel which is hardened to 50 to 52 Rockwell "C." A two-piece diepot insert is sometimes used for complex workpiece shapes.

Punch Guide Design. The guide ring minimizes the column loading on the punch by guiding the punch above the diepot. The spring-loaded guide sleeve pilots the punch into the diepot and maintains concentricity between them. The guide ring can also act as a stripper. The proper use of guide sleeves permits higher penetration ratios.

TOOL DESIGN FOR FORGING

Forge Plant Equipment

Forgings in general are seldom used in the as-forged condition but are finished by various machining processes. A typical forging is the engine connecting rod (Fig. 4-67). As can be seen by a comparison of the illustrations, very little machining is required to bring this vital component to its usable form. What then is forging? It is the plastic deformation of a metal

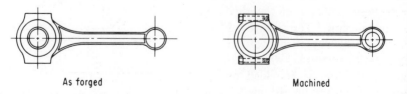

Fig. 4-67. Designs of engine connecting rod.

Chap. 4 Bending, Forming, and Drawing Dies   285

caused by the stresses imposed upon it through hammering or squeezing operations. All forgings are not made as close to their final configuration as in the case of the connecting rod. The shape of the forging is governed by the equipment and dies used in its production. Forging equipment, which will be considered first, may be divided into three broad classifications. First is the *hammer,* which imparts stress on the material by impact. Second is the *forging machine* or *upsetter.* This unit delivers its stress to the material by pressure or squeeze; it operates in a horizontal position. The third category of forge equipment is the *forging press,* which operates in a vertical position like the hammer but imparts pressure to the workpiece instead of impact. These three pieces of equipment constitute the main forging units; they do not cover the various pieces of equipment required to bring a particular forging to its final form. These will be discussed under "Auxiliary Forging and Finishing Machinery."

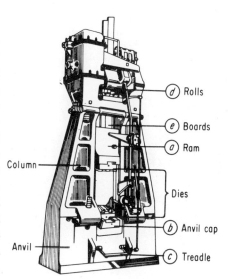

Fig. 4-68. Board-type drop hammer.

Hammers. The basic unit of the forging shop is the drop hammer (Fig. 4-68), the function of which is to form metal heated to a plastic state into a desired shape by the blows of the falling weight of the ram (*a*). There are three principal designs of drop hammers; board, air-lift-gravity-drop, and steam piston. The board-drop hammer is used for illustration in this discussion. Forging dies into which impressions have been cut are keyed into the movable ram and the stationary anvil cap (*b*). The dies must be perfectly aligned, or the forgings produced will not be in match, i.e., the impressions in the upper and lower dies will not be perfectly opposed and will produce an off-center forging. Ways in the columns and in the sides of the ram assure that with each blow the dies will strike in the same place.

The operation of securing the dies in the hammer is known as the setup. Setting up dies and tools in a hammer and its auxiliary equipment may take from several hours to a day, depending upon the complexity of the forging to be made and the size of the hammer to be used.

When the dies have been set up, they are preheated to prevent breaking under the impact of forging. A steel bar or billet which has been cut to the proper length is brought to the forging temperature, usually between 2100°F and 2400°F, and placed over the first impression in the die. The hammerman then steps on the treadle (*c*) which allows the ram to drop. After the blow has been struck, the rolls (*d*) engage the boards (*e*) keyed to the ram and raise it for the next blow.

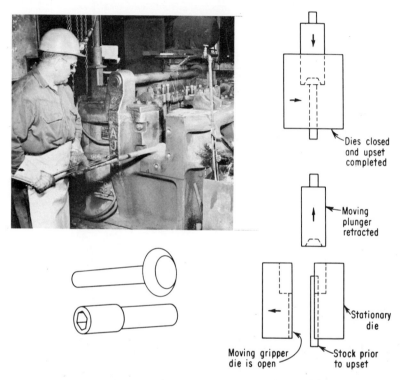

Fig. 4-69. Upsetter in operation. Two typical upset parts are shown, also a schematic of steps in the operation.

A drop hammer is classified by the weight of the ram; a hammer having a ram weighing 2500 pounds is known as a 2500-pound hammer. A well-equipped gravity-drop hammer shop has hammers ranging in size from 1000 to 5000 pounds. A set of dies for producing connecting rods is shown in place in the hammer illustration.

Forging Machines or Upsetters. Forgings that require a symmetrical gathering of stock on the end of a shaft are usually forged in an upsetter (Fig. 4-69). This machine is fitted with two gripping dies which come together horizontally to hold the forging bar in a rigid position, while the third die, called the plunger, upsets the stock on the end of the bar into an impression. This equipment is occasionally used to supplement hammer forging but generally it is used as a finished forging unit.

Forging Presses. These forging units are capable of varied forging operations which can be accomplished by means of exerting high pressures upon a heated material. Their greatest capability lies in the type of forging that is symmetrical in design. The construction compares closely with that of the crank-type presses used in stamping and drawing. In the very high tonnage ratings of over 2000 tons, forging presses are generally of the hydraulic type, whereby a high-pressure fluid acts on a piston and transmits power to the dies.

Fig. 4-70. Helve hammer for drawing out stock.

Auxiliary Forging and Finishing Machinery

The machinery discussed in the preceding section is generally used to move and shape the metal into a die cavity which contains an impression of the part to be forged, but since the initial volume of stock is seldom distributed evenly throughout the cavity, it is necessary to perform other operations to obtain a finished piece. For a clear understanding of these auxiliary units, a brief explanation of the equipment will be given, followed by an example of the forging produced as well as the basic design of the dies.

Helve Hammers. The helve hammer (Fig. 4-70) is frequently used as auxiliary equipment. Its function is to draw out stock by striking sharp, rapid blows between rounded or square dies. This operation is performed at the forging heat preparatory to forging in the hammer. Probably the best example of the use of this unit is the forging of gear shift levers. The size of steel bar used is determined by the cross-sectional area of the large boss, and is considerably larger than that required for the handle section. A bar cut shorter than the finished lever is used as raw material, and the thinner section is reduced and lengthened to fit the forging impression.

Helve hammer, forging and trim dies for gear shift levers are shown in Fig. 4-71.

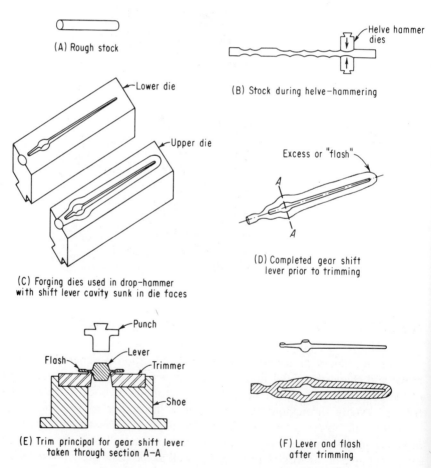

Fig. 4-71. Forging practice from rough stock to finished part.

Chap. 4 Bending, Forming, and Drawing Dies   289

Presses. Presses are used for trimming, punching, bending, and straightening operations which can often be performed at the forging heat. The presses are usually the crank type (Fig. 4-72). As can be seen from the illustration of the helve-hammered stock, a material saving in stock has been accomplished by drawing a short bar out to a rough shape which approximates the die cavity. In addition to the saving of a considerable amount of material, a further benefit is gained in finishing the forging since fewer blows are required in the hammer. This is a result of throwing out less flash around the die cavity. Since the stock cannot be proportioned to

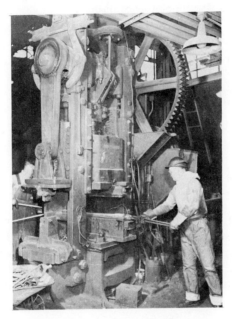

Fig. 4-72. Heavy-duty crank-type forging press.

fill the die cavity exactly, the excess flows out between the die faces and results in a thin flash. The flash cools very rapidly, and if it is not kept to a minimum, excessive pounding will be necessary to complete the forging. The result therefore is low production and decreased die life. Thus, in this forging operation as well as most others, the die designer must consider various operations to get the stock proportioned as close as possible to the ultimate shape of the piece before placing it into the final die cavity. Because stock cannot be provided with an exact volume to fill the die cavity, it is generally understood that the die cavity will throw a flash outside of its contour. Therefore, the majority of forgings must be trimmed by a press of the proper tonnage. Material at the forging temper of 2100–2400°F will require a smaller press than if the same part were trimmed cold. Conservatively speaking, it can have a capacity one-third that of a cold trimming press. Thus, if the periphery of a mild steel part was 20 in. and the flash was ⅛ in. thick, a pressure of approximately 75 tons would be required to trim the forging.

$$F = \frac{PTS}{2000} \qquad (4\text{-}18)$$

where F = force requirement in tons
P = periphery in inches of forging
T = thickness of flash
S = shear strength of material in psi.

Low carbon steel has a shear strength of 60,000 psi.

$$F = \frac{PTS}{2000} = \frac{20 \times 0.125 \times 60,000}{2000} = \frac{150,000}{2000} = 75 \text{ tons}$$

If the trimming is done at forging temperature, the shear strength S should be divided by three and the same formula applied.

$$F = \frac{PTS}{2000} = \frac{20 \times 0.125 \times 20,000}{2000} = \frac{50,000}{2000} = 25 \text{ tons}$$

In addition to the trimming operations, the crank-type press is used for a variety of other operations as previously mentioned. For example, the forged connecting rod in Fig. 4-67 will have a web of material in the hole after forging and, since it is desirable from a finish machining standpoint to have this web removed, it must be punched out. Figure 4-73 illustrates a punching die.

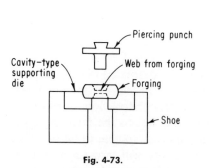

Fig. 4-73.

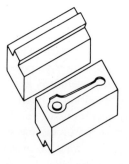

Fig. 4-74.

Fig. 4-73. Punching die.
Fig. 4-74. Restrike die.

Chap. 4   Bending, Forming, and Drawing Dies   291

Often a forging is of such an intricate nature that in normal processing by the hammer and through the trimming and piercing operation, the forging will become distorted. This can be rectified by mounting a set of restriking dies in the press and sizing the forging all over. A set of restrike dies is shown in Fig. 4-74. Often presses are used for bending operations such as illustrated in Fig. 4-5.

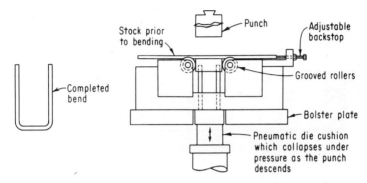

Fig. 4-75. Bending die.

Fig. 4-76. One-thousand-ton coining press.

Coining Presses. It is possible, by using a heavy duty coin press (Fig. 4-76), to obtain very close tolerances on the thickness dimensions of some forgings. A tolerance of ±0.005 in. is not uncommon, and some dimensions can be held as close as ±0.002 in. Coining dies (Fig. 4-77) are designed to allow a flow of metal so that the pressure applied will force any excess metal on the surfaces being coined to move to adjacent surfaces on which no pressure is exerted. The dimensions which are coined, therefore, must be those between opposed surfaces, and there must be adjacent surfaces to which the excess metal may move. Coining has effected savings to many forging users by replacing expensive machining.

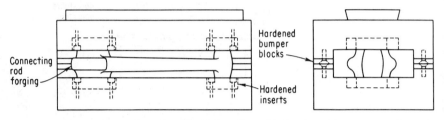

Fig. 4-77. Set of cold coining dies with hardened inserts at points of greatest wear.

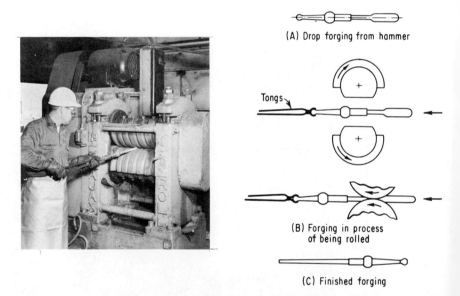

Fig. 4-78. Forging roll drawing out long slender shaft for gear shift lever.

Forging Rolls. A forging roll (Fig. 4-78) is used to draw stock from a large diameter down to a straight or tapered rod of smaller diameter and greater length. Its function is similar to that of the helve hammer; it is also used for finishing operations after a portion of the forging has been processed in a hammer. The rolls are semicylindrical to permit clearance for the preforged section. An excellent example of the work that can be accomplished on this machine is the tapered shaft of a gear shift lever. The ball end of this lever is forged in a hammer, and the shaft is later drawn out to its finished dimensions on a forging roll.

THE FORGING PROCESS

The elasticity of the metal or its size usually governs whether or not the operation of forging is to be accomplished in the hot or cold state. A good example of cold forging is the common nail, in which a small rod is gripped between moving dies and a small amount of the stock is upset on the end, forming a head. Most often, metals are forged hot because the material will shape more readily in this condition and the necessary fiberlike flow line structure is imparted. In the grain flow lies the inherent strength and toughness of forgings; the die designer must make the dies in such manner as to take full advantage of this highly desirable quality. A cast ingot

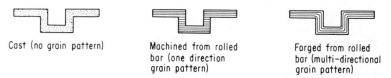

Cast (no grain pattern)   Machined from rolled bar (one direction grain pattern)   Forged from rolled bar (multi-directional grain pattern)

Fig. 4-79. Three methods of producing a crank.

at the steel mill is rolled down by successive passes to ultimately produce bars and billets with a fibrous grain which runs the length of the bar. These high-quality bars are then cropped to the proper length to make a particular forging. In Fig. 4-79 a comparison is made between three methods of producing a crank. In addition to the proper placement of grain flow pattern, the die designer must also take into account all functional surfaces of the part to help eliminate costly machining operations wherever possible. In some cases streamlining is necessary for appearance.

Above all else, overall economy must be considered in die designing for the forged part. In a costly aircraft or missile forging, quality is often of the utmost importance. Dies must be designed to impart the maximum grain flow benefit to the finish-machined part. On a hook to be used in overhead lifting, where human lives and expensive fabricated components are at stake, the ultimate in grain pattern should be developed in the dies. A small lever with a high factor of safety in its design can occasionally be produced with no regard for grain flow but a high regard for its final cost. Figure 4-80 illustrates the grain flow patterns which are developed on various forgings made by different forge die designs.

A tensile specimen pulled to destruction will exhibit a higher ductility when the grain pattern runs parallel to the pulling force than a specimen of equal dimensions and hardness having the grain pattern transverse to the pulling force. The tensile strength is affected to a lesser extent on a straight no-shock load than when impact loads occur. In this high-shock area of application grain flow plays an important part in lengthening the service life of the part. Such established data make it highly desirable for the forging die designer to work closely with the engineers who have designed a component. The forgings shown in Fig. 4-80 will all look the same when

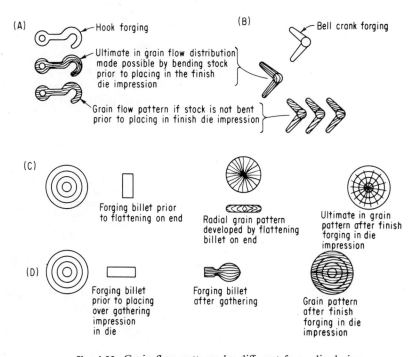

Fig. 4-80. Grain flow patterns by different forge die designs.

completed, but the grain pattern may be detrimental to the strength of the part. The hook and the bell crank can be produced more economically without bending, and the gear is produced more economically by not flattening the stock on its end prior to forging. However, if the part is to be highly stressed in the areas of transverse grain structures, it would be better to perform the bending or flattening operations. The illustrations of Fig. 4-81 depict the poorer grain patterns.

Another prime consideration in the design of tools is economy in manufacture. In this area the tool designer has two things to consider: the effect of the tool design on the part price and the cost of the tools. Some very fine tools can be constructed to accomplish an operation in a minimum of time.

Chap. 4 Bending, Forming, and Drawing Dies   295

One user's requirements may be high enough to justify the expense of these tools, whereas another may be interested in obtaining a low tooling cost with not much regard to the part price. If customer *A* needs 10,000 pieces and the tools will cost $3000 and the pieces produced from these tools 30 cents each, the total cost will be $6000 or 60 cents per part. Customer *B* needs 1000 pieces of an almost identical part; if produced from the same tool design they will cost $3.30 apiece. Therefore, it would be to customer *B*'s advantage to get tooling at $1000 and parts from these tools at 75 cents

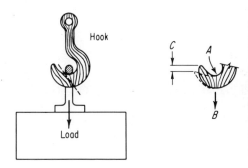

Fig. 4-81. Examples of poor grain patterns.

each. Thus the cost per part would be only $1.75. Conversely, if customer *A*'s forgings were produced from the $1000 tools the cost per part would be 85 cents. This can be depicted graphically as shown in Fig. 4-82. The flattening out of the curve along the axes shows the tool designer two basic facts: with little or no tooling the piece price rises, and with too extensive a tool design no benefit on piece price is gained. The designer should acquaint himself with all available equipment and processes and in all cases evaluate his tool designs with overall economy in mind. Occasionally the nature of the part is such that a saving can be made by upsetting a large section prior to finish impression forging rather than endeavoring to roll or draw a smaller section down from a large bar or billet. Possibly a hole could

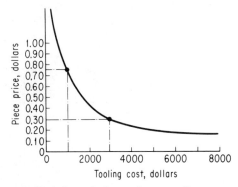

Fig. 4-82. Relation of piece price to tooling cost.

be drilled more economically than a set of piercing dies could be constructed for the quantity involved. Coining or sizing a part by exerting extreme pressures on certain contours may eliminate a costly machining operation.

FORGING DESIGN

The tools necessary to produce a given forging cannot be made until the shape of the final forging has been determined. Therefore, it is essential that the tool designer has an understanding of the underlying principles of the forging design. These will be considered in the following order: forging draft, parting planes, fillets and corner radii, shrinkage and die wear, mismatch of dies, tolerances, and finish allowances.

Forging Draft

Draft is the angle or taper that must be imparted to the surfaces of the forging that are contained in the die and are not self-releasing. For maximum production it is necessary to free the forging from the die cavity quickly. Forgings that are made in hammers and presses generally require a draft angle of 3° to 7° for external surfaces and 5° to 10° on internal surfaces. The application of draft can best be seen from the example shown in Fig. 4-83. As the dies part, owing to the action of the hammer or press, the forging will stick in whichever die offers the maximum frictional resistance to release of the part contained within its cavity. Draft as provided to all *B* surfaces permits the forging to release itself from either die. The *A* surfaces are self-releasing and require no draft addition. In the case of upsetter work (Fig. 4-69) the application of draft can be materially reduced since the stock is firmly held by the gripping dies as the plunger retracts.

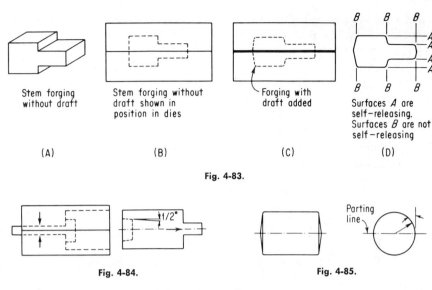

Fig. 4-83.

(A) Stem forging without draft
(B) Stem forging without draft shown in position in dies
(C) Forging with draft added
(D) Surfaces A are self-releasing. Surfaces B are not self-releasing

Fig. 4-84.   **Fig. 4-85.**

Fig. 4-83. Application of forging draft.
Fig. 4-84. Draft on head of an upset forging.
Fig. 4-85. Draft formed by arc of a circle.

Draft on the head of an upset forging can be reduced in many cases to as little as $\frac{1}{2}°$ (Fig. 4-84). In addition to flat surfaces as at A in Fig. 4-83D, a self-releasing surface which is often encountered is the cylindrical type as shown in Fig. 4-85. As can be seen, the arc of the circle forms its own draft and will not present a sticking problem during forging. Generally the draft on external surfaces can be less than for internal surfaces. The reason is that as the metal is being forged, it is undergoing a shrinkage due to cooling, and all surfaces internal as well as external are shrinking towards the center of the part. Thus, when a hole or other depression is being forged into a part, the forging tends to grip the plug in the die which forms the depression. On external surfaces, the reverse is true (Fig. 4-86).

Fig. 4-86. Surfaces a are pulling away from the die, whereas surfaces b are pulling toward the die. Greater draft angle is required on b surfaces than on a surfaces.

Parting Lines

A parting line is the dividing plane between the two halves of a pair of forging dies. This line may be perfectly flat or multidirectional. A flat parting plane is more economical to produce but very often the part which results from using a flat parting plane is not an economical forging. A straight parting line is shown in Fig. 4-83C on the stem forging. However, another type of forging may not lend itself to this type of parting. Figure 4-87 indicates an extreme amount of stock addition (shaded at *a*) which would result if a straight parting line were used. This would result in additional weight and an expensive machining operation to remove this excess; the only benefit that would be gained would be a slight reduction in the time necessary to machine the die. A lock-type parting line would be preferable and would produce a part which would have the desired shape. A lock-type parting line for the same piece is seen in Fig. 4-88. Other types of lock-type

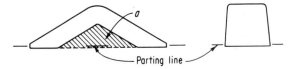

Fig. 4-87. Stock addition resulting from a straight parting line.

parting lines are shown in Fig. 4-89. In Fig. 4-89*A* the crank arm shown could have been parted through the center line *x-x*, but then draft would have to be added to faces *a* of the two large bosses. This would complicate the drilling operation which must subsequently be performed because the drill would enter a slanted surface and would tend to bend in the direction of deflection (Fig. 4-90). The boss is to enter a $1\frac{3}{16}$-in. wide clevis slot in final assembly and the remaining draft after drilling would make this impossible. A further consideration in this parting line design was economy in forging. Had the forging been parted through center line *x-x*, a separate

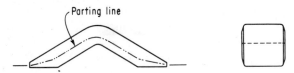

Fig. 4-88. Use of lock-type parting line for the piece shown in Fig. 4-87.

bending operation would have to be performed on the stock before placing it into the die cavity as shown in Fig. 4-91. In Fig. 89*B* a gear blank is shown with a straight parting line at *x-x* and also an elevated parting line at *y-y*. It was necessary to raise the parting plane to *y-y* in the hole to reduce the amount of stock to be removed in machining. The stock saving is indicated by the shaded areas. The plug in the top die which must form the hole is considerably stronger than if it were extended down to parting plane *x-x*. From the considerations for the designs of Figs. 4-87, 4-88, 4-89 *A* and *B*, the apprentice tool designer should readily understand the location of the parting planes shown in Fig. 4-89 *C* and *D*.

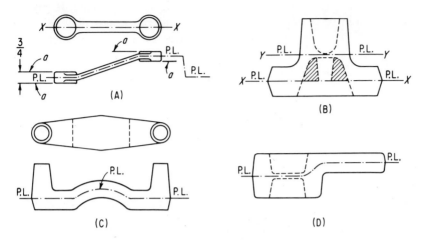

Fig. 4-89. Various types of lock-type parting lines.

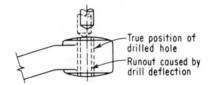

Fig. 4-90. Drilling a hole in a casting.

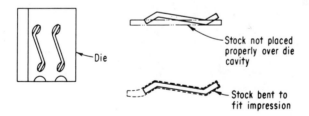

Fig. 4-91. Bending operation required because of faulty parting line.

Fillets and Corner Radii

A fillet can mean any rounding of the apex of an internal angle and a corner radius the rounding of the apex of an external angle. Fillets are employed in practically all forging work because they not only increase the life of the die but also result in sounder forgings. As can be seen in Fig. 4-92 the material will progressively flow into the die cavity. The fillet R has allowed the material to start its downward flow without nicking. If too small a fillet is used or if none is provided the material would flow as seen in Fig. 4-92B. The material has filled the cavity but has folded back against itself forming a defect. This condition is called a *lap* and is generally caused by using inadequate filleting in the die.

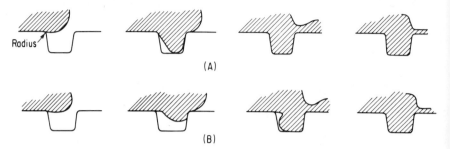

Fig. 4-92. Progressive flow of metal into a forging die.

Corners must be rounded on forgings because the material flowing into the die cavity will trap air and oil at these points, making them very difficult to fill. Scale will adhere in the sharp corner of a die to a greater extent than when it is rounded and will halt the progressive flow of material, resulting in an unfilled cavity. Table 4-5 can be used as an approximate guide in establishing the necessary corner or fillet radii:

TABLE 4-5. CORNER RADIUS DESIGN

Net wt. of forging, lb	Corners, in.	Fillets, in.
1 to 5	$1/16$	$1/8$
5 to 10	$3/32$	$3/16$
10 to 30	$1/8$	$1/4$
30 to 50	$5/32$	$5/16$

Shrinkage and Die Wear

Most steel forgings are formed at temperatures from 2100°F to 2400°F. In this state the material is greatly expanded, and during the forging operation is in the process of cooling and consequently shrinking. The die impressions are put in using a shrink scale which amounts to $3/16$ in. per foot when the material to be forged is steel. It is difficult to control both heating and cooling of a material within close limits during forging. Consequently it is difficult to have the forging leave the die impression at exactly the temperature which is equal to $\frac{3}{16}$ in. per foot shrinkage. Shrinkage variances usually amount to ±0.001 to 0.003 per inch of width or length on a given forging. *Die wear* is that quantity of dimensional difference which comes about through normal abrasion of the impression. This usually amounts to between $\frac{1}{64}$ and $\frac{1}{32}$ in. on external surfaces of forgings weighing up to 10 lb and a like amount on internal surfaces such as plugs. On forgings between 20 and 50 lb this may increase from $\frac{1}{16}$ to $\frac{3}{32}$ in.

Mismatch

When a forging is parted in such manner as to include a portion of the part in the top die and a portion in the bottom die, it is difficult to maintain alignment of the half impressions because either the upper or the lower die may shift during forging. This shift may occur sideways or endways as illustrated in Fig. 4-93. Forgings produced from a shifted die will be mismatched. This shift is generally kept to a minimum as the forging die designer will add a finish allowance if the mismatch is detrimental to the workability of the part. On forgings up to 10 lb the dies may shift 0.015 in. during a production run, and forgings from 10 to 20 lb may shift 0.020 in. A 0.005-in. shift allowance should be added for each additional 10 lb.

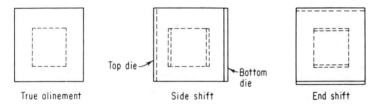

Fig. 4-93. Shifting of dies.

Tolerances

Thickness tolerances apply to the forging dimensions which cross the parting plane of the dies. They are applied both as a plus and as a minus value. If an excessive flash is thrown between the die faces and becomes too cold, the flash will tend to keep the dies from fully closing. To prevent this, the flash thickness in the die is kept as thick as possible.

Commercial plus tolerances on thickness dimensions are approximately $+\frac{1}{32}$ in. on forgings up to 1 lb, $+\frac{1}{16}$ in. on forgings from 5 to 20 lb, and $+\frac{3}{32}$ in. from 20 to 50 lb.

Commercial minus tolerances on thickness dimensions are −0.010 in. up to 1 lb, −0.023 in. from 5 to 20 lb., and −0.031 in. from 20 to 50 lb.

The forging die designer can control the plus variations by providing ample flash thickness in the die, establishing a trimming operation to remove the majority of flash before finish forging, or by designing a set of coin dies to force the excess material down to size. The minus variations are ordinarily caused by the faces of the die pounding down and thus causing an undersize condition when the dies are closed. This can be corrected by providing ample striking surface on the faces of the dies. Assuring a smooth surface finish to the die face is another control that can be applied to reduce face pound-down. Scale falling into the die cavity can also cause an undersize condition, but the die designer has little control over this factor. It is customarily controlled by the inspection department during forging.

Finish Allowances

A finish allowance is an additional amount of material added to surfaces which cannot be controlled close enough by forging and must be subsequently machined. The finish allowance for a given forging is established by considering the effects of shrinkage, straightness, mismatch, the minus tolerance on thickness, and surface decarburization. Decarburization is the loss of carbon from the outer surfaces of a forging caused by heating. On some grades of steel the loss of carbon may be very high; this should be checked with the steel supplier. If a heat treatment is to be given a forging, the outer layers will not respond to the treatment because the carbon content has been lowered. Therefore, to get full hardness even at the outer layers, it is essential to add sufficient stock to assure that virgin material is being subjected to the heat treatment. An example of the application of finish would be as seen in Fig. 4-94 where surfaces A and B require a hardened and ground finish.

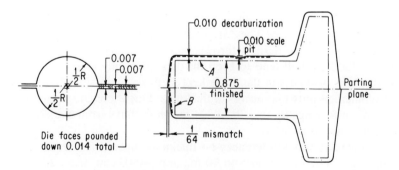

Fig. 4-94. Application of finish allowance.

Figure 4-94 shows that the forging designer has applied $\frac{1}{16}$-in. finish allowance all over specifying the forged stem (*A*) diameter 1.000 in. nominal or 0.125 in. total finish.

less	0.020 in.	scale pitting
less	0.020 in.	decarburization
less	0.014 in.	minus thickness tolerance
less	0.003 in.	shrinkage
	0.057 in.	

The nominal diameter of 1.000 in. less 0.057 in. equals 0.943 in. diameter or 0.034 in. per side. This remaining 0.034 in. has been added to guarantee a full clean-up in the event the stem forging is not completely straight as located for turning and also to allow approximately 0.010 in. for grinding after turning. The same approach was used in determining the $\frac{1}{16}$-in. allowance for surface *B*, except that thickness tolerance is not considered but end shift (mismatch) is a possibility.

DROP FORGING DIES AND AUXILIARY TOOLS

Forging dies are made from high carbon, nickel-chrome-molybdenum alloy steel blocks which have been forged to achieve the utmost grain refinement and resistance to shock. In addition to the qualities developed by hot working, a maximum resistance to wear is insured by a double heat treatment.

Before the forging impressions are sunk, shanks are shaped for locking the dies in the hammer, and the blocks are squared for perfect alignment (Fig. 4-95). The striking faces are planed, or if the forging to be produced necessitates a lock, that is, if the striking surfaces must be in two or more planes, the faces are shaped accordingly.

A blueprint or model of the part to be forged is translated into metal templates which, in turn, are used for laying out the blocks. Six types of

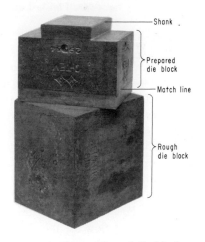

Fig. 4-95. Preparation of die blocks.

impressions may be incorporated in a die for shaping the material progressively from bar form to the finished forging (Fig. 4-96 *A* and *B*). Impression *a*, usually referred to as the *swager*, is used to reduce and draw

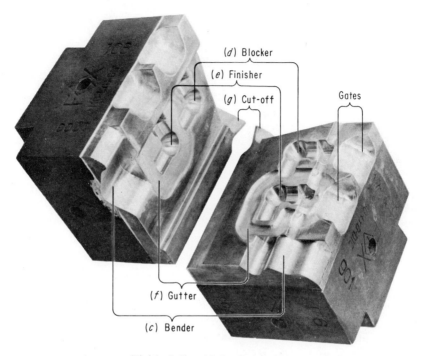

Fig. 4-96A. Finished die with bender, blocker, and finisher impressions, and cutoff.

out stock when differences in cross sections of the forging make this necessary. This operation is similar to that performed on the helve hammer (Fig. 4-70). The swager is ordinarily used for forgings having only a short section requiring reduction of area, whereas the helve hammer is used for forgings for which most of the bar length must be reduced. Impression *b*, commonly called the *edger* or *roller*, distributes the stock so it will fill the next impression without excessive waste. A *bender impression, c*, is included when curves or angles in the forging make it necessary to bend the stock before it will fit properly in the finishing impressions. Impression *d*, known as the *blocker*, gives the forging its general shape and allows the proper gradual flow of metal necessary to prevent laps and cold shuts. Although this impression has the same contour as the finished part, large fillets and radii are added to permit the easiest flow of metal. The *finisher impression, e*, brings the forging to its final size. A *gutter, f*, is cut into the block around this impression to provide space for the excess metal, or flash, which is forced out of the finishing impression when the forging is hammered

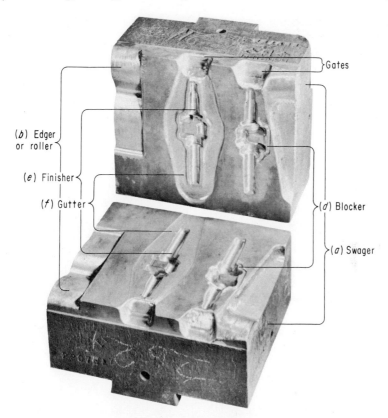

Fig. 4-96B. Finished die with roller, swage, blocker, and finisher impressions.

to size. The *cutoff, g,* cuts the forging from the bar by shearing off the tong hold. It is principally used for small forgings on which no further processing is to be performed at the forging heat. It permits rapid and economical production.

The blocking and finishing impressions are cut in the block by highly skilled men who use a milling machine especially designed for sinking dies. Cutters of various types are used in accordance with the shape of each section of the impression, but much of the accuracy of the die depends upon hand work performed after it is sunk.

When the forging die impressions are completed, the blocks are clamped together in the position in which they will meet in the forging operation, and a lead-antimony alloy is poured into the finishing impression. The resulting lead cast is used to check the accuracy of the forging dimensions, and is sent to the customer for his approval.

Since steel shrinks in cooling from its forging temperature, and the lead alloy does not, it is necessary to allow for this shrinkage in checking the lead cast. The correction amounts to $\frac{3}{16}$ in. per foot, or $\frac{1}{64}$ in. per inch.

After the lead cast has been approved, the dies are finished by machining the gutter, flash relief, and roller, and the cast is used as a model from which auxiliary tools can be made accurately. These tools include trimming, punching, and restrike dies.

Figure 4-97 shows the forging dies and working steps involved in producing a large connecting rod. The essential steps encountered in the production of this forging will explain the method for producing a wide variety of forgings.

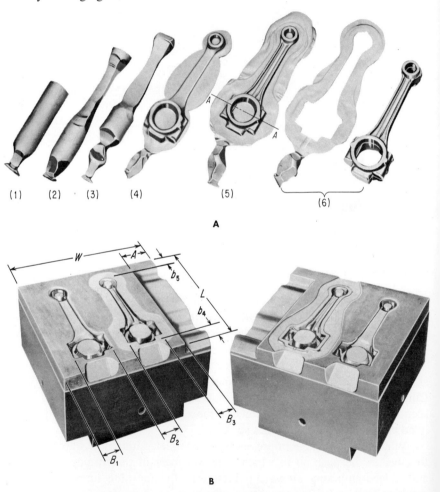

Fig. 4-97. (*A*) Working steps in producing a connecting rod forging: (1) original bar; (2) drawn in a helve hammer; (3) result of rolling; (4) rough form after blocking; (5) finished forging as it leaves the hammer; (6) the flash and the trimmed forging. (*B*) Forging dies.

Before the impressions were sunk into the die faces, it was determined that the forging would best resist the stresses imposed upon it if the grain flow ran parallel with the length. It was, therefore, decided to place the impressions in the die with the length running from front to back, and the bar was to be placed in like manner.

The next consideration was that of establishing a basic parting plane which in this case lends itself to the preferred straight-line type of parting. Subsequent machining can be materially reduced by including plugs in the upper and lower dies to form the crankpin, piston pin holes, and the depressions which form the channel section. It is, therefore, proper to place the parting plane along the thickness of the connecting rod rather than the width.

After these factors have been established, critical surfaces that require a high degree of perfection to be usable must have finish allowances added. In this case a $3/32$-in. allowance was added to the top and bottom faces of the crank and piston ends. The hole in the crankpin end was elongated $3/16$ in. to provide material for a slitting saw to part the rod from the cap portion and still end up with a round hole for the crankshaft bearing. An allowance of $3/32$ in. was added to the sides of the two holes as well as to the ends of the bolt bosses where a good seat must be established for the heads of the bolt and nut which assemble the cap to the rod portion.

Draft must now be added to all vertical surfaces in the die so that the forging will be self-releasing. All external surfaces around the periphery have a 7° angle and the plugs which form the holes have a 10° angle. Before the preforming operations of rolling and swaging can be laid out on the die face it is necessary to know the size of the original bar. From the forging design with finish, draft, and approximate flashing, the largest cross section of the rod perpendicular to the direction of stock placement is first determined. In this case it is the plane a-a in Fig. 4-96 at 5. This section is shown in Fig. 4-98A. By approximation of areas, length times width, the area of stock to fill this section would be computed as follows:

left flashing:	$1\frac{1}{2}$ in. $\times$ $3/16$ in. = 0.282 sq. in.
right flashing:	$1\frac{1}{2}$ in. $\times$ $3/16$ in. = 0.282 sq. in.
plug flashing:	2 in. $\times$ $3/16$ in. = 0.375 sq. in.
crankpin section L:	1 in. $\times$ 2 in. = 2.000 sq. in.
crankpin section R:	1 in. $\times$ 2 in. = 2.000 sq. in.
	4.939 sq. in. total

This area is slightly less than that of a $2\frac{1}{2}$-in. diameter bar. It is apparent that the $2\frac{1}{2}$-in. bar could not be laid across the length of the forging and hammered until the cavity was filled, as the volume of material required on the balance of the rod is considerably less. The I-beam section may take as little as $1\frac{1}{8}$ in. round at the small end when calculated as above. The rod forging will, therefore, require extensive reduction from a large bar diameter to a smaller bar diameter at various sections along its length. A swager such as is seen in Fig. 4-96B or a helve-hammer operation as seen in Fig. 4-70 can make this reduction. On this particular forging it was decided to use the helve hammer to reduce the steel, since a very small shipment of forgings was needed and to add a swager to the die would increase the die width approximately 3 inches. This would necessitate ordering larger, costlier blocks and would add to the block preparation cost by increasing the time to face the blocks and also the time to sink the swaging impressions in the face. A set of standard helve-hammer dies was used which did not allow the stock to be reduced beyond the $1\frac{1}{8}$-in. round minimum requirement. After the swaging operation has been elim-

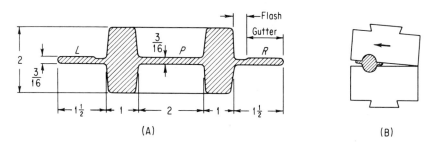

Fig. 4-98. (A) Cross section of finished forged connecting rod; (B) shift in mating halves of a forging.

inated from the sequence of operations that must be performed on the forging die blocks, it is now possible to decide on the size of blocks required.

In addition to the widths and lengths of a forging, empirical data relating to minimum striking surface are necessary to establish the die block size. The first consideration is the width required. As shown in the inset of Fig. 4-97, the roller is placed on the right side of the die. The width, a, required is approximately equivalent to the required steel size plus $\frac{3}{4}$ in. on sizes up to 2 in. round, steel size plus 1 in. on sizes from 2 in. to 3 in. round, steel size plus $1\frac{1}{4}$ in. on sizes from 3 in. to 4 in. round, and steel size plus $1\frac{1}{2}$ in. for diameters from 4 to 5 in. The next considerations are the distances b, b_1, b_2, etc. required between or around impressions. The area covered by the b allowances serves a dual purpose, the foremost being striking surface and the other being a provision for the excess metal or flashing. If too little striking surface is provided between impressions, rapid upsetting of the die face will result, causing an undersize condition on the forging thickness. The striking surface between cavities is reduced considerably by the flashing and guttering of the dies.

Flashing is that portion of excess metal adjoining the forging at the parting line. In die designer's terminology it can also be considered the portion of the die which forms the excess. The amount of flashing that must be provided varies from one type of forging to another. However, on steel sizes to $1\frac{1}{2}$ in. diameter, the dies can be flashed approximately $\frac{3}{16}$ in. wide around the periphery of the forging and about $\frac{1}{32}$ in. thick. On sizes between $1\frac{1}{2}$ in. diameter and 2 in. diameter, a $\frac{7}{32}$-in. wide by $\frac{3}{64}$-in. thick flash will usually suffice. Sizes from 2 in. diam to $2\frac{1}{2}$ in. diam: $\frac{1}{4}$ in. by $\frac{1}{16}$ in.; $2\frac{1}{2}$ to 3 in. diam: $\frac{5}{16}$ by $\frac{3}{32}$ in.; 3 to 4 in. diam: $\frac{3}{8}$ by $\frac{1}{8}$ in. In addition to the flash extension, a further provision must be made in the die for any excess material. This is called the gutter and may be incorporated in the die by use of Table 4-6.

TABLE 4-6. GUTTER DESIGN SIZES

Steel size, in.	Depth of gutter, in.	Width of gutter, in.
$\frac{1}{2}$–$1\frac{1}{2}$	$\frac{1}{8}$	1
$1\frac{1}{2}$–2	$\frac{3}{16}$	1–$1\frac{1}{4}$
2–$2\frac{1}{2}$	$\frac{3}{16}$	$1\frac{1}{4}$–$1\frac{1}{2}$
$2\frac{1}{2}$–3	$\frac{3}{16}$	$1\frac{1}{4}$–$1\frac{1}{2}$
3–4	$\frac{1}{4}$	$1\frac{1}{2}$–$1\frac{3}{4}$

From the data just given it is now possible to determine the width of block required in six steps as follows:

1. Roller width = $2\frac{1}{2}$ in. + 1 in. = $3\frac{1}{2}$ in.
2. Blocker width = widest part of forging = 4
3. Finisher width = widest part of forging = 4
4. (b_3) flash + gutter + $\frac{1}{2}$ in. = $\frac{5}{16}$ + $1\frac{1}{4}$ + $\frac{1}{2}$ = $2\frac{1}{16}$
5. (b_2) flash + gutter + $\frac{1}{2}$ in. = $\frac{5}{16}$ + $1\frac{1}{4}$ + $\frac{1}{2}$ = $2\frac{1}{16}$
6. (b_1) taken same as b_2 or b_3 = $\underline{2\frac{1}{16}}$
$17\frac{11}{16}$ in.

Therefore, an 18-in. wide block should be used and the impressions laid out on the die face in accordance with the six steps. It should be noted that in steps 4 and 5, an additional $\frac{1}{2}$ in. must be added to provide ample striking surface of the die faces. A blocking impression was added to this die to distribute the stock more evenly in the varying contours of the finish impression. If a part has very few sectional changes and is nearly symmetrical about its axis, the blocking impression can be eliminated from the die. However, generous filleting between sections is then necessary.

The overall length of this particular connecting rod is 18 inches. The distances b_4 and b_5 are also taken the same as b_2 and b_3. Thus the length of block will be approximately 22 in. (18 in. + $2\frac{1}{16}$ in. + $2\frac{1}{16}$ in.). Die block manufacturers generally stock large blocks in 2-in. increments; thus it is desirable to keep the block requirements within this category.

The dies are laid out with the finish impression as close to the center of the block as possible so that the least amount of shifting can take place as the dies come together. If it were placed on the outer edge of the die, there is a tendency to pull the dies in that direction and thus cause a shift in the mating halves of the forging (Fig. 4-98B). After the finish impression is sunk into the die face the blocking impression is sunk. This impression is very similar to the finish impression but differs in many respects. In the case of the connecting rod and similar forgings, the I-beam section is left solid but made with sufficient volume to fill the I-section when the stock is transferred to the finisher. This reduces the possibility of defects when finished. All fillets in the blocker are made two to three times larger than the same fillets in the finisher. This is done to distribute the stock to the proper areas and also to provide less resistance to the flow of material to deep sections.

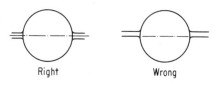

Fig. 4-99. Methods of flashing.

The blocking impressions are seldom flashed or guttered, whereas the finish impression is usually flashed in both dies and guttered only in the top die. By flashing both dies, the final forging has a neater, more symmetrical appearance (Fig. 4-99). By guttering the top die only, the forging will sit flush in the trimming die which must be used to trim the excess flash.

The roller can next be sunk into the die. The purpose of the roller is to gather steel prior to insertion in the blocking impression and also to break off scale which is formed on the bar during heating. Scale is detrimental to die life and should be kept out of the die impressions by all means possible.

As the steel comes from the helve hammer, it will be reduced in the center section only and will not be worked at the large ends of the bar. The roller is so designed as to make possible the use of a smaller bar than the $2\frac{1}{2}$ in. previously figured. This is accomplished by filleting in from both sides of the large section and thus increasing the bar size slightly at the point where the most steel is needed. On the small end of the rod, less material is needed and here the roller should be designed to elongate and reduce the steel. The side view of the roller for the connecting rod is seen in Fig. 4-100. After all impressions are sunk in the die, it is given a final dressing to fillet all sharp corners at the flash line to permit ease of metal flow (Fig. 4-101). The radius is approximately $\frac{3}{64}$ in. on $\frac{1}{16}$-in. thick flashing, $\frac{1}{16}$ in. on $\frac{3}{32}$-in. flashing, and $\frac{3}{32}$ in. on $\frac{1}{8}$-in. thick flashing.

Chap. 4 Bending, Forming, and Drawing Dies   311

In addition to forging dies, auxiliary tooling is required to complete the forging. This tooling consists of a hot punching die and hot trimming die mounted side by side in the same press. The forging is punched first so that handling time is reduced. If it were trimmed first, the forging would fall through the trim die and would have to be picked up from the base of the

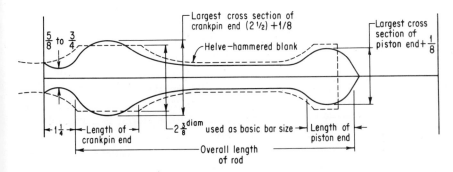

Fig. 4-100. Side view of roller for gathering the material in a connecting rod.

shoe and relocated for punching. The small punch holder is merely bolted to the top die. Any rotation about its axis due to capscrew hole clearances would not affect the punched hole since it is a full diameter. However, the large hole punch is bolted and doweled in the top die. Since it is oval, any rotation would affect the concentricity of the hole with the outside wall. The trimming punch is also keyed in the top die to prevent shift in any direction.

The trim punch is made flat where it strikes the bosses, whereas the I-beam section is contoured directly into the face of the punch. This gives full bearing all over as the forging is pushed through the trimmer. Distortion is thus eliminated and the I-beam section will hold its true shape. When push-

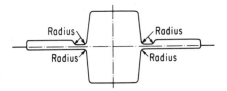

Fig. 4-101. Filleting a die.

ing a forging through a trimmer, particularly where the operation is performed hot and the metal is still plastic, it is desirable to contour the punch so that no nicks or flat spots occur in the finished forging. Both the trim and the punching die are relieved for the flash. Even though the gutter side is up, handling often results in bent flashing. This must be provided for in order that the forging rest properly in the impression. The built-up edge which remains also provides grinding stock if the trimmer needs reconditioning. If the forging is trimmed prior to punching, the piercing die need not be relieved. For trim punch design, see Figs. 4-102, 4-103A.

A hot trim die is made from unhardened medium carbon alloy steel for heavy trimming jobs and mild steel for light duty service. The cutting edge is built up with stellite, a material which maintains very high hardness characteristics at elevated temperatures. Stellite is available in the form of welding rod and is applied to the trimmer by this process. After the bead of stellite is placed around the impression, it is ground to fit the contour of the forging at the parting line. Stellited trimmers can be used for cold trimming if the base material is a medium carbon alloy. Pressures encountered in cold work are considerably higher and thus a substantial base material is needed for the support of the stellite.

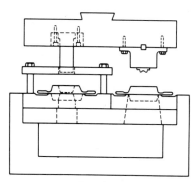

Fig. 4-102. Punch-and-trim die.

Cold trimming dies are also made from tough alloy tool-steels which can be hardened to 52–54 R/C and still maintain their cutting edges. Cold trimming punches are made from tough alloy tool-steels and are hardened 45–47 R/C. The punch is made softer so that if it comes in contact with the edge of the trimmer, its edge will shear off rather than the trimmer. Hot trim punches are made from medium carbon alloys, hardened 38–40 R/C.

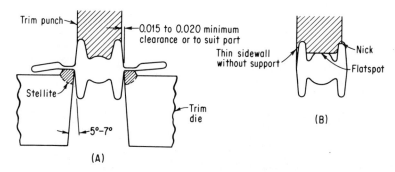

Fig. 4-103. Trimmer punch design: (A) correct; (B) incorrect.

Initial clearance is provided between the punch and the die of about 0.015 to 0.020 in. The trimming die is relieved on a 5° to 7° angle to give adequate clearance for the forgings to drop through freely. As the draft angles wear, the trimmer is opened up so that a wide trim flat is not left on the forging. However, if it is desired to hold a close trim dimension at the

Chap. 4 Bending, Forming, and Drawing Dies   313

parting line, the trimmer must be watched very closely and, if worn beyond the acceptable tolerance, must be re-stellited and ground to size. The trimming die for the connecting rod is made in two pieces to facilitate shaping of the stock and also to provide a means of adjustment (Fig. 4-104). Three- and four-piece trimmers are also illustrated to show this principle.

Heavy mild steel plates, approximately $1\frac{1}{2}$ to 2 in. thick, are used to back up a trim die, and a hole is usually burned out to a size slightly larger than the forging which the plate must accommodate.

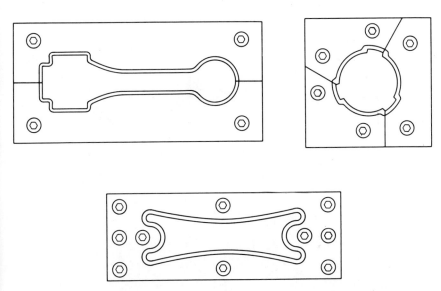

Fig. 4-104. Trimming dies made in two, three, or four sections.

Hot piercing dies are made of medium carbon alloy steels (36–40 R/C) and the punches are medium carbon alloy with a stellite facing around the periphery of the punch and on its face. Punching dies must be provided with a stripper plate (Fig. 4-105) to release the forging from the punch. The stripper posts must be set far enough apart to admit the forging, and if punched prior to trimming must be wide enough to admit the forging and the flash. Where production warrants the expense of combination dies, the trimming and punching can be combined in one die set (Fig. 4-105).

The various die components called out are: (1) shank, (2) trimmer punch, (3) trimmer, (4) bottom plate, (5) forging locator, (6) positive knockout, (7) piercing punch, (8) stripper plate, (9) legs (2), (10) knockout arms (2), (11) brackets (2), (12) knockout guides (2), (13) top shoe.

Figure 4-105 shows the die partially closed with the forging in position and about to be trimmed. As the press ram descends further, the punch will remove the web from the inside of the forging and the slug will drop through. On the return stroke the stripper plate will strip the flash from the punch and the knockout on the locator will push the forging out of the trimmer. The stripper plate and positive knockout are actuated by the

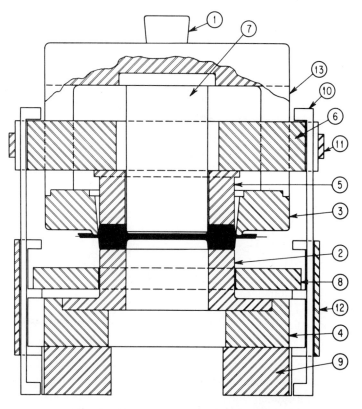

Fig. 4-105. Combination die, including trimming and punching tools.

knockout arms. This die principle can be utilized on many types of forgings but, again, the production must warrant the more complex tooling.

When small thin sections adjoin heavy sections there are cooling stresses set up which tend to distort the forging. There are also problems of distortion of light sections in normal handling operations. Consequently, a cold restrike die is required to remove the major deformations encountered after forging and subsequent handling.

Hot restriking dies are made from medium carbon alloy steels and

Chap. 4   Bending, Forming, and Drawing Dies   315

hardened to 36–40 R/C. They have half impressions sunk in the die faces about the same as the finish impression of the forging die but the impression is not flashed. A set of restrike dies for a connecting rod was shown earlier in Fig. 4-74. Cold restrike dies are made from a good grade of alloy tool-steel hardened to 50–52 R/C.

Forgings vary in size, shape, and operations. The connecting rod example required forging dies, trimming and punching dies, and restrike dies. Some connecting rods are furnished very close to size on the thickness so that a simple grinding operation is all that is required to finish the sides of the two bosses. Here a coining die is constructed such as the one shown in Fig. 4-76.

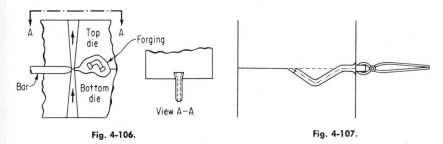

Fig. 4-106.   Fig. 4-107.

Fig. 4-106. After the forging is cut off, the bar end is then started through another forging cycle. This can continue until the bar is no longer at forging temperature.

Fig. 4-107. Incorporating a bender in a die.

Two or more small forgings may be made progressively from a long bar of steel so long as the forging temperature is adequate to keep the material plastic. In this case a cutoff is added to the die (Fig. 4-96A). The cutoff consists of two sharpened edges which pinch the steel in half as the dies come together. These cutting edges can be milled into one corner of the die or separately inserted into the side of the die. In the photograph the cutoff is milled into the die proper. Figure 4-106 illustrates an inserted type of cutoff.

The bending of stock was illustrated in Fig. 4-91 and a bending impression on the die face in Fig. 4-96A. To incorporate a bender in a die, it is necessary to leave a horn on one die which will pocket in the other die (Fig. 4-107). The horn is reduced in size and the pocket increased to allow for the stock which must fit between it.

The forging tool designer is faced with a new problem on practically every die design, and many of the problems cannot be foreseen until after the work is set up in the forging equipment and the dies tried out.

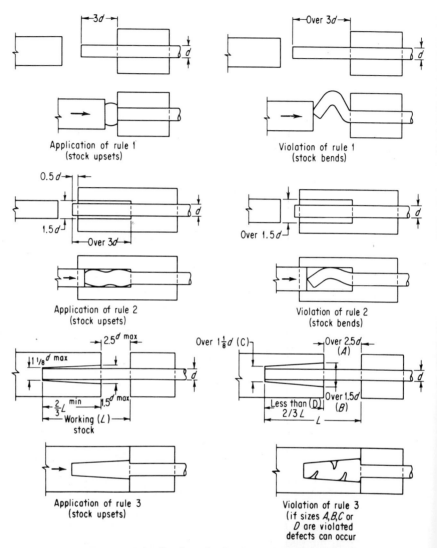

Fig. 4-108. Application of rules for gathering of stock.

UPSET OR FORGING MACHINE DIES

A wide variety of work that can be accomplished in no other way is made possible by the use of the upset principle. Forging machines can form forged head configurations on short or extremely long bars. The heads so formed will exhibit the same desirable qualities of grain flow that are accomplished by press or hammer forging. The heads vary in size and shape and

Chap. 4 Bending, Forming, and Drawing Dies   317

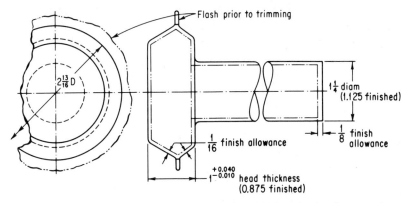

Fig. 4-109. Bevel gear forging designed for production by upsetting.

it is not always possible to form them in one blow. Occasionally the stock must be gathered in two or more blows or *passes* as they are generally called. The volume of a particular head may require the use of a great length of the original bar, and definite rules must be followed to eliminate the possibility of defects. Three basic rules which pertain to the gathering of stock are as follows:

1. The limit of length of unsupported stock that can be gathered or upset in one pass without injurious buckling is *not more than three times* the diameter.
2. Lengths of stock more than three times the diameter of the bar can be successfully upset in one blow by displacing the material in a die cavity *no greater than 1½ times the diameter* of the bar, provided the stock extending beyond the die face is no greater than ½ of the diameter.
3. An upset requiring more than three diameters of stock in length and extending up to 2½ bar diameters beyond the die face can be made if the material is confined by a conical recess in the punch which *does not exceed 1½ bar diameters* at the mouth and 1⅛ bar diameters at the bottom, provided the heading tool recess is not less than ⅔ the length of working stock or not less than the length of working stock minus 2½ times its diameter.

These rules cover the absolute limits, whereas in actual practice the dies are designed with broader limits. Figure 4-108 illustrates the application and violation of these rules.

Figure 4-109 illustrates a simple bevel gear forging designed for manufacture on an upsetting machine. Upset dies are subject to practically the same die considerations as hammer or press forgings, i.e., mismatch, die wear, die closure (thickness tolerance), scale pitting, etc. When computing the thickness tolerance of the upset forging, the weight is taken as only that portion of the forging which is actually formed by the dies. In this case it is the beveled head only and the $1\frac{1}{4}$-in. shank is excluded from the weight. If the volume of the head, in cubic inches, is calculated and multiplied by 0.284, the weight will be approximately 1.4 lb. This would require a thickness tolerance of $^{+0.040}_{-0.010}$, since 0.284 lb is the weight of *one cubic inch* of steel.

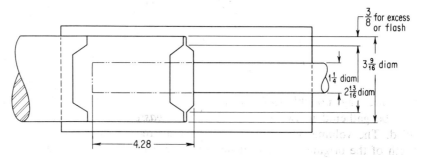

Fig. 4-110. Layout for one-blow forging of a bevel gear blank.

In designing the set of upset dies, the first step is to find the theoretical weight of the head and add the plus variations of the head thickness tolerance to this weight. This amounts to an additional 0.1 lb. Thus 1.4 lb plus 0.1 lb = 1.5 lb.

The next step is to determine the length of stock ($1\frac{1}{4}$ in.) which will be required to fill the head. Therefore the stock length is: 1.5 lb divided by 0.35 lb per in. = 4.28 in. (0.35 lb is the weight of $1\frac{1}{4}$-in. diam steel stock for a 1-in. length).

A layout of the completed forging (Fig. 4-110) is now made with a set of hypothetical dies which would be required to make the forging in *one* blow.

The die designer can now consider the three rules for upsetting. If none of the rules is violated, the die design can be completed, since the head of the forging *can* be made in one blow. Rechecking the rules:

$$\text{Rule 1.} \quad \frac{4.28}{1.25} = 3.42$$

This head exceeds the 3 diameters of rule 1. If this ratio had been between 2 and $2\frac{1}{2}$ the forging dies could be completely designed on the basis of one pass to forge. However, at times the 3.42 ratio will not result in a defective forging if rule 2 is not violated.

$$\text{Rule 2.} \quad \frac{3.562}{1.250} = 2.85$$

The 2.85 ratio exceeds the limit of rule 2 which requires that the $3\frac{9}{16}$ dimension of the dies could not exceed $1\frac{1}{2}$ in. $\times$ $1\frac{1}{4}$ in. or $1\frac{7}{8}$ in. If the cavity for the plunger were, say, $1\frac{5}{8}$ in. diam the die could be completely designed on the basis of one pass to forge and the only consideration would be to limit the stock extension beyond the die face by no more than $\frac{5}{8}$ in. or $\frac{1}{2}$ diam.

It is not necessary to proceed with the checking of rule 3 since rule 2 has already been violated.

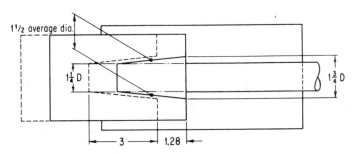

Fig. 4-111. Gathering steel by use of a conical punch.

A subsequent pass or passes must be added to the original one-pass die and the resulting die design checked against the three rules for upsetting. To best arrive at this determination, begin with rule 3 and consider a die design which will not violate this rule. The time-tested procedure called the *progressive taper method* is utilized. Gathering steel in a conical-type punch by this method can be seen in Fig. 4-111.

The volume of stock contained in the punch must equal the original weight of 1.5 lb. The punch is designed as shown with an average taper of $1\frac{1}{2}$ in. diam at the midpoint. Two conditions have now been satisfied; the upset does not violate any part of rule 3, and the weight of steel which the cone can contain is not less than the weight needed for the head.

The stock has now been increased in size so the die designer can return to the hypothetical one-pass die design and determine whether the conical pass just completed will meet all conditions. Again rechecking the rules:

Rule 1. $\dfrac{3.00 \text{ (overall length of cone)}}{1.50 \text{ (average bar diam)}} = 2$

If rule 1 is violated, additional cone passes should be designed until the stock is increased in size to the point where 3 diameters is not exceeded. Rules 2 and 3 are not applicable unless rule 1 has been violated. The dies and completed forgings from each pass are shown in Fig. 4-112.

A trimming pass has been added so as to finish the forging in the same machine without additional equipment or handling.

320   Bending, Forming, and Drawing Dies   Chap. 4

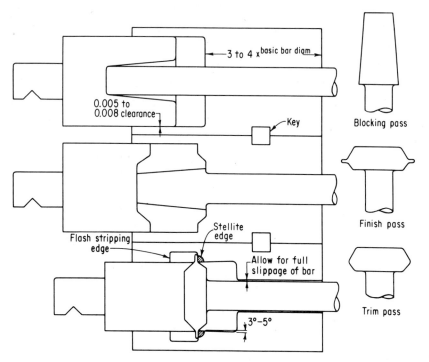

Fig. 4-112. Dies and completed forgings from each pass.

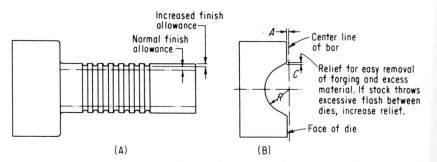

Fig. 4-113A. Finish allowance on machined part.
Fig. 4-113B. Design of grip section.

Clearance between punch and die should be as shown. This space will prevent seizing of the dies due to unequal expansion, scale formations, and the like. An important consideration in upset work is the grip which must be provided on the stock to withstand the force of the heading punches. The length of grip in the dies should be equal to a minimum of 3 bar diameters and, when heating the bar, this part of the stock should be kept as cold as possible. If the stock must be heated throughout, or the grip length is less than 3 diameters, corrugations can be bored in the dies to prevent slippage. The finish allowance on the machined part must be increased by the depth of the grooving if the die is corrugated (Fig. 4-113*A*).

Table 4-7 and Fig. 4-113*B* show the design which is necessary throughout the grip section when corrugations are not required.

TABLE 4-7. DIE DESIGN DIMENSIONS
(see Fig. 4-113B)

Stock size	R	A	Total grip, 2A	C
½	0.250	0.002	0.004	¹⁄₆₄
⅝	0.3125	0.0025	0.005	¹⁄₆₄
¾	0.375	0.003	0.006	¹⁄₆₄
⅞	0.437	0.0035	0.007	¹⁄₆₄
1	0.500	0.004	0.008	¹⁄₆₄
1¼	0.625	0.005	0.010	¹⁄₃₂
1½	0.750	0.006	0.012	¹⁄₃₂
1¾	0.875	0.0065	0.013	¹⁄₃₂
2	1.000	0.007	0.014	¹⁄₃₂
2½	1.250	0.0075	0.015	³⁄₆₄
3	1.500	0.008	0.016	³⁄₆₄

The detailed analysis of the bevel gear forging shows the method used in designing a typical set of upset dies. By not exceeding any of the three rules, interesting variations of the upset principle can be made on a variety of other forgings which lend themselves to this method of manufacture. Two variations are in Fig. 4-114: the sliding die upset and the bend and upset.

The die steels necessary for upset work vary with the severity of service and whether low or high production limits are required. Heavy dies with little change in section can be made from 0.50 per cent carbon, 2.00 per cent tungsten oil-hardening steels. If the die has light sections, warpage will become a problem during quenching and an air-hardening 5.00 per cent chromium hot work die steel should be used. Long production runs warrant the use of the 5.00 per cent chromium steel in almost all cases. This steel exhibits high resistance to thermal fatigue cracking and can be heat treated to a high strength and toughness. High thermal fatigue resistance constitutes a major requirement in the selection of upset die steels since in service they must be cooled constantly with water or water-oil mixtures.

If extreme wear on long runs is present, hardened inserts can be incorporated into the die design to reduce die replacement costs. All dies, punches, and inserts should be hardened within a 2-point range on the Rockwell C scale. The following ranges for each type of tool can be used as a guide in determining the heat treatment. However, dependability in service is still the criterion, and changes in specifications may be necessitated after the tool is used.

Material	Part	Hardness R/C
0.50C, 2.00W or 5.00CR	Gripper dies	45–50
0.50C, 2.00W or 5.00CR	Punches	42–48
or 5.00CR	Inserts	45–50

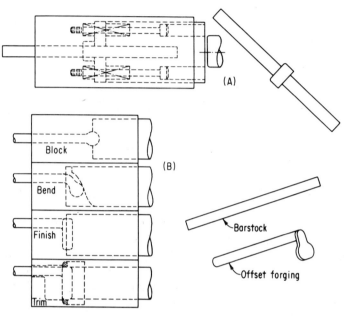

Fig. 4-114. Two types of upset: (A) sliding-die; (B) bend-and-upset.

PROBLEMS

Bending Dies

1. What is the basic principle involved in a bending die?
2. What causes spring-back?
3. Calculate bending pressures for the following bends (assume S is 30 tons):
    (a) 90-deg bend in $5/16$ metal, $1/2$ IR, 12 in. long.
    (b) 60-deg bend in $1/4$ metal, $3/4$ IR, 16 in. long.
    (c) 30-deg bend in $3/16$ metal, $3/16$ IR, 9 in. long.
4. When are edge-bending dies used? Why?
5. Calculate various blank lengths.
6. Design a simple V-bending die.
7. Design an edge-bending die.

Chap. 4 Bending, Forming, and Drawing Dies   323

Forming Dies

1. How does forming differ from bending?
2. Name five types of forming dies.
3. How does a solid form die differ from a pad-type die?
4. What are the two types of power sources for pressure pads?
5. What happens to metal during an embossing operation?
6. Name three types of embosses formed by an embossing die.
7. What is the smallest permissible radius that can be curled?
8. How are curls strengthened?
9. What is the bulging medium in a bulging die?
10. What is the difference in metal behavior in coining and swaging dies?
11. What is the range of swaging pressures?
12. What is the critical height of a hole flange?

Draw Dies

1. Will a smaller draw radius allow *less* or *more* material to flow into the draw form?
2. What is the primary use of an air cushion on a die in a drawing operation?
3. What is the primary function of a knockout die?
4. What is the minimum corner radius of a drawn cup?
5. State the formula for determining blank diameters for drawing operations.
6. Why must a pressure pad or blankholder always be used when drawing thin stock?
7. What determines the shut height of a die?
8. Why must both punch and draw dies be vented?
9. Determine the drawing pressure given a $\frac{1}{16}$-in. thick shell of 3-in. diameter of steel having a yield strength of 30,000 psi.
10. What is the range of blankholding pressure compared to drawing pressure?

Progressive Dies

1. What is a progressive die?
2. What operations can be performed in a progressive die?
3. When should a progressive die be used?
4. List the items that should be considered when designing a progressive die.
5. Why is piloting required?
6. Where are pilots located in the strip?
7. What is segmental-type die block construction? Give an example.
8. Sketch four types of shanks used on die sets.

Extruding Dies

9. Sketch the various pin and bushings available.
10. What is the feed level of the die?

Extruding Dies

1. Discuss the basic principles of impact extrusion.
2. What affects the pressures required to extrude common metals?
3. Discuss why lubricant plays an important role in successful impact extrusion.
4. Discuss the three basic methods of impact extrusion.
5. Discuss why coining is desirable at times before cold extrusion.
6. Why is tool design considered the most important factor of successful impact extrusion?
7. Discuss the important factors of punch design.
8. Discuss common characteristics of punch and pressure anvil.
9. What is the function of punch guides?
10. Discuss design considerations of diepots.

Forging Dies

1. What are the three classes of forging equipment (machines)?
2. What determines whether hot or cold forging is to be done?
3. Why are forgings inherently strong and tough?
4. What is forging draft?
5. Why are fillets and rounded corners specified for forged parts?
6. What kind of steel should be used for forging dies?
7. What is the limit of length of unsupported stock for an upset forging?
8. What is the size limit of a die cavity?
9. What is the first step in designing a set of upset dies?
10. What is the second step in designing a set of upset dies?

REFERENCES

1. Verson, J., "Tooling for Cold Extrusion," *Am. Machinist,* Oct. 7, 1957.

2. Quadt, R., "When to Use Cold Extrusion of Aluminum," *Design Engineering,* Nov., 1956.

3. "Computations for Metal Working in Presses," E. W. Bliss Co., Canton, Ohio.

5

DESIGN OF TOOLS

FOR INSPECTION AND GAGING

A visual representation of a workpiece may be used as a guide for its manufacture only to the extent that it completely defines and limits the size, shape, and composition of the workpiece by dimensions and specifications. Tolerance must be applied directly or indirectly to every dimension or specification.

Tolerance represents the total amount by which a given dimension may vary from a specified nominal size. In providing tolerances on drawings or blueprints, the designer recognizes that in the absolute sense no two workpieces, distances, or quantities can be exactly alike. The designer specifies an ideal unattainable condition (nominal dimension), and then states what degree of error can be tolerated.

Every dimension of every workpiece must in some way be specified as being between two limits. No proper dimension can ever be given as a single fixed value, because a single value is unattainable. The designer guards against this pitfall in several ways. Critical dimensions specified on a drawing or blueprint have the tolerance given as part of the dimension (e.g., 0.125 ± 0.002 in.). Less critical dimensions are provided tolerance by a general note in the title block (e.g., unless otherwise specified, all dimensions are ± 0.010). Tolerance may also be specified indirectly in the bill of material. If the designer calls for the use of purchased material or parts and does not further specify tolerance, it must be assumed that the vendor's manufacturing tolerances are acceptable.

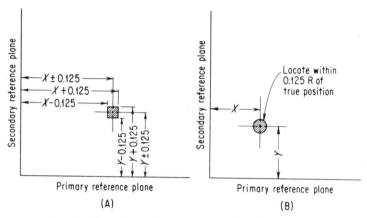

Fig. 5-1. Coordinate dimensioning and positional tolerance: (A) the element so dimensioned may be located anywhere within the shaded rectangle; (B) the element so dimensioned may be located anywhere within the shaded circle.

In establishing tolerances, the designer may consider a number of factors including function, appearance, and cost. The workpiece will have an intended function with which the specified tolerance must be compatible. Appearance is sometimes a factor in that a workpiece dimension having no functional significance may have aesthetic requirements. If for no other reason than material cost, the size of a workpiece must be limited.

Cost is perhaps the governing factor in the specification of tolerances and as such often supersedes function. As tolerances become smaller, the cost of achieving them increases rapidly. The designer often must weigh the functional necessity of a close tolerance against its cost, and must also consider the capability of the machines that will produce the workpiece. If the function of the workpiece requires extremely close tolerances to assure proper mating allowances, it may be economically advantageous to combine easily held tolerances with selective assembly. It may be possible in this way to obtain the necessary mating allowances for function without incurring the cost of close tolerances.

In a conventional blueprint or drawing the workpiece is shown in a fixed relationship to mutually perpendicular reference planes. The location of each element of the workpiece is defined by stating its distance (dimension) from each reference plane. Each dimension has a tolerance. The reference planes in a single view may be compared with the X and Y axes of a Cartesian coordinate system. The location of the element is defined as being anywhere within a rectangular area formed by the minimum and maximum values of the X and Y dimensions as established by the applied tolerances. Engineering intent can often be expressed more precisely by stating the nominal location of the element and by stating how far from the true position the element may vary. In this case a positional tolerance applied to the nominal location takes the place of the X and Y tolerances. Figure 5-1 illustrates coordinate dimensioning and positional tolerance.

WORKPIECE QUALITY CRITERIA

Surface Roughness. The surface of any solid object is composed of a pattern of irregularities. Just as, it is impossible for any two parts to be exactly alike, it is impossible to have every element of a surface coincide with the theoretically accurate design (nominal surface) usually shown and dimensioned on a drawing or blueprint. A surface is judged acceptable or unacceptable not because of the presence of irregularities, but because of their magnitude. Apparent absence of irregularity proves only that its magnitude is beyond the range of detection of the inspection device employed.

Surface roughness is judged relative to the tolerances required for the finished workpiece. A rough flame-cut or as-cast surface, with irregularities of ± 0.062 in., might be acceptable in rough construction with assembly tolerances of ± 0.125 in. The same workpiece with its surface irregularities reduced to ± 0.032 in. by a machining operation might be acceptable where assembly tolerances are ± 0.100 in. After several machining and grinding operations, with the irregularities reduced in magnitude to approximately ± 0.0003 in., the workpiece might be acceptable where assembly tolerances are ± 0.001 in.

Flatness. Flatness, when specified, is the requirement that every element of a surface be within a stated distance from a designated nominal surface plane. Figure 5-2 illustrates a common method of specifying flatness. With a liberal tolerance, flatness may be checked by applying a straightedge to the surface and by probing with feeler gages to measure the deviation. Closer tolerances may require inspection with an indicator-type gage. Extremely close tolerances might require a gaging method that incorporates amplification.

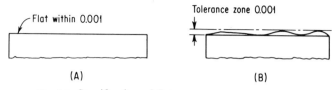

Fig. 5-2. Specification of flatness.

Straightness and *alignment* are often used interchangeably to mean in-line. Tools similar to those used for checking flatness are used to verify or test for straightness. The simplest testing device is the straightedge, a rigid bar or beam manufactured to close tolerance. The testing technique is to shut out light. Other inspection devices are: a taut wire, dial indicators, optical tools, and other devices employing amplification. Figure 5-3 illustrates use of a telescope and target (autocollimator) to check the straightness of a machine bed segment.

Parallelism. Two plane surfaces that have been individually generated in accordance with a flatness specifi-

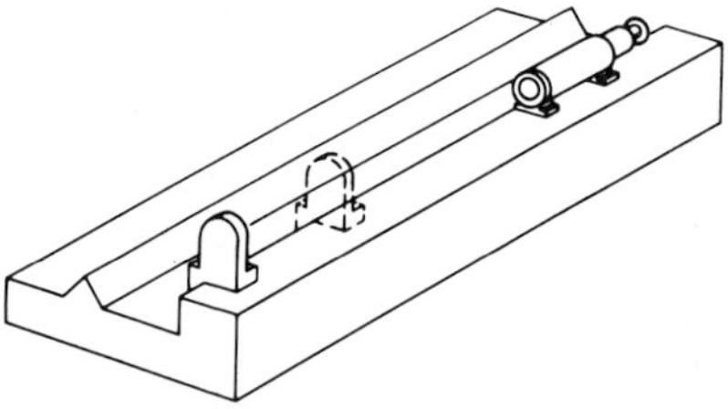

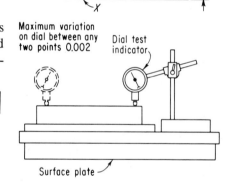

Fig. 5-3. Optical inspection of alignment.

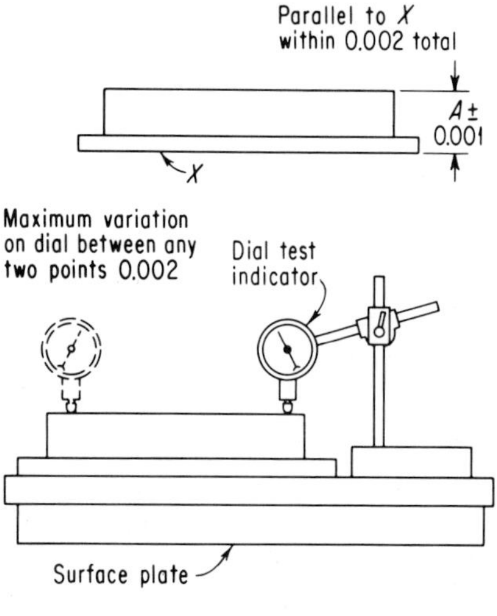

Fig. 5-4. Specification and inspection of parallelism.

cation may be subject to an additional specification, e.g., that they be parallel. Figure 5-4 illustrates a method of specifying parallelism together with one method of inspection. The inspection consists of measuring the variation of the distance between the surface plate and the indicator probe. The indicator is zeroed with the probe in contact with the workpiece. As the workpiece is moved about, variation is shown by the indicator. The distance can also be measured (qualified) by presetting the indicator to an established standard.

Distance. A drawing or blueprint must specify the relative location of every element of the workpiece. This is usually done by stating the linear distances between the elements or by dimensioning the significant points with respect to a common point or line (coordinate dimensioning). Production of a workpiece involves generation or arrangement of the elements in accordance with the blueprint. Inspection involves affirmation that the elements are located as specified.

Distance is measurable only by comparison with a known standard. Accuracy of measurement is limited by the accuracy of the standard employed. In measuring distances to establish geometric relationships, the

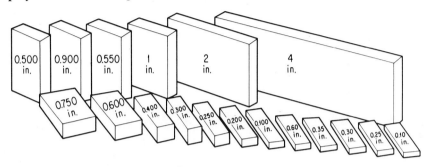

Fig. 5-5. Gage blocks.

measuring or comparing devices are called *gages* and the accepted industrial standards are the linear inch, and the meter. One method by which the standard inch is given physical form and made usable either as a direct unit of measurement or as a calibration standard is the use of *gage blocks* (Fig. 5-5). These precisely made rectangular pieces of steel, bearing a figure designating their dimensions, constitute the only practical means whereby the standard inch may be subdivided and multiplied and used in the shop as a tool.

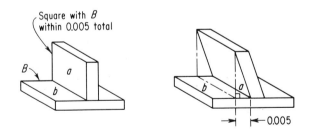

Fig. 5-6. Specification of squareness and perpendicularity.

Squareness and Perpendicularity. Geometrically, lines or planes can be absolutely perpendicular or normal to each other. In actuality, however, the perfect relationship is as unattainable as a perfect surface or exactly identical parts. Because perfection is impossible, the drawing or blueprint must specify what degree of tolerance will be acceptable. Figure 5-6 illustrates a method of prescribing tolerance as applied to squareness. Often a drawing will have a general tolerance statement as part of the title block, in which case the designer need not specify individual tolerances for every geometric relationship.

Angularity. If the geometric relationship of the lines or planes used in visual representation of a workpiece is other than one of squareness or perpendicularity, the angle between the elements must be specified. It is again impossible to obtain the exact angle required, therefore the acceptable degree of imperfection (tolerance) must also be specified. Figure 5-7 illustrates a method of specifying tolerance for an angular relationship.

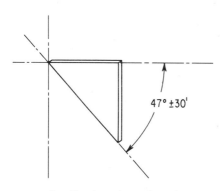

Fig. 5-7. Specification of angular tolerance.

Figure 5-8 shows a standard dial indicator application on a special gage to check angularity. The face that is finished is dimensioned by angle X, the tolerance of which is generally given in degrees, minutes, or seconds. This tolerance in arc is converted into length of chord in inches and is the permissible variation or tolerance that may be shown on the indicator. The gage consists of a soft steel base plate, two hardened, ground, and lapped tool-steel locator pins, and a standard dial indicator. It is also

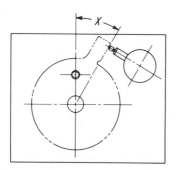

Fig. 5-8. Inspection of angularity.

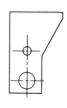

Special set master

necessary to design a special master set block to be used with the gage for setting the indicator at zero. The master is made of tool-steel, hardened, ground, and lapped.

Concentricity. A workpiece may have several surfaces of revolution. The surfaces may be generated by (1) establishing an axis through the workpiece; (2) revolving the workpiece about its axis, and (3) applying

Chap. 5 Design of Tools for Inspection and Gaging   331

a cutting tool to the workpiece at a fixed distance (radius) from the axis. Two or more surfaces so generated are concentric if they have a common axis. Representative examples include a hole with a counterbore, or a stepped drive shaft. Perfect concentricity being unattainable, the designer must specify what degree of imperfection is acceptable.

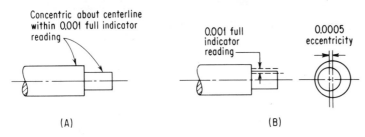

Fig. 5-9. Specification of concentricity tolerance.

Figure 5-9*A* illustrates a common method of specifying concentricity tolerance. The initials FIR mean full indicator reading. This is sometimes denoted as TIR or total indicator reading. Figure 5-9*B* illustrates the permissible variation allowed from the common axis as specified in view *A*. In most cases when a concentricity tolerance is specified it is relatively small and should be gaged with an indicator. In the shaft illustrated it can most accurately and easily be checked by rotating the part on centers and taking a direct full indicator reading. These centers are normally standard equipment in most plants and it is unnecessary to design a special gage. If the part does not have centers, one diameter is nested in a V block and the direct indicator reading is taken on the other diameter.

The part shown in Fig. 5-10 illustrates a concentricity condition that might require a special gaging fixture. All three radii are to be concentric with the center hole. The gage to check this concentricity requirement consists of a mild steel base plate, a ground and lapped center locating pin made from hardened tool-steel to a tight fit, and three standard dial indicators. The specifications for each

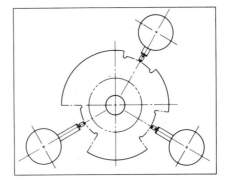

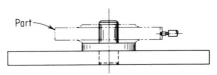

Fig. 5-10. Gaging concentricity with a special fixture.

particular dial indicator used should be consulted to determine the manufacturer's recommended method of mounting.

BASIC PRINCIPLES OF GAGING

Gage Tolerances. Since it is not possible to produce multiple parts with exactly the same dimensions, working tolerances are necessary. For the same reason, gage tolerances are necessary. Gage tolerance is generally determined from the amount of workpiece tolerance. The 10 per cent rule is in general use for determining the amount of gage tolerance for fixed limit-type working gages. When no gage tolerance is specified the gagemaker will use 10 per cent of the working tolerance as the gage tolerance for a working gage. Working gages are those used by production personnel during the manufacturing process. The amount of tolerance on inspection gages, those used by the inspection department, is generally 5 per cent of the work tolerance. Tolerance on master gages, those used for checking the accuracy of other gages, is generally 10 per cent of the gage tolerance. Where tolerances are liberal, gages used by the inspection department do not differ from the working gages.

Four classes of gagemakers' tolerances have been established by the American Gage Design Committee and are in general use. These four classes establish maximum variations for any desired gage size. The degree of accuracy required determines the class of gage to be used. Table 5-1 shows these four classes of gagemakers' tolerances.

Class XX gages are precision lapped to the very closest tolerances practicable. They are used primarily as master gages and for final close tolerance inspection.

Class X gages are precision lapped to close tolerances. They are used for some types of master gage work, and as close tolerance inspection and working gages.

Class Y gages are precision lapped to slightly larger tolerances than Class X gages. They are used as inspection and working gages.

Class Z gages are precision lapped. They are used as working gages where part tolerances are liberal and the number of pieces to be gaged is small.

Proceeding from Class XX to Class Z, tolerances become progressively greater and the gages are used for inspecting parts having progressively larger work tolerances.

To illustrate the use of the 10 per cent rule in conjunction with Table 5-1, assume a gagemaker is to choose the correct tolerance class for a working plug gage that is to be used on a 1-in. diam hole having a working tolerance of 0.0012 in. One-tenth of the work tolerance would indicate a gage tolerance of 0.00012 in. or, as noted in the table, a Class Z gage.

If the work tolerance were only 0.0006 in. on the 1-in. hole, then a Class X gage would be indicated, with a tolerance of 0.00006 in. If the work tolerance, however, were 0.015 in., then the gage tolerance indicated by the 10 per cent rule would be 0.0015 in. As this exceeds the maximum tolerance class, a Class Z gage would be indicated and the gage tolerance would be 0.00012 in.

The smaller a gage tolerance must be held, the more expensive the gage becomes. Just as the production cost rises sharply as the working tolerances are reduced, the cost to buy or to manufacture a gage is much higher if close tolerances are specified. Gage tolerances should be realistically derived from the work to be gaged.

TABLE 5-1. STANDARD GAGEMAKERS' TOLERANCES
(all dimensions are in inches)

Nominal Size		Gagemakers' Tolerance Classes			
Above	To and including	XX	X	Y	Z
0.029	0.825	0.00002	0.00004	0.00007	0.00010
0.825	1.510	0.00003	0.00006	0.00009	0.00012
1.510	2.510	0.00004	0.00008	0.00012	0.00016
2.510	4.510	0.00005	0.00010	0.00015	0.00020
4.510	6.510	0.000065	0.00013	0.00019	0.00025
6.510	9.010	0.00008	0.00016	0.00024	0.00032
9.010	12.010	0.00010	0.00020	0.00030	0.00040

Allocation of Gage Tolerances. After determining the tolerance for a specific gage, the direction, plus or minus, of that allowance must be decided. Two basic systems, and many modifications of them, are used in making this decision.

The Bilateral System. In the bilateral system the *go* and *no-go* gage tolerance zones are bisected by the high and low limits of the workpiece tolerance zone. This is illustrated by Fig. 5-11*A*, which shows the black rectangles representing the gage tolerance zones are half plus and half minus in relation to the high or low limit of the work tolerance zone.

Referring to Table 5-1 let us assume that the diameter of the hole to be gaged is 1.250 ± 0.0006 in. The total work tolerance in this case is 0.0012 in. since the hole size may vary from 1.2506 to 1.2494 in. Using 10 per cent of the total work tolerance as our gage tolerance, our gage tolerance is then 0.00012 in. From Table 5-1 we see that for this diameter this is a Class Z gage tolerance. The dimensions on the *go* plug gage for this example would be 1.2494 ± 0.00006 in., and the diameter of the *no-go* gage would be 1.2506 ± 0.00006 in.

One disadvantage in using this system is that parts that are not within the working limits can pass inspection. Using the above example, if the hole to be gaged is reamed to the low limit (1.2494 in.), and if the *go* plug gage is to the low limit (1.2493 in.), then the *go* gage will enter the hole and the part will pass inspection, even though the diameter of the hole is outside the working tolerance zone. A part passed under these conditions would be very close to the working limit, and the tolerance on the mating part should not be such as to prevent assembly. Plug gages using the bilateral system could likewise pass parts in which the holes were too large. A common misconception is that gages accept good parts and reject bad. With the bilateral system, however, parts can be rejected as being outside the working limits when they are not.

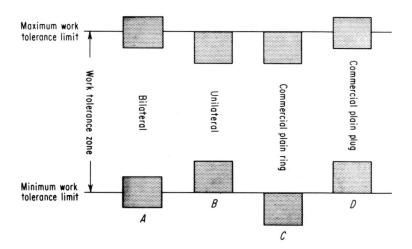

Fig. 5-11. Different systems of gage tolerance allocation.

The Unilateral System. In the unilateral system (Fig. 5-11B), the work tolerance zone entirely encompasses the gage tolerance zone. This makes the work tolerance smaller by the sum of the gage tolerance, but guarantees that every part passed by such a gage, regardless of the amount of gage size variation, will be within the work tolerance zone.

If the diameter of the hole is 1.250 ± 0.0006 in., again using 10 per cent of the working tolerance as the gage tolerance, the *go* gage dimension would be 1.2494 $^{+0.00012}_{-0.00}$ in., and the tolerance on the *no-go* gage would be 1.2506 $^{+0.00}_{-0.00012}$ in.

This system of allocation of gage tolerance, like the bilateral system, may reject parts as being outside the working limits when they are not, but all parts passed using the unilateral system will be within the working limits.

The unilateral system has found wider use in industry than the bilateral system for plain plug and ring gages.

One partial solution to the problem of gages rejecting parts that are within working limits is to use working gages with the largest unilateral gage tolerance practical, and inspection gages with the smallest unilateral gage tolerance practical. Thus no piece can pass inspection that is outside tolerance, and the possibility of the inspection gage turning down acceptable work is minimized because of its small tolerance.

Figure 5-11 shows various methods of allocating gage tolerances starting with the unmodified bilateral (A) and unilateral (B) systems. The next diagram (C) shows one commercial practice of allocating the plain ring gage tolerances negatively with reference to both the maximum and minimum limits of the workpiece tolerance. In another practice the *no-go* gage tolerance is bisected by the maximum limit of the workpiece tolerance, and the *go* gage tolerance is held within the minimum limit of the workpiece tolerance (D).

The final results obtained with this variety of allocation systems will differ greatly. The choice of the system to be used, modified or unmodified, must be determined by the product and the facilities for producing it. The objectives in choosing an allowance system should be the economical production of as near 100 per cent usable parts as possible, and the acceptance of the good pieces and rejection of the bad.

Gage Wear Allowance. Perfectly dimensioned gages cannot be made. If one did exist, it no longer would be perfect after just one checking operation. Although the amount of gage wear during just one operation is difficult to ascertain, it is easy to measure the cumulative wear of several checking operations. A gage can wear beyond usefulness unless some allowance for wear is built into the gage at the start.

The wear allowance is a dimensional increment added to the nominal diameter of a plug gage and subtracted from that of a ring gage. It is used up during the gage life by wearing away of the gage metal. Wear allowance is applied to the nominal gage diameter before gage tolerance is applied.

The amount of wear allowance is not necessarily decided in relation to the amount of work tolerance, although a small work tolerance will tend to restrict wear allowances. The material of which the gage and work are made, the quantity of the work, and the type of gaging operation to be performed must be considered in specifying a wear allowance. It is important to establish a specific value as the wear allowance. When the gage has worn the established amount, it should be removed from service without question. This averts any controversy as to whether a gage is still accurate or not.

One method, which uses a percentage of the working tolerance as the wear allowance, can be explained with the following example.

For a 1.500 ± 0.0006 in. diam hole the working tolerance is 0.0012 in. The basic dimension of the *go* plug gage would be 1.4994 in. Using 5 per cent of the working tolerance as the wear allowance, and adding this to the basic dimension, the basic dimension would then be 1.49946 in. Using the unilateral system of gage tolerance allocation the *go* plug gage would be dimensioned 1.49946 $^{+0.00012}_{-0.00}$ in. The gage tolerance would be 10 per cent of the working tolerance. Figure 5-12 shows the wear allowances and gage tolerances used by the Ordnance Department for inspection gages.

Some manufacturing companies do not build a wear allowance into their gages but set up a standard to determine when a gage has worn beyond its usefulness. The gage is allowed to wear a certain percentage above or below its basic size before being retired from service. Gages should be inspected periodically for wear. No set policy for wear allowance or gage inspection is practical for all industries. In operations where an extremely high degree of accuracy must be maintained, the amount of allowable wear is smaller and inspection must be more frequent than in operations where tolerances are more liberal.

A gage maker normally makes the gage to provide maximum wear even if no wear allowance is designed into the gage. That is, he makes the *go* end of a plug gage to its high limit, and he makes the *go* component of a ring gage to its low limit.

The *no-go* gage will slip in or over very few pieces and so wear very little, and as any wear on a *no-go* gage puts that gage farther within the product limits, no wear allowance is applied. When the *no-go* gage begins to reject work actually well within acceptable limits it must be retired.

Gage Materials. For medium production runs, hardened alloy steel is used for wear surfaces of gages. For higher-volume production runs, gage wear surfaces are often chromium plated. When a high degree of accuracy must be maintained, the production run is long, and wear is excessive, tungsten carbide contacts are often used on gages. Worn gaging surfaces can be chrome plated and reground and restored to service, provided that the wear does not exceed 0.0003 in. below the low limit. Gages can be reworked after excessive wear for use on other jobs.

Gaging Policy. Gaging policy is the standardization of the methods for determining gage tolerances and their allocation, and of fixing wear allowance. It is a guide to determining when gages are required, when and how they are to be inspected, and what types of gages should be used.

There is no one policy in universal use. Users of gages should have their own policy, the one best suited to the nature of their work. A gage policy is helpful in eliminating controversy over the method of gaging, gage tolerances and allocation, and gage wear.

The policy should establish a general rule for determining the amount of

Chap. 5 Design of Tools for Inspection and Gaging

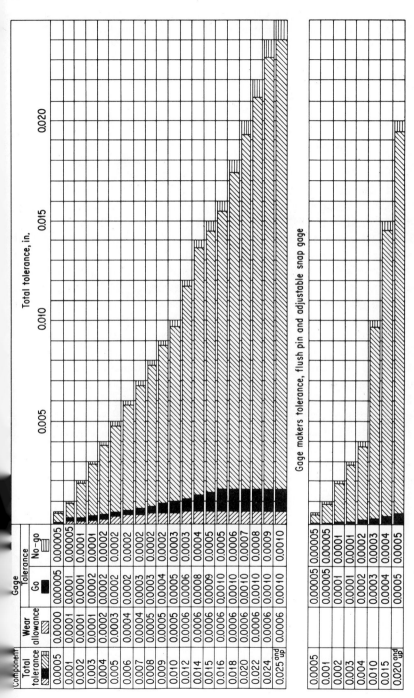

Fig. 5-12. Wear allowances and tolerances used as standard practice by plants in U.S. Ordnance production.

gage tolerance. The 10 per cent rule, used in conjunction with the four standard classes of gage tolerances, is widely used in industry.

The magnitude and application of wear allowance is also part of the gage policy. How often gages are inspected will depend upon how often they are used, what material they are used to check, and how close the tolerances are. Where a high degree of accuracy must be maintained, the gage policy will necessarily be more stringent.

GAGE TYPES AND APPLICATIONS

Measurement is the act of comparing quantity or distance with a known standard. Dimensions are verified by end measurement, e.g., the end point of a specified distance is measured for its coincidence with the end point of the standard; the last drop of a specified quantity of liquid brings the fluid level up to a mark on the side of the vessel. Virtually all dimensional measurement is end measurement. End measurement is predicated on coincidence of origin, that is, both the standard and the distance or quantity being measured must start at a common point.

Every structure must in its creation or measurement be organized with reference to three mutually perpendicular planes. Every dimension has as its ultimate origin one of the three planes. Given a horizontal reference plane, two mutually perpendicular vertical reference planes can be constructed, and any structure can be defined by dimensions originating at the reference planes.

Surface Plate. The surface plate is the physical representation of the primary horizontal reference plane. As such it is the primary tool of layout and inspection. Surface plates are usually made of cast iron or granite, with one surface finished to a high degree of flatness. When accuracy of 0.00001 in. or finer is required, toolmakers' flats or optical flats are used instead. A number of accessories such as squares, straightedges, parallels, sine bars, vernier height gages, and dial indicators are used in conjunction with the surface plate.

The workpiece being inspected is carefully placed on the surface plate with its primary reference plane coincident if possible with the plate surface. If the workpiece has elements extending below its established reference plane, it may be necessary to mount the workpiece on adjustable supports (jacks) or parallels to achieve parallelism between the two reference planes. Vertical reference planes are erected on the surface plate as required.

To avoid cumulative error, all dimensions specified on a drawing or blueprint should originate from a reference point or plane rather than from the end point of another dimension. If, however, a workpiece feature is dimensioned from a point other than a reference plane, its location should be verified in relation to that point rather than the reference plane.

Chap. 5 Design of Tools for Inspection and Gaging

Templates. A template is a physical representation of a specified profile, or may be a guide to the location of workpiece features with reference to a single plane. A straightedge may be used as a template to check flatness. To control or gage special shapes or contours in manufacturing, special templates are used for visual comparison to insure uniformity of individual parts. These templates are made from thin, easily machinable materials, some of which may be later hardened if production requirements demand extended usage. Figure 5-13 shows an application of the contour template to inspect a turned surface. Templates of this type are also used widely in the sheet metal industry and where production is limited. Templates are satisfactory where the part tolerance will permit this type of visual inspection.

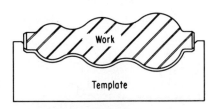

Fig. 5-13. Gaging the profile of a workpiece with a template.

To inspect or gage radii or fillets visually, a standard commercial type of template is obtainable from various gage manufacturers. These templates or gages are used by inspectors, layoutmen, toolmakers, diemakers, machinists, and patternmakers. The five different basic uses are shown in Fig. 5-14. These gages are usually made to nominal sizes in increments of $\frac{1}{64}$ in. Special gages of this type might readily be designed and made for specific jobs.

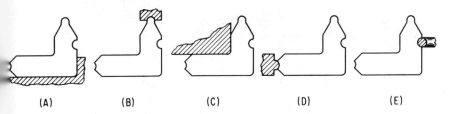

Fig. 5-14. Commercial radius gage and applications: (A) inspection of an inside radius tangent to two perpendicular planes; (B) inspection of a groove; (C) inspection of an outside radius tangent to two perpendicular planes; (D) inspection of a ridge segment; (E) inspection of roundness and diameter of a shaft.

A *screw pitch gage* is used to determine the pitch or number of threads per inch. This is also accomplished by visual comparison with a template or gage (Fig. 5-15). These gages are obtainable for most of the commonly used thread forms and sizes. A screw pitch gage is seldom designed for special use since it does not check thread size and will not give an adequate check on thread form for precision parts.

Plug Gages. A plug gage is a fixed gage usually consisting of two members. One member is called the *go* end and the other the *no-go* or *not-go* end.

The actual design of most plug gages is standardized, being covered by American Gage Design (A.G.D.) standards. However, there are many cases where a special plug gage must be designed.

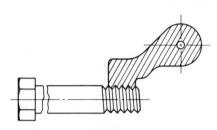

Fig. 5-15. Screw pitch gage and method of application.

A plug gage usually consists of two parts: the gaging member and a handle with the size, *go* or *no-go*, and the gagemaker's tolerance marked on it. There are generally three types of A.G.D. standard plug gages available. First in order of usage is the *single-end plug gage.* This type of gage consists of two separate gage members, a *go* and a *no-go,* each having its own handle (Fig. 5-16A). The second type is called the *double-end plug gage.* This type consists of two gage members, a *go* and a *no-go,* mounted on a single handle, one gage member on each end (Fig. 5-16B). The third type, called the *progressive gage,* consists of a single gage member mounted on a single handle (Fig. 5-16C). The front two-thirds of the gage member is ground to the *go* size and the remaining portion is ground to the *no-go* size. The *go* and *no-go* sizes are therefore incorporated in the same gage member.

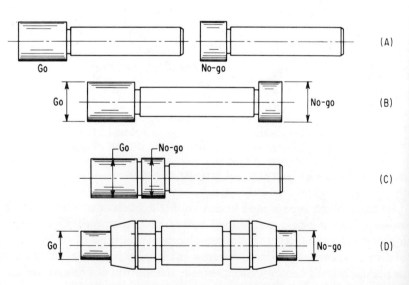

Fig. 5-16. Standard plug gages used to inspect the diameter of holes.

Standard A.G.D. plug gages generally have three methods of mounting the gage members on the handle. The smaller sizes are usually called *wire-type* gages. The gage member is simply a straight blank or nib with no shoulder, taper, or threads. The gage member on this type is held in the handle by a setscrew or with a collet-type chuck built into the handle (Fig. 5-16D). The advantage of this type of mounting is that the gage members are reversible when one end becomes worn, thus increasing the life of the gage.

Another type of mounting is called the *taperlock* design. In this type the gage member is manufactured with a taper ground on one end. This taper fits into a tapered hole in the handle much the same as a taper-shank drill fits into a drill-press spindle. This type is not reversible.

The third type is called the *trilock* design. This type is usually for larger-size gages. The gage member has a hole drilled through the center and is counterbored on both ends to receive a standard socket head screw. Three slots are milled radially in each end of the gage member. The gage is held to the handle by means of a socket head screw with the three slots engaging three lugs provided on the end of the handle. This type of gage is also reversible.

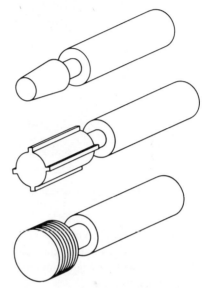

Fig. 5-17. Special plug gages used to inspect the profile or taper of holes.

It is sometimes necessary to design special plug gages. There are many types, depending on the individual job requirement. A square, hexagonal, or octagonal hole requires a special plug gage. A hole with serrations or involute gear teeth also requires a special plug gage. Figure 5-17 shows several special plug gages.

Ring Gages. Ring gages are usually used in pairs, consisting of a *go* member and a *no-go* member (Fig. 5-18). They are fixed-type gages and their design is covered by American Gage Design standards.

In sizes up to 1.510 in. the design is a plain ring, knurled on the O.D., and lapped in the I.D. to a close tolerance. The *no-go* member is identified by a groove around the O.D. of the gage.

In sizes above 1.510 in., the gage has a flange to reduce weight and increase rigidity. The *no-go* in this type is also identified by a groove around the O.D. of the gage.

Special ring gages are required at times. An example would be the inspection of a shaft having a keyway. A key-shaped segment added to a ring gage would allow simultaneous qualification of shaft size and key-slot width and depth.

Snap Gages. A snap gage is a fixed gage arranged with inside measuring surfaces for calipering diameters, lengths, thicknesses, or widths. Snap gages can be divided into three types as shown in Fig. 5-19.

A *plain adjustable snap gage* is a complete external-caliper gage employed for size control of plain external dimensions. It has an open frame, in both jaws of which gaging members are provided. One or more pairs of the gaging members can be set and locked to any predetermined size within the range of adjustment.

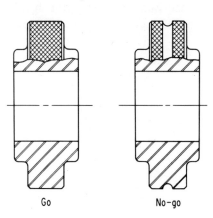

Fig. 5-18. Ring gage set used to inspect the diameter of shafts.

A *plain solid snap gage* is a complete external-caliper gage employed for size control of plain external dimensions. It has an open frame and jaws, the latter carrying gaging members in the form of fixed, parallel, nonadjustable anvils.

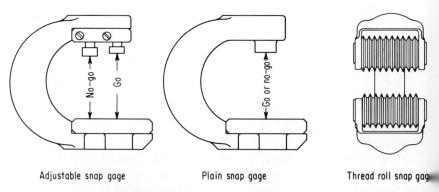

Fig. 5-19. Snap gages.

A *thread roll snap gage* is a complete external-caliper gage employed for size control of thread pitch diameter, lead of a thread, and thread form. It has an open frame and jaws in which gaging members are provided. One or more pairs of the gaging members can be set and locked to a predeter-

Chap. 5 Design of Tools for Inspection and Gaging   343

mined size within range of the thread to be checked. Gaging members vary with the different diameters and pitch of thread.

Many companies have master drawings of snap gages. The designer need only fill in the dimension required to gage a part. These master drawings are then sent to gage manufacturers for filling the order. There are special cases where a standard snap gage cannot be used, for example, checking the outside diameter of a groove. Figure 5-20 shows a grooved

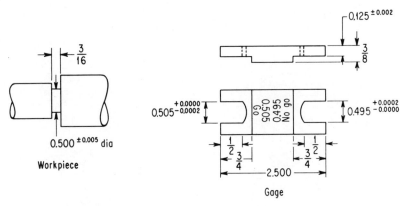

Fig. 5-20. Special snap gages.

shaft. Width of the groove and shaft diameter in the groove must be qualified. Also shown is a special double-end snap gage designed for the required inspection.

A snap gage may at times be superior to a ring gage as an inspection tool. Figure 5-21 illustrates such a case where a ring gage may accept out-of-round workpieces that would be rejected by a snap gage.

Flush Pin Gages. The flush pin gage (Figs. 5-22 and 5-23) is a simple mechanical device used to measure a linear dimension. The essential parts of the gage are the body and a sliding pin or plunger. The indicating device is a step ground on the plunger equal to the total tolerance of the dimension. With the gage applied to a workpiece the position of the plunger can be checked visually or by fingernail touch.

Figure 5-22 shows a slotted workpiece being qualified with a flush pin gage. Also shown are the relative positions of the plunger at the high and low limits of the depth dimension. The flush pin principle applied in this way is simple in operation, is rugged and foolproof, does not require a master for presetting, and is economical when compared to micrometer or dial gaging methods. The dimension could be checked with a depth micrometer. This, however, would require a greater degree of operator skill, and the instrument can easily be misread.

344　　　　　　　　　Design of Tools for Inspection and Gaging　　Chap. 5

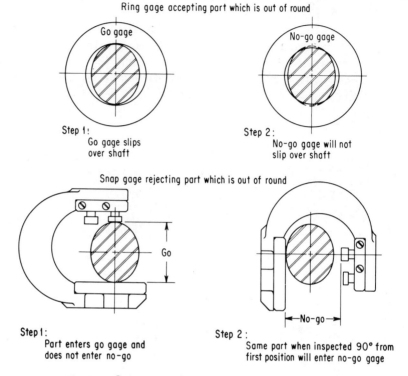

Fig. 5-21. Gage comparison.

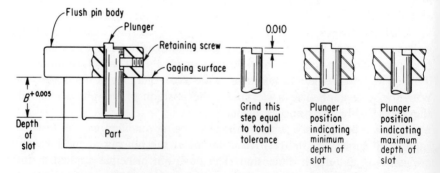

Fig. 5-22. Basic application of flush pin gage indicating various positions of plunger.

Figure 5-23 shows an inspection fixture containing two flush pin gages. Also shown is the workpiece being qualified. The first dimension being checked (X) is the distance from the center of a spherical radius to a flat surface. The second dimension (Y) is between the center of the same spherical radius and over the outside of a roll (wire). A standard tooling ball in the base of the fixture locates the radius of the workpiece and provides the origin for both dimensions.

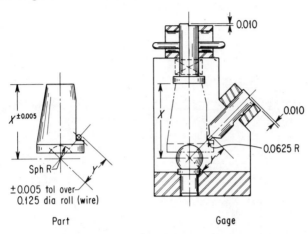

Fig. 5-23. Workpiece with flush pin-type inspection fixture.

AMPLIFICATION AND MAGNIFICATION OF ERROR

A surface can never be exactly flat. Two parts can never be exactly alike. Measured distances or quantities can never be exactly equal. The inability of an inspection device to detect error proves only that the capability of the inspection device is limited. Error or deviation is ever present, and unless specified or necessary limits are exceeded is not in itself a disqualifying factor. The designer specifies what magnitude of error can be tolerated. The inspector confirms that the magnitude of the error in the workpiece is within tolerance. Accuracy of measurement is limited by the accuracy of the known standard used for comparison, and by the skill of the person making the comparison.

As tolerances become smaller, the primary gaging methods are not precise enough to detect the magnitude of error. An inspector may, for instance, find that qualification of a workpiece no longer depends on whether or not a gage enters a hole, but rather on how much pressure is required to insert the gage or whether the gage enters with too little pressure. Inspection becomes more dependent on human judgment and consequently less reliable.

As variations in dimension become too small to be conveniently measured, it becomes necessary to amplify or magnify them prior to measurement. This can be done mechanically, pneumatically, optically, or electrically. The principle of amplification is illustrated in Fig. 5-24 which shows a simple mechanical indicator suitable for in-plant fabrication, which can be incorporated in many gaging fixtures. The indicator is zeroed to a master, after which any deviation from the nominal dimension is amplified tenfold and may be read directly from scale B.

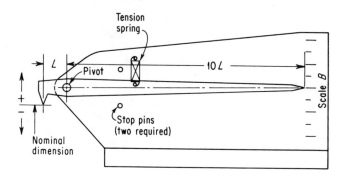

Fig. 5-24. Mechanical indicator.

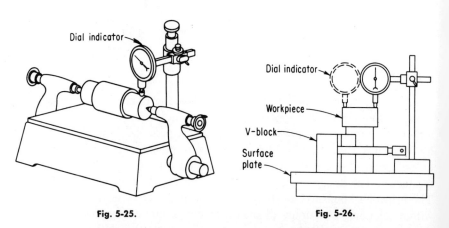

Fig. 5-25. Fig. 5-26.

Fig. 5-25. Concentricity gaging fixture composed of standard inspection tools.

Fig. 5-26. Squareness gaging fixture composed of standard inspection tools.

Dial indicators. The dial indicator (see Fig. 5-25) is perhaps the most widely used instrument for precise measurement. Basically it consists of a probe, rack, pinion, pointer, dial, and case. The probe, which is attached to the end of the rack, bears on the workpiece. A change in workpiece size

Chap. 5 Design of Tools for Inspection and Gaging

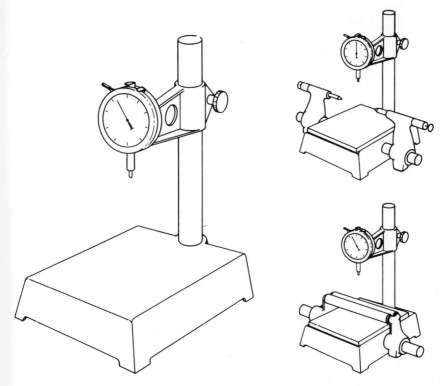

Fig. 5-27. Gaging fixture with accessories.

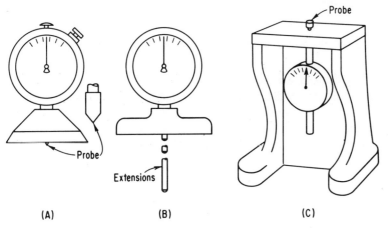

Fig. 5-28. Indicator-type depth gages: (*A*) special indicator for checking shallow recesses; (*B*) depth gage with probe extensions; (*C*) bench mounted depth gage.

alters the position of the probe, which in turn moves the rack. The rack movement turns the pinion, which through a gear train causes the dial pointer to move. The graduated dial is calibrated for direct reading of variation from the nominal dimension. Several amplification factors are involved.

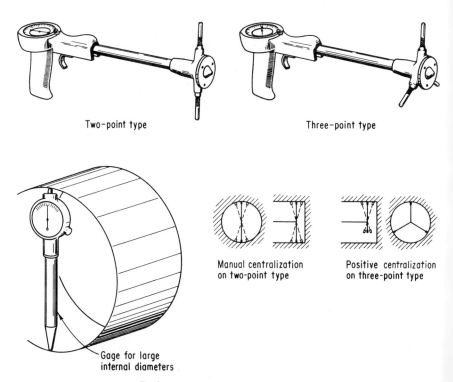

Fig. 5-29. Indicator-type plug gages.

Dial indicators are commercially available from a great number of sources. Standard models offered vary greatly in size, amplification ratio, mounting facilities, and precision. Inexpensive models are available for production use as well as very precise models for use in the gage crib.

An indicator gage has one primary advantage over a fixed gage, in that it shows how much a workpiece is oversize or undersize. When using an indicator as part of a gaging device, a master block made to the nominal dimension to be checked must be used to preset the indicator to zero. Then, in applying the gaging device, the variation from zero, the nominal dimension, is read from the dial scale.

Extremely precise dial indicators are part of the standard inspection equipment found in a gage crib, together with surface plates, parallels,

Chap. 5 Design of Tools for Inspection and Gaging 349

V blocks, and vernier height gages. Figure 5-25 shows standard inspection tools being used to check concentricity of a workpiece. As the part is manually revolved, the indicator shows any variation in concentricity between the two diameters. Figure 5-26 shows standard inspection tools being used to check squareness of a workpiece. In checking squareness or concentric-

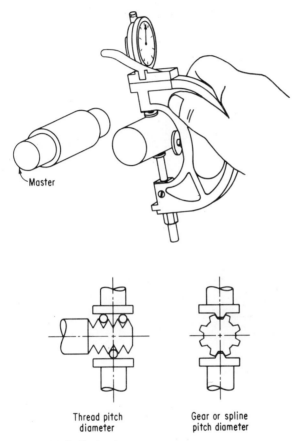

Thread pitch diameter

Gear or spline pitch diameter

Fig. 5-30. Indicator-type snap gages.

ity, the indicator is zeroed with the probe bearing on the workpiece. If the same standard components are used directly as a height gage, the gage must first be zeroed to a known standard such as a gage block or master.

If inspection is frequently required, it may not be economical to tie up quality control equipment. Figure 5-27 shows a commercially available gaging fixture complete with surface, indicator, and stand. Also available are attached centers and rolls for checking concentricity of centered or uncentered parts.

Dial indicators are often used with or as an integral part of commercially available plug and snap gages. Also available are tools for special purposes. Figure 5-28 shows three different types of depth gages commercially available. Figure 5-29 shows several indicator-type plug gages, together with a gage for measuring large interior diameters. Figure 5-30 shows a typical indicator-type snap gage, together with special tooling for the measurement of threads, gears, or splines. Figure 5-31 shows an indicator-type plug gage used as part of a special gaging fixture.

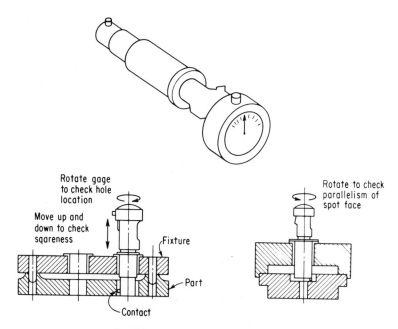

Fig. 5-31. Special gaging fixture using indicator-type plug gage as component.

Figure 5-32 is an example of a lever-type amplifying comparator, specially designed for production and easily made. By suitably proportioning the scale and pointer, the device will measure 0.0001-in. increments.

A ground steel rod is fitted to the reamed hole in the measuring column, and a steel ball is brazed into the end of this rod. A knurled setscrew allows vertical adjustment of the rod to accommodate different sizes of work.

All four arms must be drilled at exactly the same spacing, with free rotation at all eight swivel points. A tension spring keeps the measuring head in the up position until needed. A ground glass plate provides an accurate base for the part being measured.

The pointer should be set to a nominal dimension with gage blocks. When setting, the arms should always be at right angles to the column to minimize parallax, and gage blocks should be used for laying out the graduations on the scale.

Pneumatic Amplification Gages. There are two general types of air gages. One type is actuated by varying air pressure, and the other is operated by varying air velocity at constant pressure.

Figure 5-33 illustrates a pressure-type gage. Compressed air is filtered and reduced to a regulated pressure before it enters the measuring circuit. The air then flows into two channels and into opposite sections of a differential pressure meter.

Air in one of the channels is allowed to escape through a gaging plug. In the other channel, air is allowed to escape through a zero setting valve. Escape of air through the latter is balanced with the escape of air between the gaging plug and the master.

Any change in pressure caused by variation in the sizes of the gaged workpieces is detected by the differential meter. The system illustrated uses one master gage for setting to zero, after which the actual size of the part can be read on the indicating dial.

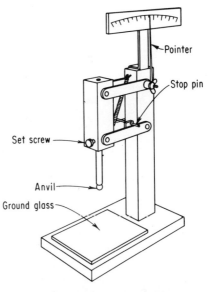

Fig. 5-32. Lever-type comparator.

This type of gage can also be used for checking external dimensions by replacing the plug gage with either a ring gage or a probe (Fig. 5-34). The ring gage operates exactly like the plug gage.

The probe operates in a different manner, but on the same basic principle. As the plunger within the probe rises or lowers, owing to part variation, it blocks the flow of air escaping through a jet in the side of the probe. Zero setting occurs when the escape jet is blocked halfway so that the mean amount of air escaping is read as the mean pressure reading.

In the flow-type air gage (Fig. 5-35), compressed air from the regular plant supply enters the gage through an automatic compensating pressure regulator, passes through a vertical transparent indicator tube and out through the orifice in the gaging spindle.

A float in the indicator tube is free to move up and down in response to the changes in velocity of the air passing through the tube and around the float. The greater this velocity, the higher the float rises in the tube. Its position gives a direct reading on a scale marked with the appropriate units of measurement.

Any practical degree of magnification is available merely by changing the size and weight of this indicator float, which is readily accessible from the top of the gage.

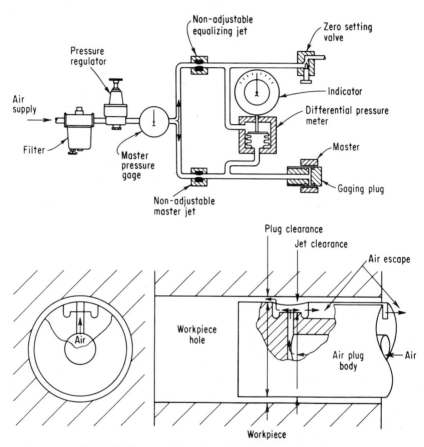

Fig. 5-33. Pressure-type air gaging system.

The velocity of air flowing at any given instant during the gaging operation depends only on the clearance between the gaging spindle and the sides of the bore being gaged. The greater the clearance, the higher the velocity of air that passes between the sides of the spindle and the internal surface of the bore.

As this gage is a limit type, two masters must be used to set high and low limits. Setting these limits is done with the air pressure on. A minimum master ring is slipped over the gaging spindle, and the air pressure is adjusted so that the indicator float rises to a point in the central zone of the tube. One of the sliding marker points is then set opposite the float position. Substituting the maximum master ring sends the float higher in the tube. The second marker is then used to mark this second position. The length of tube between the two markers now represents the difference in diameter between maximum and minimum master rings or the tolerance of the work to be gaged.

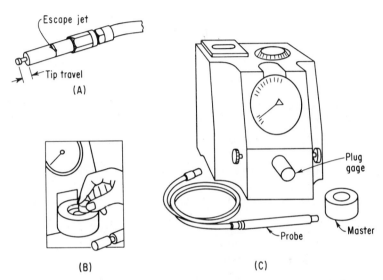

Fig. 5-34. Pressure-type air gage: (*A*) probe for inspecting internal diameter; (*B*) ring gage for inspecting outside diameter; (*C*) complete gage with accessories.

The length of tube between the two markers may be divided or calibrated on the adjacent scale in any number of divisions by the use of additional master setting rings. For classification inspection, the position of the float shows the exact diameter of any individual piece.

Flow and pressure type gages, excluding probes, are excellent for applications to parts having a high surface finish because the plugs or rings float in and out of the part without contact.

Electrical Amplification Gages. In this type of gage (Fig. 5-36), one extreme of the needle's swing closes an electrical circuit to a red signal light. The other extreme closes a circuit to a green signal light. The actual position of the contacts, which fix the sweep of the needle, is controlled by two micrometer adjusting screws, one for each terminal. The sweep of the needle between contacts is proportional to the dimensional tolerance for which the gage is set.

In setting an electric gage head for any given dimension, a master fixed-size gage or a stack of precision gage blocks equivalent to the minimum tolerance limit is placed in gaging position on the anvil. By means of the adjusting knob on the spindle the needle is moved toward the terminal that controls the red signal light. Its movement may be viewed through the window in the gage head case.

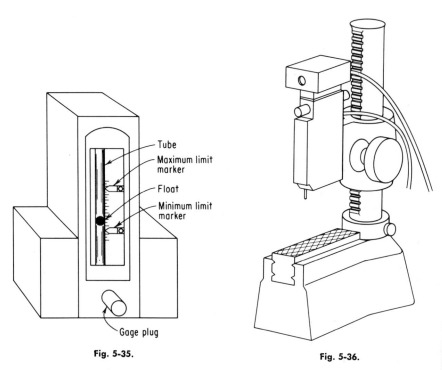

Fig. 5-35. Flow-type air gage.
Fig. 5-36. Reed-type electrical gage.

By manipulating the micrometer adjustment screw on the side of the head, the terminal is moved so that the signal light just changes from red to amber. Thus, any workpiece smaller than the minimum master gage will allow the needle to swing far enough to break the green-light circuit, leaving the red light on, and thus show that the part is undersize.

In the same way the opposite terminal is set with a maximum master limit gage. When a workpiece is larger than the maximum master gage, the needle makes contact and breaks the circuit which lights the red light, allowing the green light to remain on. The needle swinging anywhere between these points fails to break either circuit and keeps the signal light amber to indicate that the part being checked falls within its prescribed tolerance range.

This type of electrical gage is a limit gage, and can determine whether or not a workpiece is within tolerance. The gage will indicate which limit is beyond tolerance, but does not disclose by what amount. Accuracy in checking limits is 0.000005 in. Tolerance range is from 0.00005 to 0.012 in.

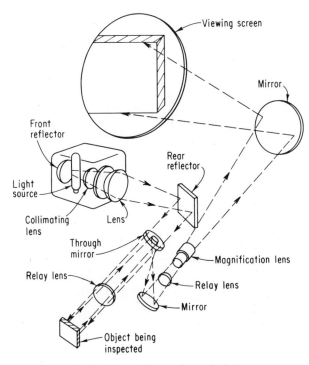

Fig. 5-37. Optical projection gaging principle.

An electronic gage is available which will indicate, by means of a calibrated dial, the exact variation of the part from its basic size. This gage can be used where selective inspection is required. The electronic gage needs but one master setting block to set it to zero.

Optical Projection Gaging. Optical projection is a method of measurement and gaging employing a precision instrument known as an *optical projector* or *comparator*. Optical gaging is gaging by sight rather than feel or pressure. Measurement or gaging is accomplished by placing a workpiece in the path of a beam of light and in front of a magnifying-lens system, thereby projecting an enlarged silhouette shadow of the object upon a translucent receiving screen (Fig. 5-37).

Measurement of a workpiece normally requires a chart gage with reference lines in two planes. In gaging applications a precisely scaled layout of the contour of the part to be gaged, usually containing tolerance limits, is drawn on the translucent receiving screen. Figure 5-38 shows a workpiece with the chart gage used for its inspection.

Optical gages are virtually unaffected by wear. There is no wear to a light beam, and any fixture wear can be compensated by repositioning to the setting point. Little operator skill is required; no tactile skill or sensitivity is required. Dimensions can be altered on the translucent screen

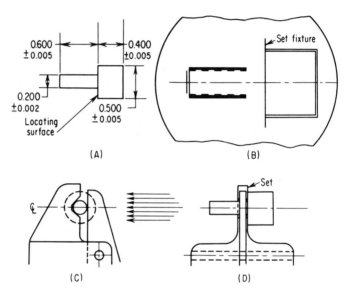

Fig. 5-38. Optical gage setup: (*A*) workpiece; (*B*) chart gage; (*C*) side view of holding fixture; (*D*) front view of holding fixture.

easily and quickly. The chart provides exact duplication. Several dimensions can be checked simultaneously eliminating excessive handling of the part.

Most optical projectors have standard magnification of 10 up to 100 power. To establish the effective specimen area that can be projected through a given lens system, divide the screen diameter by the magnification of the lens being used. For example, a comparator with a 14-in. diam screen, using 10-power magnification, will project a complete area of 1.400-in. diam specimen at one setting.

A *chart gage* is a very accurately scribed, magnified outline drawing of the workpiece to be gaged, containing all the contours, dimensions, and tolerance limits necessary for the purpose. The chart is made of glass or plastic, or for very short and quick checks drafting paper can be used, glued to a sheet of glass. Paper should not be used consistently because of shrinkage with the changes of weather. Chart layout lines should be dense black, sharply defined, and from 0.006 to 0.010 in. wide for best legibility. Dimensions extend normally to the center of the lines. When maximum and minimum tolerance lines are used, the magnification should be high enough to maintain a minimum of 0.020 in. spacing between the lines. For closer tolerance checking, special lines or bridge arrangements, based on gaging to the edge of the lines, are often used. Figure 5-39 shows two portions of a chart gage. The first portion (*A*) uses maximum and minimum tolerance lines. The second portion (*B*) has a close tolerance bridge arrangement.

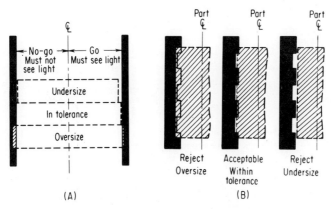

Fig. 5-39. Chart gage segments: (*A*) maximum and minimum tolerance lines; (*B*) close tolerance bridge arrangement.

Workholding equipment includes vises for flat workpieces, and staging centers for round workpieces such as shafts and cylindrical parts with machined centers. V blocks, in a range of sizes, are mounted on bases to clamp to projector work table slots. Diameter capacities run from virtually zero to 5-in. diameter.

GAGING POSITIONALLY TOLERANCED PARTS

Figure 5-1 illustrates the concept of position tolerances as compared with coordinate dimensions. Dimensions are usually labeled "TP" (true position) or "basic" on positionally toleranced parts. Tolerances of position and form are usually expressed as diameters, widths, or contours. The tolerances or tolerance zones are applicable throughout the depth, length, or circumference of the workpiece feature unless otherwise specified.

The diameter (cylindrical) tolerance zone specified in Fig. 5-40 is also a squareness tolerance since the actual centerline of the 0.510 to 0.520-in. diam hole must lie entirely within the tolerance cylinder to meet the specification.

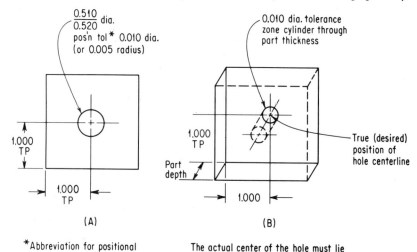

*Abbreviation for positional tolerance

The actual center of the hole must lie within the tolerance zone shown through depth of part

Fig. 5-40. Squareness specified by positional tolerance: (*A*) drawing or blueprint; (*B*) resultant workpiece.

The width tolerance zone specified in Fig. 5-41 is equally disposed about the true center of the 3.000 to 3.010-in. width. The actual center of the slot must lie entirely within the rectangular zone as shown by the broken line.

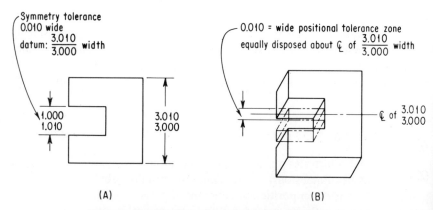

Fig. 5-41. Specification of squareness by a width tolerance zone: (*A*) drawing or blueprint; (*B*) resultant workpiece.

The contour tolerance zone specified in Fig. 5-42 is equally disposed about the true (desired) contour. The tolerance zone path is normal or square to the true surface and usually of constant width. Contour tolerance zones may be unequally disposed about true contours or all on one side of the true contour.

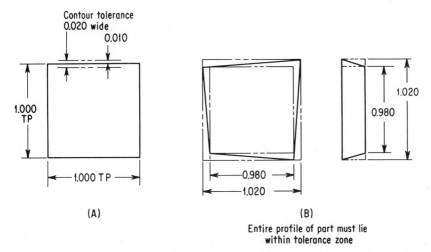

Fig. 5-42. Implications of contour tolerances: (*A*) drawing or blueprint; (*B*) resultant workpiece.

Maximum Material Condition (MMC)

The most critical specified interchangeable size of a part or part feature is called its *maximum material condition,* the maximum (biggest) size limit of an external feature and the minimum (smallest) size limit of an internal feature. When the abbreviation MMC is specified after the locational tolerance to a part feature, the specified locational tolerance applies only when the feature is at its MMC size, and the locational tolerance may increase in direct proportion as the actual feature is finished, or made away from its MMC size.

MMC Applied to Internal Features. Figure 5-43 illustrates two identical parts containing clearance holes, assembled with two 0.500-in. diam bolts. Each part could be dimensioned and toleranced as shown, with MMC specified after the hole location tolerance. Also shown is a hole relation gage for each of the parts. Hole relation gages check hole-to-hole relationship, not hole location to some other part feature. The gage could contain fixed pin gage features in lieu of the separate gage pins shown which fit tight into nominally loacted bushings in the gage.

The locational and squareness tolerance is the actual difference in size between the gage pin and the clearance hole feature. Since feature size is bound to vary from hole to hole, from part to part, and from process to process, the true tolerance is a variable since the 0.500-in. diam bolt or gage pin and the interchangeable design requirement remains a constant. The positional tolerance is 0.010-in. in diam only at MMC, in this case when the hole diameter is 0.510 in.

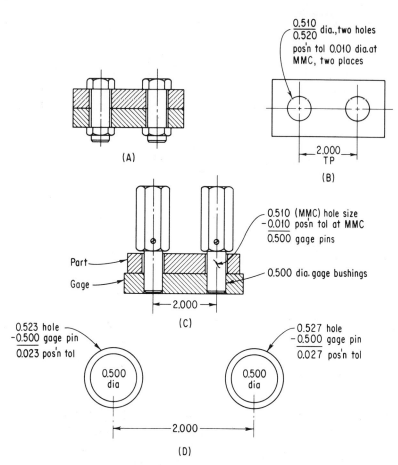

Fig. 5-43. MMC as applied to internal features: (*A*) two identical parts assembled with 0.500-in. diam bolts; (*B*) workpiece drawing or blueprint; (*C*) hole relation gage required; (*D*) hole location tolerance varies with hole size.

The omission of the MMC callout or the specification of the abbreviation RFS (Regardless of Feature Size) makes gaging quite difficult since the gage pin must always be 0.010 in. smaller than the actual hole size. Eleven or more gage pins would be required for each hole (0.500-in., 0.501-in., 0.502-in., . . .) and if a particular hole measured 0.512 in., the 0.502-in. gage pin would be used, etc.

Table 5-2 shows how the true tolerance varies with hole size. All tolerances greater than 0.010 in. in diameter are bonus tolerances and are not specifically allowed in any system other than the positional tolerancing system.

TABLE 5-2. VARIABLE TOLERANCES ALLOWED BY MMC
(all dimensions are in inches)

Feature diameter	Positional and squareness tolerances diameter
0.510 (MMC or most critical size)	0.010 at MMC (tightest tolerance)
0.511	0.011
0.512	0.012
0.513	0.013
0.514	0.014
0.515	0.015
0.516	0.016
0.517	0.017
0.518	0.018
0.519	0.019
0.520 (least critical size)	0.020 (largest tolerance allowed)

MMC Applied to External Features. Figure 5-44 illustrates the tolerances and gages for the manufacture and assembly of two different parts—one part containing clearance holes and the other part containing holes tapped for 0.500-in. diameter bolts. Illustrated also is the feature relation gage required for the tapped part. Gage thickness or bushing height must be at least the maximum thickness of the untapped part to guarantee that the bolts will be properly located and square for assembly. The two *go* thread gages simulate the bolts at assembly. Gage bushing size is determined by adding the 0.010-in. diameter positional tolerance specified for the tapped features to the bolt or tap size. Stepped gage pins with *go* threads may be used to take advantage of standard bushing size as long as a 0.010-in. difference in size is maintained between the gage bushing and that portion of the gage pin that lies within the bushing. Since there is little if any tolerance of size on the pitch diameter of a thread, the MMC callout is usually not applied to tapped features.

Basic Design Rules. Two basic rules govern the design of gages for positionally toleranced parts. These principles apply regardless of the number of features that make up an interchangeable pattern.

1. For parts with *internal features,* the nominal gage feature size is directly determined by subtracting the total positional tolerance specified at MMC from the specified MMC size of the feature to be gaged for location.

2. For parts with *external features,* the nominal gage feature size is directly determined by adding the total positional tolerance specified at MMC to the specified MMC size of the feature to be gaged for location.

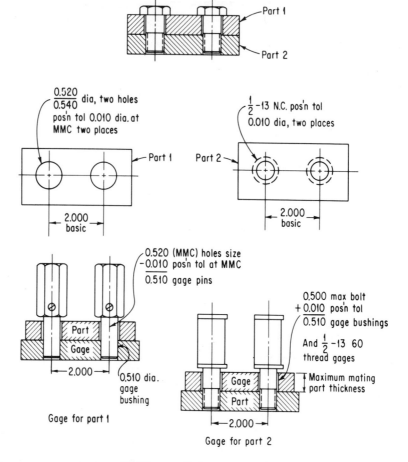

Fig. 5-44. MMC as applied to external features.

Practical Applications

Gaging Critical Part Datums. Figure 5-45 shows a workpiece with four holes that must be located from the specific center (datum) of the workpiece regardless of the actual workpiece size (RFS). The design specification drawing includes the exact pickup points so that the same center can be repeatedly found. Also shown is a hole relation gage which uses four dial indicators to determine and correctly position the datum for the hole gaging operation.

Gaging Noncritical Part Datums. 1. Design Requirement: The pattern of interchangeable features (holes in Fig. 5-45) need be located from the center of the datum diameter only when the datum is at MMC (4.010 diam

Chap. 5 Design of Tools for Inspection and Gaging   363

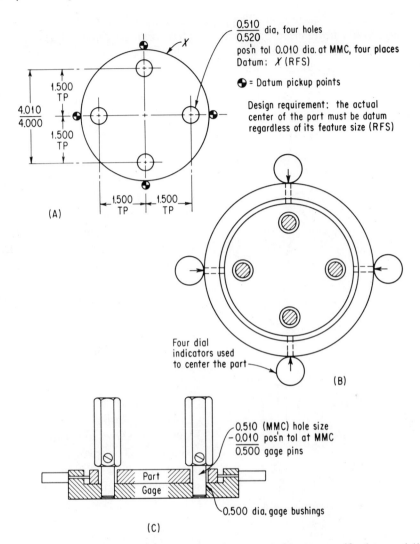

Fig. 5-45. Gaging workpiece features relative to a specific datum: (*A*) workpiece; (*B*) plan view of gage showing centering by dial indicators; (*C*) cross section of gage showing gage pins in position.

in this case). This allows the use of a "shake" gage of fixed diameter which should usually approximate the mating part at its MMC size. Figure 5-45 illustrates the hole relation and location gage specified by the datum callout, which centers the part only when the part approaches its MMC size, and fits loosely or "shakes" on the part when the part is not at its most critical interchangeable size.

2. Design Requirement: The pattern of interchangeable features (holes in Fig. 5-46) never need be exactly located from the center of a specified datum feature. A gage fit allowance has been specified that is dirctly reflected in the size of the datum gage feature, since the gage is 0.995-in. diam and differs from the MMC size of the part datum (1.000) by 0.005. Quite a large allowance could be specified if the datum was merely a convenient starting place for manufacturing.

Gaging an End Product. A pattern of interchangeable features (holes) are the most critical features on the part shown in Fig. 5-47, and are not locationally critical in relation to any single datum feature. The 0.300-in. minimum breakout specification is the result of a stress analysis and is an end-product requirement. No single datum is specified and the 0.300-in. minimum specification can be readily gaged with a micrometer or a turning fork type *no-go* gage as shown.

Gaging Radial Hole Patterns. Figure 5-48 shows a workpiece with seven holes. The specified positional tolerance includes the location and squareness (angularity) tolerances for each radial hole. Also shown is the gage required for checking the part. In use, all seven gage pins must go through the part at one time. In designating the datum, if RFS callouts had been used instead of MMC for the diameter and width of the slot, the gage would be required to center on the two datum features; as a result, one gage pin could be used to individually qualify each hole in reference to the datum.

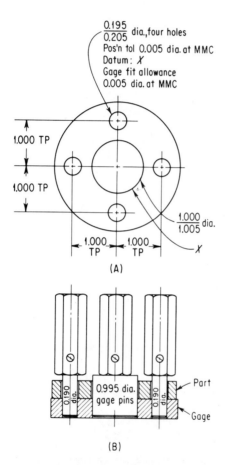

Fig. 5-46. Gaging workpiece features not specifically located relative to the datum: (*A*) workpiece; (*B*) hole location (shake) gage.

Gaging Squareness. Since the positional tolerance is also the squareness tolerance (Fig. 5-40), squareness may be specified when it is more critical than the positional tolerance. Figure 5-49 shows a workpiece so

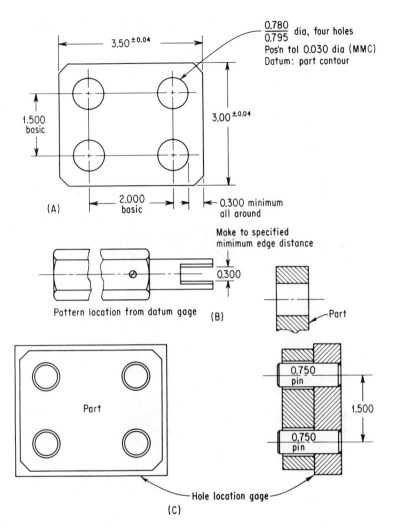

Fig. 5-47. Gaging an end-product: (*A*) workpiece; (*B*) gage for checking required edge distance; (*C*) gage for checking hole location.

dimensioned with the required squareness gage which can be used independently of any required pin location or relation gages.

Gaging Straightness. A straightness specification increases the effective interchangeable size of a workpiece feature. The pin shown in Fig. 5-50 will require a mating hole with a minimum diameter of 0.376 in. (0.375-in. diam + 0.001-in. straightness tolerance). Shown also is a gage for checking the straightness of the pin. The actual diameter of the pin must be inspected separately.

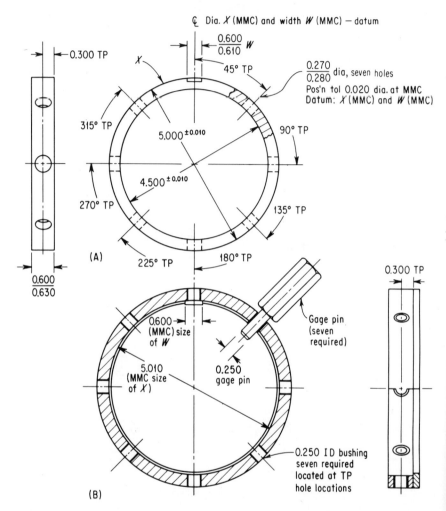

Fig. 5-48. Gaging radial hole patterns: (*A*) workpiece; (*B*) hole location and squareness gage.

Gaging Concentricity. Features that lie on the same axis may have a specified positional relationship. The gage in Fig. 5-51 allows the effective size of the part to increase to 0.750-in. diam and 1.015-in. diam at MMC. The concentricity tolerance could approach 0.016-in. if the datum was 0.749 in. in diameter and the other diameter was 1.000 in. (i.e., both diameters at their least critical interchangeable size).

Gaging Contour. The gage illustrated in Fig. 5-52 will give variables inspection data since the dial indicator will be set to read zero along a perfect part. Actual surface form variations can be measured as the indica-

Chap. 5 Design of Tools for Inspection and Gaging

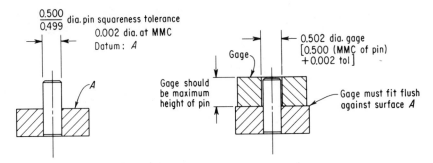

Fig. 5-49. Specification and gaging of squareness.

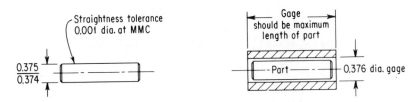

Fig. 5-50. Specification and gaging of straightness.

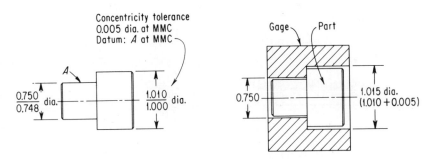

Fig. 5-51. Specification and gaging of concentricity.

tor is moved along a gage contour parallel and normal to the desired part contour.

Optical Gaging. Figure 5-53 illustrates a part that has been designed purposely for an optical gage. The contour tolerance zones include size, form, and location of the part and part features. The optical gage (or template) also shown would be drawn on a transparent sheet of very stable plastic material and would be to the increased scale (10 to 1, 20 to 1, or whatever) to which the part would be magnified in an optical comparator. When the magnified part is compared to the scaled-up tolerance zone requirements, all part features must lie within their respective contour tolerance zones.

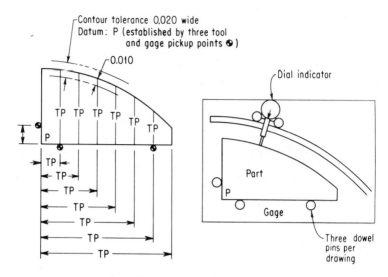

Fig. 5-52. Specification and gaging of contour.

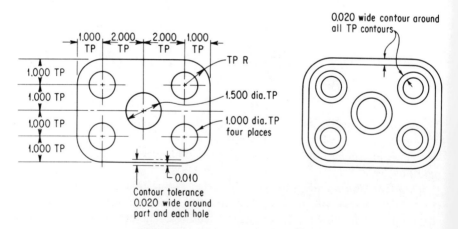

Fig. 5-53. Workpiece purposely designed for optical gaging, with optical gage (template).

Chap. 5 Design of Tools for Inspection and Gaging

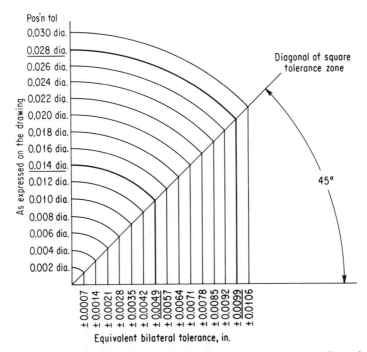

Fig. 5-54. Conversion chart: Positional tolerance to bilateral. (*The Sandia Corp.*)

Conversion Charts. Since most machine movements are made in two transverse directions X and Y, the chart shown in Fig. 5-54 may be used to determine the bilateral tolerance approximations of positional tolerance diameters.

The chart shown in Fig. 5-55 may be used to determine whether an actual hole center (as measured on an open setup) is within the positional tolerance specified on the drawing, and also to determine the actual amount of deviation (radius) from TP. For example, on a drawing a hole is located by dimensions labeled TP with a positional tolerance. The location of the actual center of the hole on the part is derived by coordinate measurements. The difference in the actual measurements and the corresponding TP dimensions are plotted coordinately on the chart using the values in the upper and right-hand borders. It can now be shown whether the location is within the required positional tolerance limits using values in the left-hand border. To determine the actual deviation from TP, a radius may be drawn through the point, intersecting the 45-deg diagonal line. The amount of deviation may now be read from the scale on the 45-deg line using the values in the lower border.

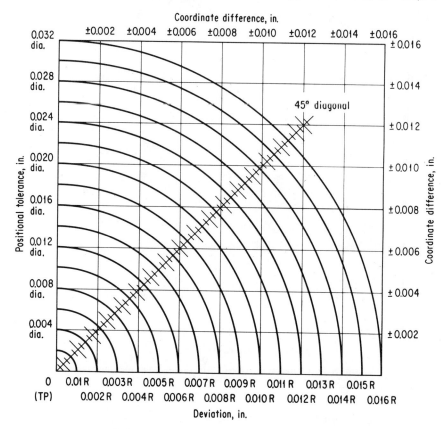

Fig. 5-55. Conversion chart: Bilateral to positional tolerance. (*The Sandia Corp.*)

PROBLEMS

1. Define tolerance.
2. Why is it impossible to obtain an exact dimension?
3. In specifying purchased components what precautions must be taken to insure proper function of the tool?
4. How do you establish the location of each element of the workpiece?
5. How is the location of each element of the workpiece defined?
6. Name seven quality characteristics of a product which might be specified.
7. What is the function of a master gage and what tolerance is used?
8. What is the disadvantage of the bilateral system of gaging?
9. Find the *go* and *no-go* gage dimension of a plug gage using the bilateral system for a hole size 0.750 ± 0.0005 in. and what class of gage is to be used. Note: Do not apply wear allowance.

10. Find the *go* and *no-go* gage dimension of a plug gage using the unilateral system for a hole size 0.750 ± 0.0005 in. and what class of gage to be used. Note: Do not apply wear allowance.
11. What determines the amount of wear allowance to be incorporated in the design of a plug gage?
12. What would happen if wear allowance were not incorporated in the design of a gage?
13. Can a tungsten carbide gage which has excessive wear be restored by chrome plating?
14. Can a flush-pin gage be used for checking a tolerance of ±0.001 inches?
15. What are the advantages of optical gaging?
16. What is MMC?
17. Find the *go* and *no-go* gage dimensions of a plug gage using the unilateral system and including wear allowance for gaging a 0.750 ± 0.0005 in. diam hole.

REFERENCES

1. Campbell, James S., Jr., *Principles of Manufacturing Materials and Processes,* McGraw-Hill Book Co., Inc., New York, 1961.
2. Kennedy, Clifford W., *Inspection and Gaging,* The Industrial Press, New York, 1951.
3. Michelon, Leno C., *Industrial Inspection Methods,* Harper & Brothers, New York, 1950.
4. *The Science of Precision Measurement,* The DoALL Co., Des Plaines, Illinois.
5. American Society of Tool and Manufacturing Engineers, *Tool Engineers Handbook,* 2nd ed., McGraw-Hill Book Co., Inc., New York, 1959.
6. *American Drafting Standards Manual,* American Standard ASA Y14.5-1957, American Standards Association, New York.
7. *Surface Roughness, Waviness and Lay,* American Standard ASA B46.1-1955, American Standards Association, New York.
8. *Gage Blanks,* Commercial Standard CS8-51, U.S. Department of Commerce, National Bureau of Standards.
9. *Screw Thread Gages and Gaging,* American Standard ASA B1.2-1951, American Standards Association, New York.

6

TOOL DESIGN

FOR THE JOINING PROCESSES

The joining processes are generally divided into two classes: *mechanical* and *physical*. Mechanical joining does not ordinarily involve changes in composition of the workpiece material. The edges of the pieces being joined remain distinct. Physical joining usually involves a change in composition at the mating edges, with the edges losing their identity in a homogeneous mass. Two pieces of wood nailed together are joined mechanically. The same two pieces of wood could be joined physically by an adhesive. At the exact center of the joint, only the adhesive would be found. The adhesive would penetrate into the pores of the wood for some distance, and the workpiece edges would no longer exist as true entities but would have become a blend of wood and glue.

Joining processes may require tooling to hold the parts in correct relationship during joining. Another function of the tooling is to assist and control the joining process. Quite often several parts may be joined both mechanically and physically. Thus, two workpieces may be bolted together to assure alignment during subsequent welding. Mechanical joining at times may be considered as the tooling method for the final physical joining.

TOOLING FOR PHYSICAL JOINING PROCESSES

Physical joining processes cannot generally be performed without tooling, because the elevated temperatures required usually make manual positioning impractical. The tooling must hold the workpieces in correct relationship during joining and it must assist and control the joining process by affording adequate support. Tooling used for hot processes must not only withstand the temperatures involved, but must in many cases either accelerate or retard the flow of heat. Hot fixtures must be so designed that their heat-expanded dimensions remain functional.

Design of Welding Fixtures

The purpose of a welding fixture is to hold the parts to be welded in the proper relationship both before and after welding. Many times a fixture will maintain the proper part relationship during welding, but the part will distort after removal from the fixture. Good fixture design will of itself largely determine product reliability. Major fixture design objectives, some basic and some special, are: (1) to hold the part in the most convenient position for welding; (2) to provide proper heat control of the weld zone; (3) to provide suitable clamping to reduce distortion; (4) to provide channels and outlets for welding atmosphere; (5) to provide clearance for filler metal; (6) to provide for ease of operation and maximum accessibility to the point of weld.

Other factors that will also have a very definite influence on fixture design are: (1) cost of tool; (2) size of the production run and rates; (3) adaptability of available welding equipment on hand to do the welding process; (4) complexity of the weld; (5) quality required in the weldment; (6) process to be employed; (7) conditions under which the welding will be performed; (8) dimensional tolerances; (9) material to be welded; (10) smoothness required; (11) coefficient of expansion and thermal conductivity of both workpiece and tool materials.

The tool designer must be familiar with the gas, arc, and resistance welding processes. Each of these processes will require individual variations of the general design factors involved. For instance, heat dissipation is not a critical factor in some of the welding processes.

Gas Welding Fixtures

The general design of a gas welding fixture must take into consideration the heating and cooling conditions. A minimum of heat loss from the welding area is required. If the heat loss is too rapid, the weld may develop cracks from cooling too rapidly. Heat loss by materials, particularly aluminum and copper, must be carefully controlled. To do this, large fixture masses should not be placed close to the weld line. If the fixture supports are too far from the weld line, however, the part may distort. The contact area and clamps should therefore be of the minimum size consistent with the load transmitted through the contact point. In welding copper and aluminum, the minimum contact surface often permits excessive heat loss,

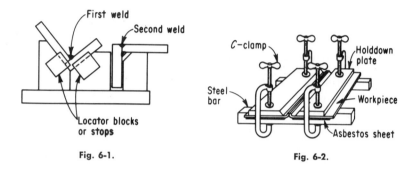

Fig. 6-1.   Fig. 6-2.

Fig. 6-1. Simple welding fixture using gravity to help locate parts.

Fig. 6-2. Workpieces with simple fixturing for gas welding operations.

and prevents good fixture welds. This necessitates tack welding the fixtured parts at points most distant from the fixture contact points, with the rest of the welding done out of the fixture. With this method, excessive distortion may result, and subsequent stress relieving of the part may be required.

One of the simplest fixtures for gas welding is a gravity-type fixture as shown in Fig. 6-1. This design eliminates excess fixture material from the weld area to minimize heat loss, while providing sufficient support and locating points. The design also permits making welds in a horizontal position, which is generally advantageous.

Figure 6-2 shows another simple form of gas welding fixture which holds two flat sheets for joining. C-clamps hold the workpieces to steel support bars. Alignment is done visually or with a straight edge. A heat barrier of asbestos is placed between the workpieces and the steel bars. Holddown plates are used to keep the workpieces flat and to prevent distortion. If the parts to be welded have curved surfaces, the supporting bars and holddown plates may be machined to match the part.

Simple parts may be properly located or positioned in a fixture visually. As workpiece shape becomes complex, or the production rate increases, positive location is desirable. The same locating methods used in workholder design (Chap. 2) can readily be adapted for the design of welding fixtures.

The selection of material for gas welding fixtures is governed by these factors: (1) part print tolerances; (2) material heat resistance; (3) heat transfer qualities; (4) the fixture rigidity required to assure workpiece alignment accuracy. The fixture material should not be affected in the weld zone and should prevent rapid heat dissipation from the weld area. Some of the fixture materials commonly used are cast iron, carbon steel, and stainless steel.

Arc Welding Fixtures

Arc welding concentrates more heat at the weld line than gas welding. The fixtures for this process must provide support, alignment, and restraint on the parts, and also must permit heat dissipation.

Some of the more important design considerations for arc welding fixtures are: (1) the fixture must exert enough force to prevent the parts from moving out of alignment during the welding process, and this force must be applied at the proper point by a clamp supported by a backing bar; (2) backing bars should be parallel to the weld lines; (3) backing bars should promote heat dissipation from the weld line; (4) backing bars should support the molten weld, govern the weld contour, and protect the root of the weld from the atmosphere.

Backing bars are usually made from solid metal. A simple backup could be a rectangular bar with a small groove directly under the weld. This would allow complete penetration without pickup of the backup material by the molten metal. In use the backup would be clamped against the part to make the weld root as airtight as possible. Some common groove shapes are shown in Fig. 6-3. Figure 6-4 shows a backing bar in position against a fixed workpiece.

The size of the backup bar is dependent on the metal thickness and the material to be welded. A thin weldment requires larger backup to promote heat transfer from the weld. A material with greater heat-conducting ability requires less backup than that required for a comparable thickness of a poor conductor.

Figure 6-5 shows backing bars designed for use with gas, which may be used to blast the weld area (A), flood the weld area (B), or may be concentrated in the weld area (C). Backup bars may be made of copper, stainless steel (used for tungsten inert gas), beryllium, titanium, or a combination of several metals (sandwich construction).

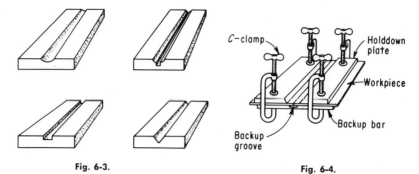

Fig. 6-3. Typical grooves in backing bars.

Fig. 6-4. Workpieces with simple fixturing for arc welding operations.

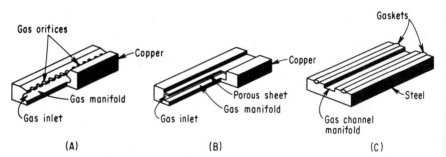

Fig. 6-5. Backing bars with provisions for (A) directed gas flow, (B) diffused gas flow, and (C) pressurized gas.

Resistance Welding

One of the simplest and most economical processes for joining two or more metal parts is resistance welding. In resistance welding fusion is produced by heat generated at the junction of the workpieces by local resistance to passage of large amounts of electric current and by the application of pressure. Figure 6-6 illustrates the elements of a resistance welding machine. Resistance welding processes used for low-cost high production are shown in Fig. 6-7 and include: spot welding, projection welding, seam welding, pulsation welding, flash butt welding, upset butt welding, and cross wire welding.

Spot Welding. Spot welding, compared to riveting, may be considerably faster and less expensive since there is no need to drill holes and insert rivets. Figure 6-8 illustrates typical spot-welded joints and electrode shapes which can be produced on standard welders. Figure 6-9 illustrates the principle of series welding where two welds are made with each stroke of

Chap. 6 Tool Design for the Joining Processes   377

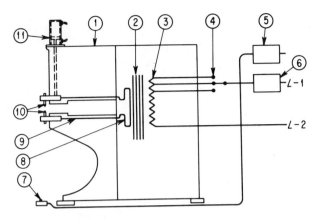

Fig. 6-6. Elements of typical resistance welder: (1) housing; (2) low-voltage, high-current transformer; (3) primary coils; (4) tap switch; (5) welding timer; (6) power interrupter; (7) foot switch; (8) secondary loop; (9) bands from electrodes to secondary; (10) electrodes; (11) cylinder which exerts electrode pressure on work.

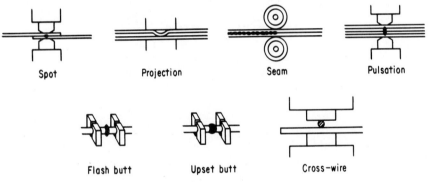

Fig. 6-7. Resistance welding methods. (*Assembly and Fastener Engineering.*)

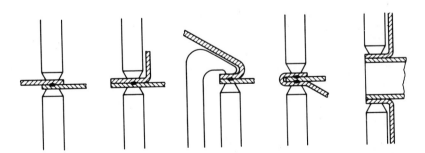

Fig. 6-8. Typical spot-welded joints. (*Machine Design.*)

the welder without any markings, indentations, or discoloration on one side of the assembly. Figure 6-10 illustrates the principle of indirect welding ordinarily used where the welding current must pass through the side or ends of parts due to design. The welding electrode tip size directly affects the size and shear strength of the weld. Tip area is a function of the workpiece material gage. For thin sheets (up to 0.250 in.) the diameter of the electrode tip can be calculated by using the formula $d = 0.1 + 2t$ where t is the material thickness. For thick material ($\frac{1}{4}$ to 2 in.) $d = t$.

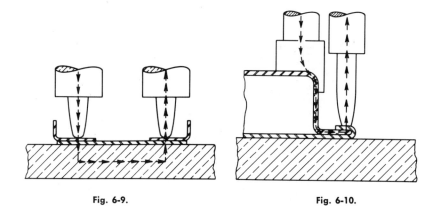

Fig. 6-9.   **Fig. 6-10.**

Fig. 6-9. Assembly showing series-welded joint. (*Machine Design.*)

Fig. 6-10. Spot-welded assembly showing a typical joint design for an indirect weld. (*Machine Design.*)

Projection Welding. In projection welding, embossments or projections are formed on one or both workpieces for localization of heat. The dimpled workpieces are placed between plain large-area electrodes. Projection welding provides increased strength with reduced electrode maintenance.

Seam Welding. In seam welding the material to be welded passes between two rotating disk electrodes. As the current is turned on and off a continuous tight seam is produced.

Pulsation Welding. In pulsation welding the current is applied repeatedly to make a single weld while the pressure is applied. This process will produce a better weld for heavier material.

Flash Butt Welding. In flash butt welding the work is clamped in dies, the current is turned on, and the two joints are brought together by means of cam control to establish flashing and upsetting followed by discontinuation of the welding current. Figure 6-11 illustrates a flash welding fixture.

Upset Butt Welding. Upset butt welding differs from flash butt welding in that pressure is applied continuously through the clamping dies after the welding current is applied so that heat is developed entirely from the resistance effect of the current.

New Welding Processes. Newer joining techniques growing in popularity are: ultrasonic welding, high-frequency resistance welding, foil-seam

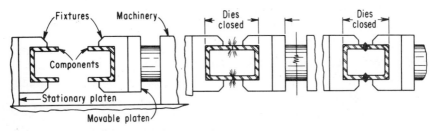

Fig. 6-11. Flash butt welding.

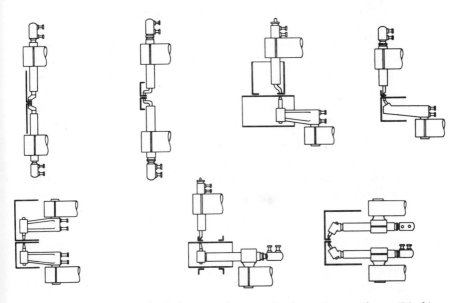

Fig. 6-12. Typical standard electrode tips and operations. (*Machine Design.*)

welding, magnetic-force welding, percussion welding, friction welding, thermo-pressure welding, diffusion-bond welding, electro-slag welding, and electron-beam welding.

Function of Electrodes. In resistance welding the parts are positioned between electrodes which exert heavy pressure, conduct the current into the materials to be welded, and dissipate the heat from the outer surface of the materials being welded. Wherever possible, the electrode tips should be water cooled. Figure 6-12 illustrates typical standard electrode tips. The design of welding electrodes and the material from which they are made is of great importance. For increased life, the design must provide sufficient strength with adequate heat conduction and cooling.

Resistance Welding Fixtures

There are two general types of fixtures for resistance welding. The first type is a fixture for welding in a standard machine having a single electrode. The second type is a fixture and machine designed as a single unit, usually to attain a high production rate.

Certain design considerations apply to fixtures for resistance welding: (1) keep all magnetic materials out of the throat of the welding machine, particularly ferrous materials; (2) insulate all gage pins, clamps, locators, index pins, etc.; (3) protect all moving slides, bearings, index pins, adjustable screws, and any accurate locating devices from flash; (4) give consideration to the ease of operation and protection of the operator; (5) provide sufficient water cooling to prevent overheating; and (6) bear in mind that stationary parts of the fixture and work are affected by the magnetic field of the machine. Workholder parts and clamp handles of nonmagnetic material will not be heated, distorted, or otherwise affected by the magnetic field.

There are other considerations that will affect the design of resistance welding fixtures and the machine if high production may be required:

1. The fixture loop or throat is the gap surrounded by the upper and lower arms or knees containing the electrodes and the base of the machine which houses the transformer. This gap or loop is an intense magnetic field within which any magnetic material will be affected. In some cases, materials have actually been known to melt or puddle. Power lost by unintentional heating of fixture material will decrease the welding current and lower the welding efficiency. This power loss may sometimes be used to advantage; e.g., if the current is burning the parts to be welded, the addition of a magnetic material in the throat will increase the impedance, lower the maximum current, and halt the burning of parts.
2. The throat of the machine should be as small as possible for the particular job.
3. Welding electrodes should be easily and quickly replaceable. Water for cooling should be circulated as close to the tips as possible. Adjustment should be provided for electrode wear. If the electrodes tend to stick, knockout pins or strippers may be specified. Current-carrying members should run as close to the electrodes as possible,

have a minimum number of connections or joints, and be of adequate cross-sectional area.
4. Provide adjustment for electrode wear.
5. Check welding pressure application.
6. Have knockout pins or strippers if there is a tendency of the electrode to stick to the electrode face. These may be leveraged or air operated.

General Fixture Design Considerations. Simple fixtures may have the part located visually with scribed lines as a guide. This is quite similar to locating parts for gas welding. For higher production, a quicker locating method is needed. A locating land may be incorporated in the fixture to accurately establish the edge position of the part to be welded (Fig. 6-13).

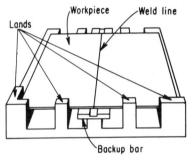

Fig. 6-13. Locating lands.

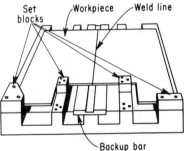

Fig. 6-14. Set block locators.

In some cases setup blocks may be used in place of a locating land (Fig. 6-14). When welding a variety of similar parts with different dimensions, setup blocks have a distinct advantage over the land method of locating. With proper design, setup blocks can be interchangeable to accommodate varying workpieces. Dowel pins may be used as locators (Fig. 6-15). Other means of locating are V blocks and adjustable clamps.

Clamping Design Considerations. Clamps used in welding fixtures must hold the parts in the proper position and prevent their movement due to alternate heating and cooling. Clamping pressure should not deform the parts to be joined. Clamps must be supported underneath the workpiece (Fig. 6-16). Owing to the heat involved, deflection by clamping force could remain in the part.

Quick-acting and power-operated clamps are recommended to achieve fast loading and unloading time. C clamps may be used for low production volumes. Power clamping systems may be direct acting or work through lever systems (Fig. 6-17).

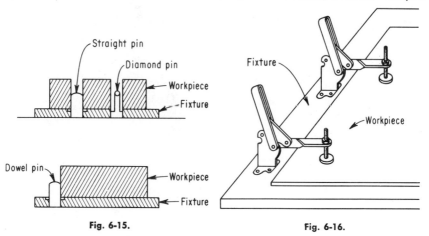

Fig. 6-15.

Fig. 6-16.

Fig. 6-15. Dowel pin locators.

Fig. 6-16. Typical clamp installation with the fixture supporting the workpiece directly beneath the clamps.

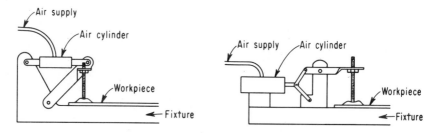

Fig. 6-17. Air-actuated clamping methods.

TOOLING FOR SOLDERING AND BRAZING

Soldering and brazing differ from welding in several respects. The metal introduced to the workpiece for the joining operation is nonferrous, usually lead, tin, copper, silver, or their alloys. The workpiece or base metal is not heated to the melting point during the operation. The added metal is melted and usually enters the joint by capillary action. Lead-tin soldering is called *soft soldering* and is conducted below 800°F. Silver soldering is called *hard soldering* and requires temperatures from 1100°F to 1600°F. Copper brazing requires temperatures from 1900°F to about 2100°F.

The success of these processes depends on chemical cleanliness, temperature control, and the clearance between the surfaces to be joined. Cleanliness is usually obtained by introducing a flux which cleans, dissolves, and floats off any dirt or oxides. The flux also covers and protects the area by shielding it from oxidation during the process. It may to some extent reduce the surface tension of the molten metal to promote free flow. The worst contamination is usually due to oxidation during the process. Many joining operations are conducted in a controlled atmosphere with a blanket of gas to shield the operation.

Temperature control, although influenced somewhat by fixture design, is dependent primarily on the heat application method. In a simple low-production process, the workman may hold a torch closer to the workpiece for a longer period of time. In a very precise high-production process, the instrumentation of a controlled atmosphere-type furnace may be adjusted.

Clearance between the surfaces being joined determines the amount of capillary attraction, the thickness of the alloy film, and consequently the strength of the finished joint. The best fitting condition would have about 0.003 to 0.015-in. clearance. Larger clearances would lack sufficient capillary attraction, while smaller clearances would require expensive machining or fitting.

Many soldering and brazing operations are conducted without special tooling. As with mechanical joining methods, the many workholding devices of Chap. 2 can be used to conveniently present the faces or areas to be joined. An electrical connecting plug can be conveniently held in a vise while a number of wires are soldered to its terminals. In many high-production assembly operations, parts are manually mated with a preformed brazing ring between them, and are then placed directly on the endless belt of a tunnel-type furnace.

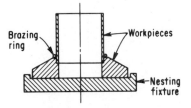

Fig. 6-18. Simple nesting fixture with work in place.

If the shape of the workpiece is such that it will not support itself in an upright or convenient position, a simple nesting fixture may be required. Figure 6-18 shows a simple nesting fixture in which two workpieces and a brazing ring have been placed. The fixture can be mounted on a table while an operator applies heat with a hand torch. The same fixture could be mounted on a powered rotating base in the flame path of a fixed torch, while a feed mechanism would introduce wire solder at a predetermined rate (Fig. 6-19). The same fixture could in quantity be attached to the belt of a tunnel-type furnace. A number of the fixtures could be attached to a rack for processing in a batch-type furnace.

Tooling for Induction Brazing. Figure 6-20 shows a nest-type fixture to hold mating workpieces within the field of an induction work coil. Figure 6-21 shows the same fixture as altered to permit use of an internal induction heating coil. If the external coil is used, the fixture designer must provide some method of moving the fixture or coil while workpieces are loaded and unloaded.

Induction coils (inductors) provide a convenient and precise way of quickly and efficiently heating any selected area of an electrically conductive part or assembly of such parts to any required depth to provide a specified brazed joint. Correct selection must be made of frequency, power density, heating time, and inductor design.

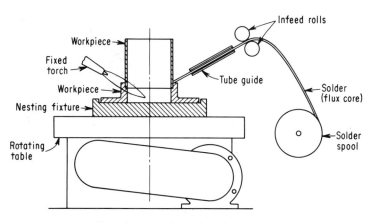

Fig. 6-19. Soldering machine using simple nesting fixture.

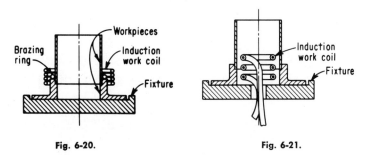

Fig. 6-20.   Fig. 6-21.

Fig. 6-20. Nesting fixture for brazing with an external inductor.
Fig. 6-21. Nesting fixture for brazing with an internal inductor.

Induction Heating Theory. The flow of current through an electrical conductor results in heating as the current meets resistance to its flow. Thus I^2R losses (I, current; R, resistance) may be low as in the case of current flowing through copper wire (resistance is low) or high when this same current reaches and flows through a heating element (resistance is high). This high loss is the resistance heating obtained with conventional electric heaters.

Induction heating is also resistance heating with current flow meeting resistance. A workpiece so heated has been made the secondary of a simple transformer, the primary being the inductor which generally surrounds the part and through which alternating current is flowing. The flow of alternating current in a primary induces by electromagnetic forces (magnetic flux) a flow of current in the secondary. Alternating current tends to flow on the surface and there is a relationship between frequency of the alternating current, the depth to which it flows, and the diameter of stock which can be heated efficiently.

Considering a piece of plain carbon steel being heated for brazing or surface hardening, the depth at room temperature in which there is instantaneous flow of current is related only to the frequency:

$$D = \sqrt{4/F} \quad \text{where } D = \text{depth in inches} \\ F = \text{frequency in cycles/sec}$$

The depth D increases with temperature. For heat to be generated, the I^2R must have a time factor t to become I^2Rt.

Additional depth results from the current following the path of least resistance (cooler underlying metal), and from heat flow by conduction. This depth may be approximated by the equation $D = \sqrt{0.0015t}$ where D is inches and t time in seconds. The higher the frequency for a given heating time, the shallower will be the depth of heat. The converse is also true. For a given frequency, the depth of heat is also directly proportional to the time (square root).

An examination of the equation $D = \sqrt{4/F}$ will give an answer to the relationship between frequency and diameter of stock which can be heated efficiently. Large diameters may be heated efficiently with low frequency. Sixty cycles is used efficiently on 10-in. diameter steel workpieces.

The theoretical depth of current penetration mentioned above increases with temperature and must never exceed the radius of the stock being heated. For practical purposes, this depth D ($D = \sqrt{4/F}$ at room temperature) should be two to three times the radius for through heating and ten times the radius for surface hardening.

Power density and heating time are closely related to the design of the inductor. It is upon the inductor design that the success of any induction heating application is largely dependent.

The purpose of the inductor is to set up a magnetic flux pattern in which the work to be heated is correctly positioned so that the required heating is accomplished. Coil design is influenced by the application, as is the selection of frequency, power density, and heating time.

Regardless of the material used in the induction brazing process, the inductor must be designed to heat the area of the joint sufficiently to cause the bonding material to flow. Temperatures can range from a few hundred degrees for soft solder and epoxies to 2100°F for copper. No attempt will be made here to discuss joint design, placement of material, type of material, or fluxing. These factors must have first been resolved and proper consideration given to their relationship to the induction heating method, which enables the precise heating of the joint area only. Neither will any discussion be given to frequency selection other than to comment that extremely small diameters usually require the high frequencies of a vacuum tube oscillator, and either oscillators or motor-generator sets are suitable for parts of medium cross section. However, when attempting to heat heavy sections and to avoid overheating of sharp corners on small or medium-size workpieces, the use of motor-generators at 3000 or 10,000 cycles is preferred.

Copper is normally used for the coils with appropriate water cooling. This can be accomplished by using copper tubing, solid copper to which tubing has been attached, or solid copper which has been drilled out or machined to provide water passages.

An inductor should have the following characteristics: (1) proper physical shape to surround the section to be heated; (2) large enough diameter to permit loading and unloading or, in the case of a continuous operation, to permit clear passage; (3) proper support or rigidity to maintain its designed shape; (4) insulation to avoid electrical breakdown (spacing or an air gap between the turns and the workpiece is usually sufficient); and (5) ability to withstand operating conditions (exposure to water, dirt, and flux must be taken into consideration).

The air gap or space between the work and the inside of the inductor should be kept low for reasons of efficiency. A gap of $\frac{1}{8}$ in. is reasonable. For irregularly shaped sections where a round inductor is used and where efficiency is not too important, gaps of $\frac{1}{4}$ in. and greater can be tolerated.

The area to be heated determines the length or width of the inductor. For most joint designs on small parts, a single turn is sufficient. On larger diameters it is necessary to heat a wider band requiring either several turns or a wider inductor. An inductor that is too narrow will simply require a longer heating time to allow the heat to flow to cover the area. An inductor that is too wide will heat more metal than necessary and, therefore, will be less efficient. A single-turn inductor should be no wider than the bore diameter; for greater coverage, multiturn inductors are used.

The electrical characteristics of the high-frequency power source determine the number of turns in an inductor. Generally, the high frequencies of a vacuum tube oscillator (450 kc) usually require multiturn coils, whereas the motor-generator sets (3 and 10 kc) can operate into either multiturn coils or single-turn coils through the use of variable-ratio transformers. Turns of the inductor should be kept as close as possible without touching. Typical inductors are shown in Fig. 6-22.

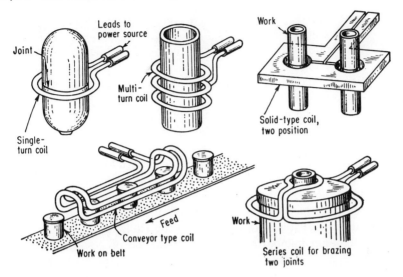

Fig. 6-22. Designs of induction-heating coils for brazing.

Number of Inductors. The number of inductors to be used at a time is, of course, influenced by production requirements. The number can be extended to as many as it may appear convenient to load. Instead of a large number, a continuous coil can be used, through which parts are moved progressively. This method is generally limited to a workpiece having a joint on or near its end. The determination of how many parts are to be processed at a time is related to heating time and power density. The time generally runs from 10 or 15 seconds on small parts to as long as two minutes on large parts where the heat must penetrate. The power density is approximately 0.5 to 1.5 kw per square inch of surface area. An average value of 1 kw/sq in. is recommended. The actual value depends on the ratio of surface area to volume.

The amount of energy (kilowatt seconds) is determined by the volume of metal to be heated and the temperature necessary to flow the bonding material. Since the exact amount of metal heated in a brazing operation depends on many factors such as conductivity of the work and heating time, only a rough approximation can be attempted. The following procedure is recommended:

1. Estimate the volume of metal which would be expected to be brought to brazing temperature.
2. Convert the volume to weight in pounds based on steel weighing 0.3 lb per cubic inch.
3. For 400°F solder, convert to kilowatt seconds by multiplying by 100.

4. For 1200°F brazing material, convert to kilowatt seconds by multiplying by 300.
5. For 1600°F brazing material, convert to kilowatt seconds by multiplying by 500.
6. For 2000°F brazing material, convert to kilowatt seconds by multiplying by 700.
7. Estimate the surface area of the joint. Based on 1 kw/sq in., divide the area into the above value of kilowatt seconds to determine the approximate brazing time in seconds.
8. Determine from production requirements (allow for handling each part) the number of parts to be processed at a time.
9. Total power required will depend on the above value and kilowatts per piece based on 1 kw/sq in. of surface area.
10. Power available is determined by the equipment to be used. The maximum number of parts which can be processed at once can be approximated by dividing the kilowatt rating of the equipment by the surface area of each part in inches (1 kw/sq in.).

Material. A solid-type inductor is preferred because of its greater rigidity. Accidental contact while loading parts will not bend it out of shape. Less expensive coils can be made of copper tubing which, except for extremely low power (1 kw per coil), should not be less than $3/16$ in. in diameter. Both round and rectangular sections are used; the rectangular section is especially adaptable when a wide, single turn is needed. Its wide flat area can be adjacent to the work for more uniform heating. A 1-in. wide inductor can be made of $1/4$ by 1-in. tubing or from a wide flat strip brazed to a round tube for cooling.

Leads. The leads to the inductor should be kept as short as possible and close together. In some situations coaxial leads may be necessary (450 kc). If the leads are not a continuation or part of the inductor and a brazed joint is used, they must not reduce the flow of current to the inductor. Silver solder is recommended, and carefully prepared joints (even mitered) are essential. Insulating materials should be used between leads if they are not self-supporting. The connection to the power source will be determined by the design of the output transformer.

Cooling. Inductors must have sufficient cooling to avoid overheating. Inductors for brazing operations can be made by forming a copper tube on a suitable mandrel. The tube must not collapse to restrict internal coolant flow.

Cooling water may be brought in from the output transformer through the connectors carrying the current or by separate insulated connections to a water supply.

Coil Support. Solid inductors are usually self-supporting, as are single-turn inductors. Support of multiturn inductors or an array of inductors can be accomplished with any nonmetallic material. Many plastic materials can be used. Coils can be attached by studs (preferably copper) brazed to the inductor.

Heating Internally. When tubular sections are to be joined it is often more convenient to heat from the inside. The factors of coil design are essentially the same as for OD inductors except that the leads must necessarily be brought to the coil axially instead of radially (Fig. 6-21). Since heating from the ID is considerably less efficient (the magnetic flux is weak outside the confines of the coil), two to three times as much energy (kilowatt seconds) may be required. This can be obtained by using more power or time or both. When high production requirements preclude the use of longer heating cycle, the ID coil efficiency can be improved by the use of an inside core which increases the magnetic flux outside of the coil. Basic design elements are as follows:

1. For vacuum tube oscillators, compacted ferromagnetic stock may be used. This material, which can be purchased in round bars, is then made a core inside the inductor. The turns can be held tightly on the surface or wound in grooves machined in the core. No insulation is required, and one lead may be brought through a hole in the center of the core.
2. For motor-generators, iron laminations are used. They are made from 0.007-in. thick strips of transformer steel, ¼ in. wide, which has been cut to the correct coil length. The strips are together and inserted in a multiturn coil to form a core. If a single-turn inductor is used, laminations are cut in a C shape and stacked tightly around the turn as shown in Fig. 6-23.

The design of inductor coils for brazing operations is relatively simple. Electrical matching to the proper power source is no problem. The physical configuration is dictated by the shape and size of the area to be heated. The inductor is a primary winding for a transformer with the workpiece acting as a secondary. The coil need only be formed to surround, or for ID heating, be adjacent to the surface to be heated. A wide latitude of shape is permitted. A square part can be heated with an inductor formed on a square mandrel (Fig. 6-24) or a round mandrel. Irregularly shaped parts can be heated with round inductors. If a corner of a workpiece overheats, the inductor can be modified to merely provide a larger air gap at the corner.

Occasionally a joint must be heated by proximity. An ID coil cannot be used and an OD coil would necessitate heating entirely too much metal. An inductor for heating such an area is shown in Fig. 6-25. Such an inductor is inefficient and can be improved by use of laminations as shown. The C shape is open toward the work.

Generally, an inductor should have the same contour as the area to be heated. With a reasonably uniform air gap the part can be loaded with no difficulty. When there are large changes in diameter in the section to be heated, the larger diameter will tend to overheat unless some compensation is made. This can be done by having fewer turns per linear inch in that area, by increasing the air gap, or by using two turns in series (electrically) on the smaller diameter (Fig. 6-26). The intensity of the magnetic field in which the surface finds itself should be as uniform as possible.

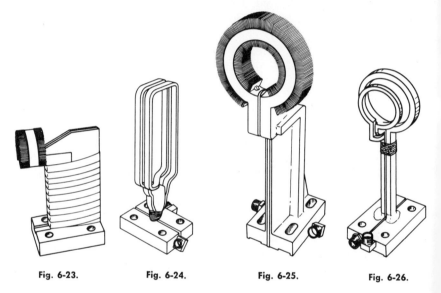

Fig. 6-23.   Fig. 6-24.   Fig. 6-25.   Fig. 6-26.

Fig. 6-23. Special-purpose inductor.

Fig. 6-24. Inductor formed on rectangular mandrel.

Fig. 6-25. Inductor with C-shaped laminations added.

Fig. 6-26. Inductor with loops in series.

TOOLING FOR MECHANICAL JOINING PROCESSES

Many workpieces can be held for mechanical joining without tooling. A workman can often manually align two workpieces and insert a fastener. This method has several limitations. The workpiece must be small or light enough to be positioned manually, and the forces incurred in the joining process must be relatively small. The complexity of an assembly so joined is also limited by the number of components that a workman can conveniently handle. An elementary workholder can often be used to advantage in even the most simple joining operation.

A universal-type vise (Fig. 2-49) can be of great value in mechanical joining. A primary workpiece can be held in any position while other workpieces are fastened to it. Several workpieces may be held in alignment between the jaws, while the workman applies a fastener. Clamping pressure can be used to counteract the joining forces, such as the torque applied to a threaded fastener.

Threaded Fasteners

Threaded fasteners are universal in application, the bolt and nut being the commonest type. Tooling for threaded fasteners is as varied as their application. There are, however, design principles that apply to all cases.

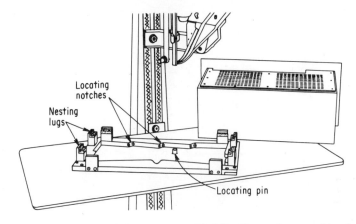

Fig. 6-27. Partial nest-type workholding fixture for assembly of cabinet. (*Detroit Power Screwdriver Company.*)

The most common method of threaded assembly is to place two workpieces in their correct relative location, drill a hole through them, insert a bolt in the hole, and torque a nut onto the bolt. The hole may have been drilled in both pieces prior to mating. In many cases no tooling will be required. If it is inconvenient for a workman to hold the workpiece together while inserting the bolt, adding the nut, then tightening the nut, a simple workholder may be advantageous. Many of the workholders shown in Chap. 2 will suffice. The workholder should locate the workpiece so that the holes are conveniently positioned, and should support the workpiece against the torque and thrust loads imposed in tightening the fasteners.

One of the two workpieces being assembled may be tapped to receive the threaded fastener, or a self-tapping fastener may be used. Power tools may be used to drive the fasteners. The fixture design principles for nut runners are: (a) *location*—the fixture must align the workpiece precisely with the nut runner; and (b) *support*—the fixture must withstand the weight of the workpiece plus the thrust and torque loads imposed.

Figure 6-27 shows a template-type nesting fixture resting on the table of a power screwdriver. It is designed to hold a metal cabinet while two side shields are assembled to it by twenty self-tapping screws.

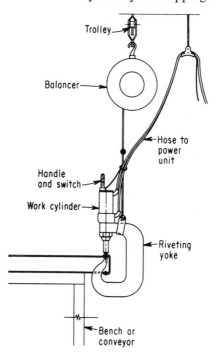

Fig. 6-28. Workpiece simply supported for riveting. (*Machinery and Steel World.*)

A round locating pin is attached to the machine table exactly under the driver. The fixture base is a template and has ten radiused notches that can be manually held against the locating pin. When the components of the cabinet are placed in the fixture, each template notch is directly beneath a predrilled hole in the components into which a screw is to be inserted. As each notch is held against the locating pin the predrilled hole is placed directly beneath the driver, and the machine is cycled to insert and drive one screw. After ten screws have been driven to secure one side shield the cabinet is inverted and another ten screws are driven to fasten the other shield.

Chap. 6 Tool Design for the Joining Processes   393

The fixture design principles are again the same. The fixture must precisely locate the workpiece relative to the driver. In this case ten precise location points are involved. The fixture must also establish the exact height of the workpiece. The fixture must support the workpiece in resistance to the torque and thrust loads imposed.

For high-production volumes, several machines may be arranged to simultaneously insert and drive screws into a single-fixtured workpiece.

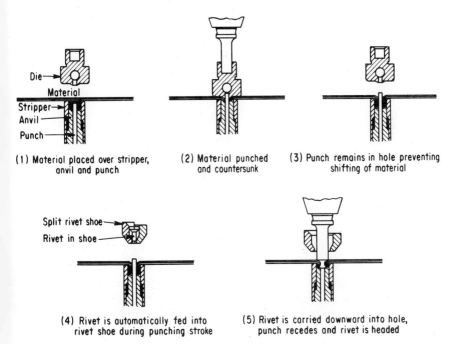

Fig. 6-29. Sequence of punching and riveting operations. (*Engineering and Research Corporation.*)

Rivets

Perhaps the most widely used pin-type fastener is the rivet—a pin headed on one end, the other end being plastically deformed after insertion to prevent retraction. The riveting process is extremely varied, and may be used to assemble the parts of a timepiece or the structural members of bridge. Rivet diameters may vary from perhaps 0.015 to 5.000 inch. The holes may be drilled or pierced before or during the operation. Workholders and tooling will have in common only the ability to support and locate the workpieces.

Figure 6-28 shows an L-shaped workpiece clamped in a fixture while a second workpiece, a channel, is being riveted to it. The portable riveting

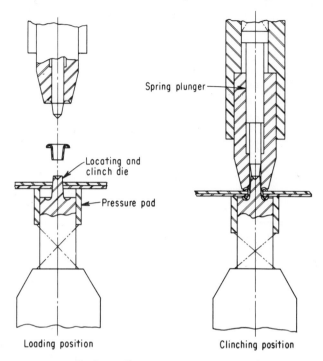

Fig. 6-30. Eyelet curling.

yoke literally squeezes the rivets to deform them. The holes are drilled and the rivets are inserted prior to the operation. Figure 6-28 is highly schematic. In practice it would be necessary to employ stops or other means to locate the channel with reference to the primary workpiece.

Figure 6-29 shows the sequence of a punching and riveting operation used in mass production. The two workpieces are placed between the tools, and the machine is cycled to automatically pierce a hole, insert a rivet, and then head the rivet.

Eyeleting. Eyelets like rivets are used extensively as low-cost fasteners for light assembly work. For high-production operations costs can be reduced to a minimum with the use of eyelet attaching machines. These machines are actually small power presses with hopper feed mechanisms. Figure 6-30 illustrates typical tooling of an automatic eyelet-setting machine in loading and clinching positions.

Tubular Riveting. A tubular rivet is a cross between a solid rivet and an eyelet. The straight end of the rivet has a center hole which permits this part of the rivet to clinch easily when it is struck by the contoured riveting tool. Figure 6-31 illustrates tooling for a typical tubular riveting operation.

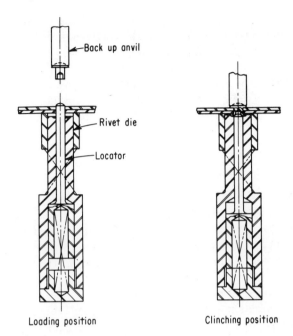

Fig. 6-31. Tubular riveting.

Spin Peening. Some materials to be assembled are brittle, while other assemblies require slender unsupported rivets or eyelets which will not withstand the single impact required to rivet or eyelet without distortion or cracking. For delicate assembly operations spin-peening machines may be used. These machines deliver innumerable light blows while the hammer spins. The peening and spinning action of these machines is either mechanical or pneumatic.

Clinch Allowance. Clinch allowance is that part of a rivet or eyelet which extends beyond the combined thickness of the assembly before the rivet or eyelet has been set. Figure 6-32 shows the relation of rivet or eyelet length to material thickness and clinch allowance. The proper rivet or eyelet length should be approximately equal to the material thickness plus the clinch allowance. The clinch allowance should be approximately 70 per cent of the shank diameter. Since both rivets and eyelets are avail-

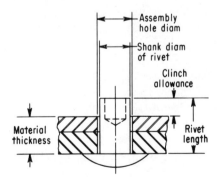

Fig. 6-32. Determination of rivet length for tubular rivets.

able commercially in $\frac{1}{32}$-in. length increments, the clinch allowance can vary from a minimum of 50 per cent to a maximum of 90 per cent of the shank diameter. The hole diameter and the method of producing the hole in the material also affect clinch diameter. The minimum clinching radius for tubular rivets or eyelets can be determined by multiplying the wall thickness by three. The clinching contour of riveting tools must be free of nicks and circular grooves. After hardening, the contour must be highly polished.

Riveting Equipment

Rivets can be deformed or set in many ways. Pressure may be applied continuously or by a series of hammer blows. The rivets may be manually or automatically inserted. The holes may be drilled prior to or during the riveting sequence. The tool used to apply the deformation pressure may be a common hammer, a pneumatic hammer (riveter or rivet gun), a portable squeezing yoke, or a stationary machine.

The final shape of the rivet becomes that of the tool (die) used to apply the deformation pressure. The shape may be flat (reflecting the contour of a hammer or a conventional bucking bar) or may be curved as shown in Figs. 6-30 and 6-31. The rivet die may be a simple bucking bar, an interchangeable rivet set placed in the nozzle of an air hammer, or a complete forming die placed in a standard hydraulic or mechanical press.

The workholders or fixtures used to hold and locate the workpieces being assembled by riveting are generally of three types: stationary, portable, and self-contained (riveting die fixture).

Pneumatic Hammer. Conventional riveting is performed by placing the rivet in a predrilled hole, holding a pneumatic hammer against the head of the rivet, holding a bucking bar against the end of the rivet, and then cycling or activating the hammer. The initiating pressure is exerted by the hammer while the deformation pressure is exerted by the bucking bar. Pneumatic hammers (rivet guns) are commercially available in a wide range of types and sizes. The nozzle of the rivet gun receives and retains the rivet set or die. Rivet sets are commercially available in a wide variety of shapes to mate with the many types of rivets commonly used.

Portable Yoke Type. This type of riveting equipment is commonly used for squeezing large rivets ($\frac{3}{16}$ to 1 in. diam), and consists of a yoke with a cylinder (air or oil) to provide the squeezing action. Equipment size can be minimized by using hydraulic cylinders and high pressures (5000 lb per sq in.). Equipment weight is minimized by using special high-strength heat-treated steel. The cylinder advances an anvil to the rivet with a primary pressure of approximately 1000 lb per sq in. When resistance is encountered, the pressure increases to 5000 lb per sq in. for the rivet upsetting portion of the action. Figure 6-28 shows a hydraulic riveting yoke.

Stationary Machines. Stationary machines are often used for riveting. The workpieces being assembled are located with reference to each other (mated) and are placed between the upper and lower elements of the machine. Small stationary machines have a spring-loaded pin locator in the rivet die. Predrilled holes through the mated workpieces engage the pin locator as shown in Fig. 6-33. The upper riveting die (backup anvil)

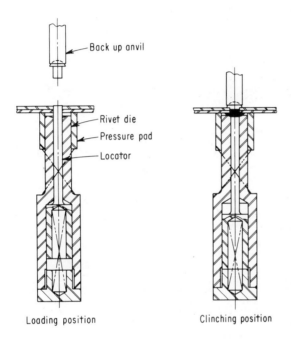

Fig. 6-33. Stationary riveting machine operation.

pushes the rivet into the holes, depresses the spring pin, and upsets the driven head on the lower anvil. The rivets are fed from a hopper to a feed track. The lower end of the feed track locates the rivet directly above the pin locator. The end of the track is split to allow the upper die to pick off the rivet for insertion.

Large stationary machines often combine both riveting and hole piercing. Figure 6-29 illustrates the sequence of operations. The workpieces need not be predrilled, but must be located with reference to each other. The locating and holding fixture may be part of the machine. Extremely large stationary machines can hold and precisely locate large aircraft sections while thousands of holes are pierced and rivets are driven. The entire

398 Tool Design for the Joining Processes Chap. 6

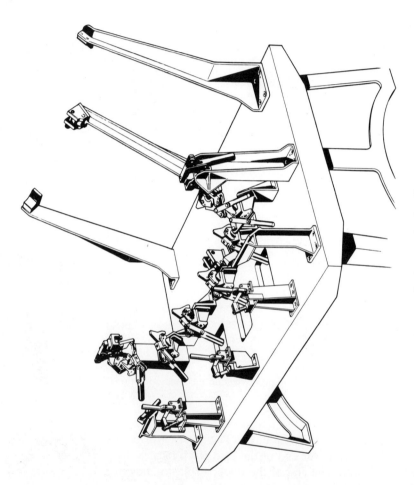

Fig. 6-34. Stationary riveting fixture.

Chap. 6 Tool Design for the Joining Processes 399

production sequence is automatic, with the machine motions governed by tape control. Closed-circuit TV is used to monitor the operation.

Stationary Holding Fixtures. Stationary holding fixtures are those in which two or more parts are located and pinned or clamped in position. They are usually free-standing fixtures where portable riveting equipment is used. Air-operated riveting hammers are used for smaller rivets such as those used in aircraft or appliance work. Larger rivets used in structural work are deformed with hydraulic or air-operated riveting yokes (rivet squeezing action). Figure 6-34 illustrates a simple holding fixture where riveting equipment is brought to the work.

Portable Fixtures. Portable riveting fixtures are those used to locate two or more parts for transport to a stationary-type rivet unit or machine. The fixtures also hold the workpieces during the riveting sequence.

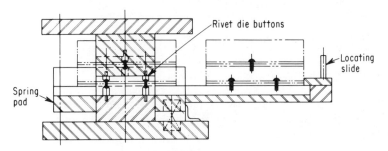

Fig. 6-35. Punch press riveting.

Riveting Die Fixture. A riveting die fixture is completely self-contained for use in a punch press. This fixture consists of locating elements for two or more parts, and rivet buttons for driving one or several rivets in one stroke or hit of a press. The locating elements may be movable so that parts can be positioned outside of the press and rivets can be placed in location. The assembled parts are placed in location in the riveting die, and all rivets are driven at once by the action of the press. Figure 6-35 shows a riveting die fixture.

Riveting Fixture Design. Beyond the primary requirements that the fixture precisely locate and hold the workpieces, ease of loading and unloading is extremely important. Many parts can be riveted without clamping if the weight of the parts alone will keep the riveting surfaces in contact with one another. Light-gage panels do require clamps because of the tendency to wrap or twist. Clamps for low-volume production are usually of the hand-operated cam or toggle type. For high production, air-operated clamps of the same type are used. Locating pins are used in preference to clamps wherever possible. The area of application must be accessible to the riveting tool. Tilting fixtures are often used for this reason.

Stapling

Stapling is a joining operation using preformed U-shaped wire staples. Staples are made in a variety of shapes, wire sizes, leg lengths, and crown sizes. Staples are available in blunt-end, chisel-point, or divergent-point styles and are cohered into strips or sticks. The sticks are loaded manually into staplers. Figure 6-36 illustrates staple nomenclature. Gun tackers and hammer tackers are used to drive wire staples. Wire staples can also be driven and clinched by low-cost stapling machines.

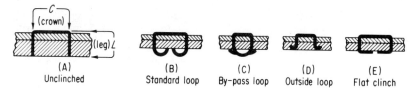

Fig. 6-36. Staple nomenclature.

Wire Stitching

Wire stitching is the process of joining two or more pieces of material with wire fed from a coil, cut to length, U formed, driven through, and clinched by a specially designed machine called a stitcher. Figure 6-37 illustrates the principle of wire-stitching machines. For low-cost, high-volume production, automatic wire-stitching machines are ideal for fastening com-

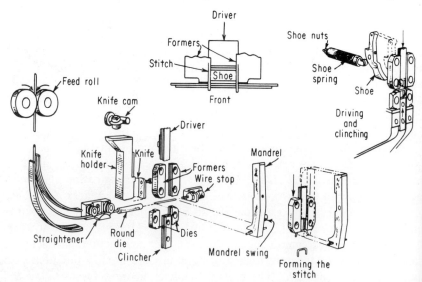

Fig. 6-37. Metal-stitching nomenclature and principles.

ponents together. These machines are available to perform the following operations: (1) carton or industrial-type stitching; (2) carding, bagging, and labeling; (3) book stitching; and (4) metal stitching. Basically these machines are the same, varying in sizes from small bench models to large floor models. Each type machine draws wire from a coil, cuts it to proper length, forms it into a stitch, drives it through the material, and clinches it with every stroke of the machine.

Metal Stitching

Metal stitching is one of the newest methods of fastening thin-gage metals, and fastening metals to nonmetals. Metal stitching will fasten the work more economically since there is no need for other operations such as punches or drilling prior to fastening. Standard stitching machines can be

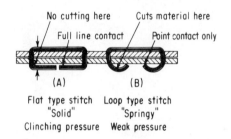

Fig. 6-38. Types of metal stitches.

tooled to form loop- or flat-type clinches. Flat-type clinches are used when the stitched joint must carry heavy loads, such as aircraft constructions where joints are subject to severe stresses. The flat clinch is formed by an upward movement of the clinching die, folding the legs flat against the bottom of the material. The loop-type clinches provide only point contacts with the material and, therefore, provide less strength than flat clinches. Loop clinching is formed by curling the legs of the wire with stationary solid dies. Figure 6-38 illustrates flat- and loop-type clinches.

Special Applications. Metal stitches are used extensively by manufacturers of electronic components. Variable-resistor control and switch shafts are assembled into mounting bushings by wire retaining or snap rings. The rings are made of round or flat spring-tempered steel wire. The specially tooled machine draws the wire from a coil, cuts it to proper length, forms it to a U shape, and drives it against a grooved die to form the round retaining ring into the circular groove of the shaft. Figure 6-39 illustrates the parts and clinching dies of C-ring retaining-ring assembly.

Fig. 6-39. C-ring retaining-ring assembly.

Staking

In staking, two or more parts are joined permanently by forcing the metal edge of one member to flow either inward or outward around the other parts. Staking is an economical method of fastening parts. The operation is completed with a single stroke of an arbor press, kick press, punch press, air, or hydraulic press. Figure 6-40 illustrates locking a ring to a shaft with center punch staking. Figure 6-41 illustrates joining together a bushing, a bracket, and a shell with a ring-staking punch. Figure 6-42 illustrates the principles of inward staking by forcing metal of a ring against the knurled portion of a shaft. To provide added rigidity and torque in assembly, spot staking may be used. Spot staking is essentially the same as ring staking except that three or more equally spaced chisel edges of the staking punch force metal into splined portions of the parts to be assembled. Figure 6-43 illustrates a typical splined design and spot-staking punch to force metal into this spline. In some cases it is desirable to use a combined spot- and ring-staking punch, as is shown in Fig. 6-44.

Chap. 6 Tool Design for the Joining Processes   403

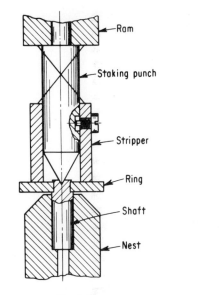

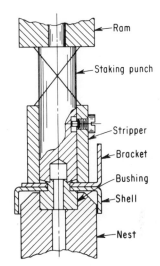

Fig. 6-40.   Fig. 6-41.

Fig. 6-40. Staking by center punch.
Fig. 6-41. Staking by ring punch.

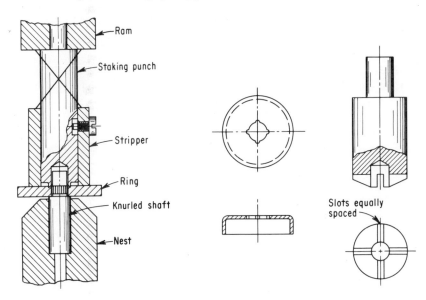

Fig. 6-42.   Fig. 6-43.

Fig. 6-42. Inward staking by ring punch.
Fig. 6-43. Spot-staking punch and workpiece with splined hole.

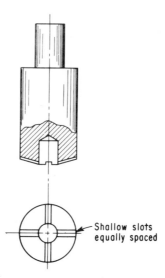

Fig. 6-44. Combined spot- and ring-staking punch.

To facilitate efficient metal flow, the tips of punches must be free of nicks and circular grooves, and, after hardening, these surfaces should be highly polished. Polishing the punch tips in the direction of metal flow is desirable instead of circular polishing.

PROBLEMS

1. What basic design principles are used in the physical joining process that are common to the mechanical joining processes?
2. What is the general purpose of a welding fixture?
3. Why is it important to provide for proper heat control in the weld zone?
4. What are some of the major design objectives for a welding fixture?
5. Before designing a welding fixture, what factors should be considered?
6. Why is it advantageous to tack weld copper and aluminum before finish welding out of the fixture?
7. When are parts that have been welded together stress-relieved?
8. What is the advantage of a gravity-type fixture as shown in Fig. 6-1?
9. What is a disadvantage of vertical gas welding?
10. What function does the holddown plate serve in a gas welding fixture?
11. What factors govern the selection of materials used in the design of a gas welding fixture?
12. What materials are commonly used in gas welding fixtures?
13. What are some of the design considerations for arc welding fixtures?
14. What is the design function of a backing bar?

Chap. 6 *Tool Design for the Joining Processes* 405

15. What is the position of the backing bars in relation to the weld line?
16. What determines the size of the backup bar?
17. What materials are used for backup bars?
18. What important design considerations should be incorporated in the resistance welding fixture?
19. What results if the clamps are not supported underneath the workpiece?
20. Would setup blocks be used for high-production multiple-part fixtures? Why?
21. What joining materials are used in brazing or soldering?
22. Why is it important to use inert gas for shielding?

7

TOOLING FOR CASTING

Casting may be defined as the production of a desired shape by introducing a molten material into a previously prepared cavity and allowing it to solidify. The metals most frequently cast are iron, steel, aluminum, magnesium, some copper alloys, and certain zinc alloys.

Although the craft is an ancient one, modern casting methods frequently involve precise science and highly skilled engineering. An example is the one-piece V8 engine block cast from aluminum or gray iron, with cylinder chambers and water jackets formed by the many intricate cores set in the mold and holding wall thicknesses within close dimensional accuracy.

Casting is the fastest method for producing complex shapes from raw materials that can be temporarily liquified by heat or chemical means. Sand casting accounts for some 95% by weight of all metal cast. The fluid material is poured or forced into a shaped cavity called a *mold* or *die*. After it has solidified, the casting is removed from the mold. In sand casting the mold is destroyed after one use, but molds for many materials are reusable.

Chap. 7 Tooling for Casting

The materials to be cast, the volume of production, and the requirements of the functional design all affect the selection of the mold material and design, and largely determine the other tooling required. For nonmetallic castings, the molds are usually made from gray iron or steel, cast, machined from blocks, or formed from sheet.

SAND CASTING

Sand molds are compacted by hand or machine depending on the tooling that can be economically justified. A simple bench mold requires green sand, a flask, molding and bottom boards, a pattern, and certain hand tools

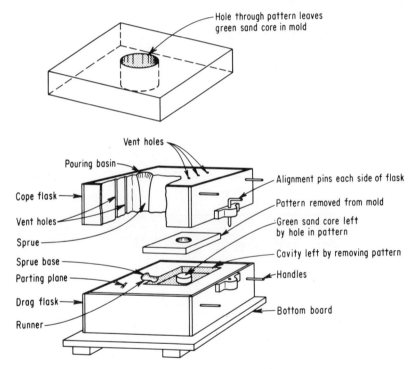

Fig. 7-1. A typical bench mold partially sectioned to show detail.

(Fig. 7-1). Machine molding requires a match plate which forms the parting plane and aligns the two halves of the pattern for making the cope and drag (Fig. 7-2). Additional mechanization would deliver correctly prepared sand to hoppers above the molding machine and carry away spilled sand which would pass through the grate floor onto moving conveyors below. *Sprues,* or pouring channels, are either hand cut with tubular sprue cutters or mounted directly on the plate. Mounted metal sprues require draft and thus give a sprue that is wider at the base than at the top. This is poor from a fluid flow standpoint and may be avoided by using a correctly designed rubber sprue which deforms on squeezing to form the correct sprue contour and relaxes for easy withdrawal.

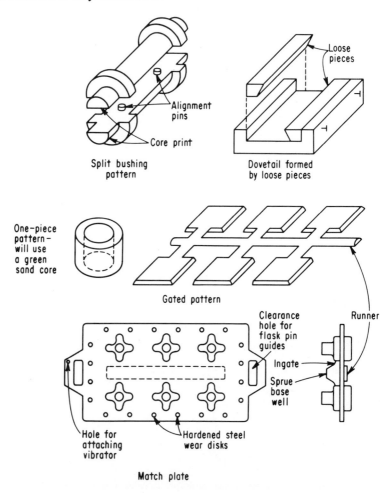

Fig. 7-2. Typical small pattern equipment.

Once the mold is rammed the pattern is rapped (moved slightly relative to the sand) or vibrated and drawn. If needed to form internal cavities, reduce weight, or form more uniform wall sections, a *core* is placed in aligning *core prints*. Long slender cores as for steam radiators require additional support from chaplets, which are spacers used to center the core in order to maintain uniform wall thickness. Chaplets must fuse with the metal to form a pressure-tight joint. Core tooling includes core boxes, core driers, and core assembly fixtures (Fig. 7-3).

Finally the cope and drag halves of the mold are aligned, assembled, and clamped or weighted. Pouring the mold includes filling the pouring basin with molten metal and maintaining it full while the metal flows down the sprue, along the runner, and through the ingates into the mold cavity.

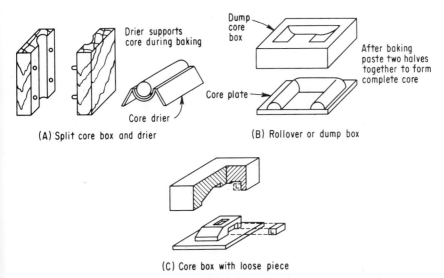

Fig. 7-3. Core boxes and accessories. (*By permission of McGraw-Hill Book Co., Inc.*[2])

Figure 7-4 shows schematically the major steps taken in the production of a sand casting of very simple configuration. The pattern, being a facsimile of the casting to be produced, is itself very simple. Castings of more complex design require somewhat different and more complex tooling and operations. They also demand higher skills in patternmaking, coremaking, and molding.

Usually castings weighing up to ten to twenty pounds are gated at the parting line. Larger floor and pit molds, especially if poured with steel or light metals, are bottom gated to minimize turbulence, but this creates unfavorable thermal balance within a casting that requires *risers*. In bottom gating the coldest metal fills the riser, thus reducing its ability to compensate for solidification shrinkage and to act as a reservoir of heat. Sound castings can result only if proper *directional solidification* is established from thin-

410 Tooling for Casting Chap. 7

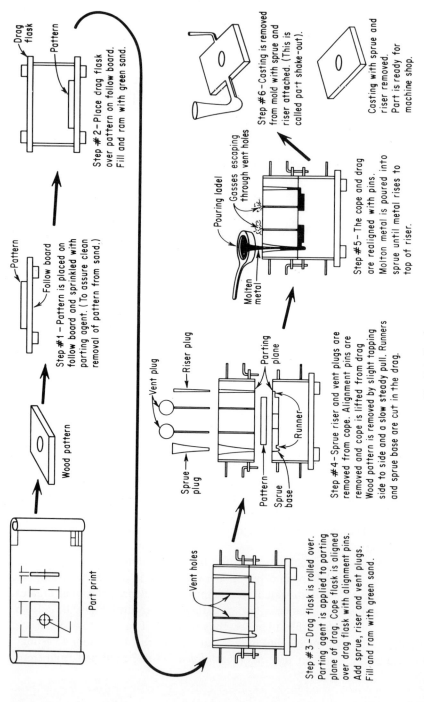

Fig. 7-4. Casting process.

ner, faster cooling sections to hotter sections leading to the risers (Fig. 7-5). Special casting techniques must be used to supplement bottom pouring:

1. Chills can promote directional solidification or reduce undesirable heat concentration.
2. Insulating riser sleeves help maintain riser heat.
3. Radiation shielding on top risers reduces radiant heat loss.
4. Exothermic riser compounds liberate much heat through chemical reaction.
5. Pouring the risers last with fresh metal places the hottest metal there.

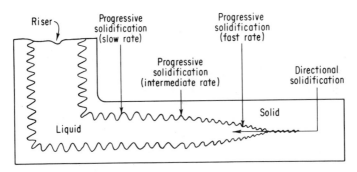

Fig. 7-5. Directional solidification within a casting.[3]

Patterns

A pattern (Fig. 7-2) forms a mold cavity such that after the cast metal reaches room temperature the product will be the expected size. Shrinkage allowances (Tables 7-1 and 7-2) are at best approximations. The size and shape of a casting and the mold wall's resistance to deformation by the liquid metal affect the shrinkage allowance significantly. For example, a round steel bar shrank $18/64$ in. per ft; with a large knob on each end, it shrank $12/64$ in. per ft; when the knobs were replaced by flanges, it shrank $7/64$ in. per ft.[1*] Thus if tolerances are important, a tool engineer must work closely with qualified foundrymen. Again a casting may shrink differently from the allowances in the tables, owing to its geometry. In a large bull gear the shrinkage across the diameter will be different from the shrinkage across the rim.

Stop-offs are portions of a pattern that form a cavity in a mold which is refilled with sand before pouring. This might be desirable to prevent breakage of a frail pattern member. The stop-off may be filled by a core later. This may simplify the location of the pattern parting line by making it

---

* Superior numbers indicate specific references at the end of this chapter.

unnecessary to carry it out of a plane position (to provide for lugs or certain cored holes).

Once a mold has been rammed around a pattern the latter must be removed from the mold cavity. This is aided by rapping or vibrating the pattern and by a taper called *draft* on all vertical surfaces of the pattern. Usually draft will vary between $1/16$ to $1/8$ in. per foot (1 to $1/2°$). Inside surfaces such as cored holes require greater draft than outside surfaces, and manual molding equipment requires greater draft than that required if mechanical drawing equipment is to be used.

Pattern Materials. Material for patterns must resist the moisture and abrasion of green sand. Which material is chosen depends on the quantity of castings to be made and the casting process. The most common pattern materials are wood and metals, but epoxy resins and plaster are used on occasion.

Softwood patterns of white pine are suitable for short runs (1–50 pieces) of medium castings (up to 6 ft long) but they wear rapidly; hardwoods (mahogany, cherry, birch) are used for runs of 50 to 200 pieces. At the latter production volume, patterns should be gated and flasks and core boxes should have metal wear strips. In some cases, additional equipment may be needed, such as core setting and pasting fixtures, drier patterns or core driers. At still higher production volumes (200 to 5000 pieces), patterns and accessories must be metal with hardened steel wear plates, and core boxes should be constructed with venting devices to permit blowing.

Wood is inexpensive and easily worked, and is therefore commonly used for low production and large patterns. But even when painted, wood patterns vary in size and shape with humidity changes, and they are easily damaged on the molding floor or on their way to and from storage. Metal patterns, at the same time more accurate and durable, are used for all match plates, aluminum being used for manual operations and gray iron or brass for machine-lifted plates.

Pattern Colors and Coatings. In 1958 a Tentative Pattern Color Code for New Patterns was adopted by the Pattern Division of the American Foundrymen's Society:

1. Unfinished casting surfaces, the face of core boxes, and pattern or core box parting faces: *clear* coating.
2. Machined surfaces: *red*.
3. Seats of and for loose pieces: *aluminum*.
4. Core prints: *black*. In the case of split patterns and where cores are used, paint the core area *black*.
5. Stop-offs: *green*.

Chap. 7 Tooling for Casting

TABLE 7-1. SHRINKAGE, MACHINING, ALLOWANCE, MINIMUM SECTION THICKNESS, AND TYPICAL TOLERANCES FOR SAND CAST FERROUS ALLOYS

(all dimensions in inches)

Pattern dimension	Shrinkage Solid	Shrinkage Cored	Machining bore	Allowance * (outside dimension)	Cope side	Minimum section side	Typical tolerance
			GRAY IRON				
Up to 6	1/8	1/8	1/8	3/32	3/16	1/8	±1/32
6–12	1/8	1/8	1/8	1/8	1/4		±1/16
13–24	1/8	1/8	3/16	5/32	1/4		
25–36	1/10	1/10	1/4	3/16	1/4		
37–48	1/10	1/12	5/16	1/4	5/16		
49–60	1/12	1/12	5/16	1/4	5/16		
61–80	1/12	1/12	3/8	5/16	3/8		
81–120	1/12	1/12	7/16	3/8	7/16		
			CAST STEEL				
Up to 1	..	..	Cast solid	..	..	3/16	..
Up to 6	1/4	1/4	1/4	1/8	1/4		1/4
6–12	1/4	1/4	1/4	3/16	1/4		1/4
13–18	1/4	1/4	9/32	1/4	5/16		1/4
18–24	1/4	3/16	9/32	5/16	3/8		5/16
25–48	3/16	3/16	5/16	3/8	1/2		3/8
49–66	3/16	5/32	3/8	3/8	1/2		1/2
67–72	3/16	1/8	1/2	7/16	9/16		5/8
Over 72	5/32	1/8	5/8	1/2	5/8		3/4
			DUCTILE IRON				
Up to 24	1/10–1/8	1/10	3/16	5/32	1/4		

			MALLEABLE IRON				
Section thickness	Shrinkage (no core)	Section thickness	Shrinkage (no core)	Minimum sec. size		Typical tolerance	
1/16	11/64	1/2	7/64	1/16		Up to	±1/32
1/8	5/32	5/8	3/32			5–8	±3/64
3/16	19/128	3/4	5/64			9–12	±1/16
1/4	9/64	7/8	3/64			13–24	±1/8
3/8	1/8	1	1/32				

* Allowance on the bore given for radius.

TABLE 7-2. SHRINKAGE AND MACHINING ALLOWANCES FOR SAND CAST NONFERROUS CASTINGS

(all dimensions in inches)

Metal	Pattern dimensions	Section thickness	Shrinkage allowances	Machining allowances	Minimum sec. size	Typical tolerance
Aluminum	Up to 24 25–48 49–72 Over 72	5/32 5/32 9/64 1/8	5/32 9/64–1/8 1/8–1/16 1/8–1/16	3/32 1/8 1/8	3/16	1/32
Magnesium	Up to 24 24–48 Over 48	11/32 11/32 5/32	5/32 5/32–1/8 5/32–1/8	3/32 1/8 1/8	5/32	1/32
Admiralty metal	Up to 24		1/8	1/4–3/8		
Brass			3/16		3/32	3/32
Bronze			1/8–1/4		3/32	3/32
Beryllium copper			1/8			
Everdur			1/4			
Hasteloy			1/4			
Nickel and its alloys			1/4			

The clear coating on most surfaces will disclose the quality of material and workmanship of the pattern. Likewise important construction details and layout or centerlines used by the patternmaker will not be obliterated and will then be available for use during repair or modification of the equipment.

There are numerous types of pattern coating materials on the market. Nitrocellulose lacquers cure rapidly but have relatively poor moisture resistance. Present-day modified lacquers have much better moisture resistance. Again, shellac modified with various synthetic resins is superior to pure shellac. Wood patterns coated with synthetic resin (plastic) have been used under the most adverse conditions, such as being rammed up with hot sand or left in the mold for many hours, without noticeable damage. Plastic coatings require considerable time and care for application but prolong the pattern life up to several times. In addition there is less tendency for the sand to cling to the pattern surface; such patterns draw more easily.

Design of Coremaking Equipment

Core boxes contain cavities to form sand cores to the desired shape. Patternmakers construct core boxes from wood, metal, synthetic resin, or other suitable materials. Generally, the core material would be the same as

Chap. 7 Tooling for Casting

that chosen for the pattern. There are several types of core boxes and accessories (Fig. 7-3).

   a. Two- or three-piece split core boxes (hand or machine rammed).
   b. Dump boxes for making a half core (later the two halves are pasted together).
   c. Blow boxes for the high production of relatively small cores (few ounces to 400 pounds).
   d. Multiple-piece, loose-piece, or special-purpose core boxes.
   e. Auxiliary equipment such as core driers and pasting jigs.

The design and construction of most core boxes and patterns are often left largely to the discretion of the patternmaker. But a tool engineer should understand the principles of core box design, vent location, vent areas, and the general break-even point at which it is better to move from wooden core boxes to metal core boxes or from hand ramming to core blowing.

Core Blowing. Small and medium-size cores are rapidly produced by the use of a core blower. It clamps the core box shut, seals the sand reservoir tube to the box, then fills and rams the core by the kinetic energy of a sand-laden air stream. Proper orientation of the blow and vent holes will promote uniform filling and ramming of the core in less than two seconds, even for the largest core boxes, which are usually of aluminum with added wear plates of steel at points of high attrition.

Blow Holes. A steel blow plate fits the bottom of the sand reservoir and provides holes from $3/16$ to $1/2$-in. diam. to direct the sand to the proper location within the core box (Fig. 7-6). The number and size of the blow holes are largely a matter of experimentation. Too few blow holes prevent the box from filling completely and promote channeling. On the other hand, too many holes cause the box to fill before the remote corners are rammed sufficiently hard.

The blow plates and sand reservoirs should be designed to be interchangeable for as many core boxes as possible.

Core Box Venting. All the air that enters the core box through the blow holes must be vented. As the air enters the box it expands, moves at a lower velocity, and deposits the entrained sand before it flows through the vents. The impact of one grain upon another rams the core.

Venting must be designed in proportion to the blow hole area to achieve adequate ramming. Less venting is required for more flowable sands because a smaller volume of ramming air is required to ram cores to suitable strength. Conversely, more vent area is required for core sands with high green properties because larger volumes of air are needed for equivalent rammed properties.

Venting is obtained by using vent screens or slotted vents (Fig. 7-6). If screens are used the venting area should be about twice the blow hole

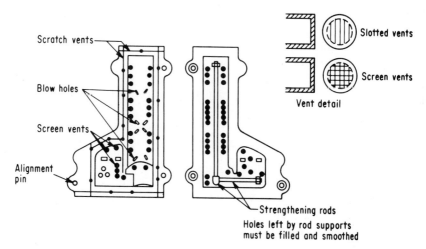

Fig. 7-6. Vents for core boxes. (*By permission of McGraw-Hill Book Co., Inc.*[4])

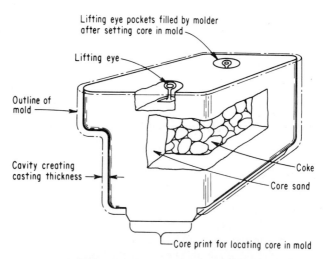

Fig. 7-7. Coking a core.

area; vent plugs require still greater area. Vent plugs do not leave surface imperfections on a core, as screens do, because the former can be contoured to the cavity. If surface finish is important, screens should be placed on core prints and at the other spots where blemishes are not critical, or else vent plugs must be used.

Chap. 7   Tooling for Casting

Core Venting. Cores must be permeable to permit the hot gases from the charring binders to escape to the atmosphere. This is usually accomplished by placing wire or rod in the core box prior to blowing the core. Subsequently the wire is removed. If possible the vents should pass through the locating prints to permit venting through the back of the core prints.

Where cores are very large, as in the case of press beds or the bases of heavy machines, coke coal is sometimes used on the inside of the core. This materially shortens drying time, and makes the core lighter. In some cases (Fig. 7-7), lifting eyes are inserted in the core for use in lowering the core into the mold.

Core Support. Larger cores require internal reinforcement with embedded wires or rods to prevent breakage while handling, or premature sagging in the mold after it is poured.

Cores with irregular contours require special core driers to support them during the curing process. In high production the drier may be the lower half of the core box itself. Such a practice would require a considerable investment in tooling, which would have to be balanced by cost saving.

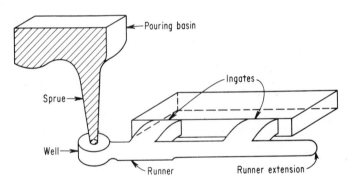

Fig. 7-8. Gating system terminology in sand casting.

Gating System Design

The gating system for a casting is a series of channels which lead molten metal into the mold cavity. It may include any or all of the following: pouring basin, sprue, sprue base, runners, and ingates (Fig. 7-8). A well-designed gating system should:

1. minimize turbulence within the molten metal as it flows through the gating system. The use of tapered sprues and proper streamlining will reduce excessive erosion and gas entrainment.

2. reduce the velocity of the molten metal in order to attain minimum turbulence.
3. deliver the molten metal at the best location to achieve proper directional solidification and optimum feeding of shrinkage cavities.
4. provide a built-in metering device to permit uniform, standardized pouring times regardless of variations in pouring techniques.

Turbulence Within a Gating System. Extensive research has shown not only that molten metal and water flow similarly, but also that gating systems can be designed using the principles of fluid mechanics.

There are several limitations. The high density of metals, up to ten times that of water, makes it difficult to force molten alloys to turn a corner as from a runner to an ingate. But, once Newton's law of inertia is applied to flowing metal, proper gating systems can be designed. The density of a metal does not affect its flow characteristics because the rate and nature of fluid flow depend on the inertia of the fluid and the forces applied to it. Both factors depend on density in the same way, so they have no effect on the fluid flow. But impact depends upon density alone; hence mold erosion increases directly with density.

Although water has a surface tension approximately one-tenth that of molten metals, recent experiments have shown that Wood's metal and mercury have nearly equal ability to entrain air. However, the greater surface tension and the natural oxide coating that envelops a stream of molten metal seem to permit it to flow in a laminar fashion at a greater velocity than that suitable for water in a given channel.

Fluids flow in either a laminar (streamlined) or a turbulent manner. In either case the degree of turbulence can be described quantitatively by the Reynolds number R_e, which is a ratio of the inertia forces within the fluid to the viscous forces acting on the fluid during flow.

The Reynolds number (Table 7-3) is sensitive to changes of viscosity and density. For example, water flowing in a 1-in. diam. pipe at a velocity of 1 in./sec has a Reynolds number of 360 at 32°F and 2180 at 212°F. But the real significance of the Reynolds number in metal casting is that it relates the size of the channel to the velocity of flow, the density of the molten metal, and its absolute viscosity at the pouring temperature.

In water, laminar flow occurs at Reynolds numbers under 2000 to 4000. The surface tension of liquid metals and their oxide coating seem to justify the assumption that detrimental turbulence in molten metals will occur only after much greater Reynolds numbers are reached. Eastman refers to a maximum $R_e = 20,000$ for aluminum alloys. Metals with low Reynolds numbers may always flow turbulently in a gating system of sufficient capacity to prevent laps or cold shuts. In practice, gating system design does not necessarily eliminate all turbulence but rather reduces it to a level that is sufficient for the required quality.

TABLE 7-3. RELATIVE REYNOLDS NUMBERS FOR CERTAIN METALS AND LIQUIDS

Liquid	Temp., °F	Relative Reynolds no. (Water = 1)
ZnCl (34%)	68	8800 *
Glycerine	68	690 *
Mineral oil	68	625 *
Water	68	1.00 *
Aluminum alloy (356)	1472	0.57
	1472	0.57
	1409	0.75
	1350	0.90
	1346	0.84
	1292	1.16
	1290	1.05
	1328	2.62
Magnesium alloy	1634	0.80
	1256	0.80
Wood's metal	212	0.13
Mercury	59	0.11
Cast iron (3.27% C)	2372	0.45 *
Copper	2192	0.40 *
Iron	2920	0.89
Steel (mild)	near m.p.	0.44 *
Steel (0.75% C)	2732	0.10 *

* See Reference 5.

Velocities Within a Gating System. The flow of molten metal in a gating system is a function of a number of other variables which can best be related by the terms of Bernoulli's theorem. This theorem states that the sum of the potential, pressure, kinetic, and friction energies at any point in a flowing liquid is a constant, or

$$Wh_1 + \frac{WP_1}{\gamma} + \frac{WV_1^2}{2g} + WF_1 = Wh_2 + \frac{WP_2}{\gamma} + \frac{WV_2^2}{2g} + WF_2 \quad (7\text{-}1)$$

Dividing by W one obtains the usual form:

$$h_1 + \frac{P_1}{\gamma} + \frac{V_1^2}{2g} + F_1 = h_2 + \frac{P_2}{\gamma} + \frac{V_2^2}{2g} + F_2 \quad (7\text{-}2)$$

where h_1, h_2 = the respective heads at stations 1 and 2 (in.)
P_1, P_2 = the respective pressures on the liquid (lb/in.^{2})
V_1, V_2 = the respective liquid velocities (in./sec)
γ = the specific weight of the liquid (lb/in.^{3})
g = the acceleration of gravity (386 in./sec ^{2})
F_1, F_2 = the respective head losses from friction (in.)

The velocity of the molten metal at any point in a gating system (Fig. 7-9) and the friction losses that occur at junctions between gating system elements (Table 7-4) can be evaluated by the use of Bernoulli's theorem. Proper streamlining will permit a significant increase in the flow rate of a gating system.

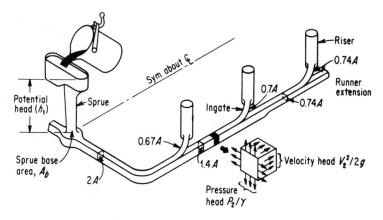

Fig. 7-9. Design for a non-pressurized gating system with a gating ratio of 1:4:4.[6]

TABLE 7-4. VALUES OF LOSS AND FRICTION COEFFICIENT [6]

	Sharp	Streamlined (rounded)
Sprue entry from pouring basin:		
Round or square tapered sprue	0.75	0.12–0.14
Round untapered sprue		0.47
Bend of sprue to runner	2.0	1.0
Right angle bend in runner:		
Square cross section	2.0	1.5
Round cross section	1.5	1.0
Junction at right angles to runner	4.0–6.0	..
Junction with 25% or more area reduction from runner to gate	2.0	0.5
Runner choke when choke area is approximately ⅓ runner area, plus bend of sprue into runner	13	..

Losses from wall friction: *
  Round channel loss   = $0.02 L/D$
  Square channel loss   = $0.06 L/D$
  Rectangular channel loss = $0.07 L(AB)2AB$

* L = length, D = diam of round or side of square, A = one side of rectangle, B = other side of rectangle.

The Vertical Elements of a Gating System. The law of continuity in fluid flow requires that the same quantity (flow rate) of material must exist at all points in a flowing stream. In the vertical part of a gating system (sprue) the acceleration of gravity increases the velocity of flow.

If a straight-sided sprue is used, the cross-sectional area of the flowing stream at the sprue base will be less than that of the sprue. Consequently, air will be aspirated from the surrounding mold until the sprue volume is completely filled. However, if a tapered sprue is designed to conform to the dimensions of the descending stream, such a condition can no longer exist and the metal quality will improve.

The Pouring Basin. Most foundrymen look with disfavor upon pouring basins because they represent increased molding and pouring costs. However if maximum effectiveness of the gating system is desired, or if higher

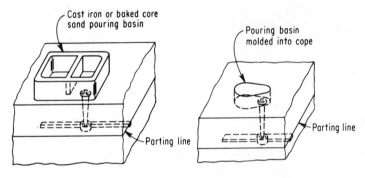

Fig. 7-10. Designs for pouring basins.[7]

quality is a necessity, then a pouring basin is required. It may be molded into the cope (Fig. 7-10), or made from cast iron for nonferrous alloys if chilling is not a problem, with sharp lower edges which may be pressed into the top of the cope (Fig. 7-10), or it may be molded in core sand. Regardless of type, the entrance to the sprue top should have a 1-in. radius and the sprue in the basin should align correctly with the cope sprue. The pouring basin depth is a function of the sprue entrance diameter. Experience shows that a sprue entrance diameter of 1 in. requires at least 2 ½-in. liquid metal depth to prevent the formation of a vortex.[8]

The Sprue Base Well. Most investigators agree that there should be a well at the base of the sprue. This serves a dual function. A molten metal pool is an excellent device for preventing excessive sand erosion where the molten metal impinges on the runner at the sprue base. Also there is considerable velocity loss in the well. Kura [8] recommends the following empirical well dimensions: for a narrow deep runner, the diameter should be 2½ times the width of the runner in a two-runner system (Fig. 7-11), and twice its width in a one-runner system. For square, round, or wide shallow runners, the well should have a cross sectional area five times the area of the bottom of the sprue, and a depth twice that of the runner.

Ratio Gating in the Horizontal Gating System. A major decision must now be made. Is the gating system to be free flowing (nonpressurized) with a gate ratio 1:2:4 or 1:4:4 (as in Fig. 7-9), or choked (pressurized) with a gate ratio 2:4:1 or 2:2:1? Ratio gating considers the relative cross-sectional area of each component of the gating system as a part of an interrelated unit. The gating ratio considers the sprue base area as unity, followed by the total runner area and the total ingate area. Thus, a nonpressurized gating system having a sprue base of 1, a runner area of 2, and a total ingate area of 4, would be expressed 1:2:4. A pressurized gating system maintains the back pressure throughout with a gate area equal to or less than the sprue base area, e.g., 1:0.75:0.5 or 1:2:1.

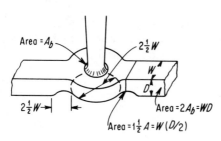

Fig. 7-11. Design for sprue base well.[7]

Foundries casting drossy alloys such as light metals and manganese bronze have traditionally used the sprue base as the metering orifice and they have designed the balance of the gating system to permit unpressurized flow in order to reduce velocity and turbulence. In contrast, the ferrous and heavy metal founders have used pressurized systems in which the ingate has been the metering orifice.

Unpressurized gating systems reduce velocity, turbulence, and aspiration but must have tapered sprues, sprue base wells, and pouring basins. In addition they deliver metal uniformly to multiple ingates only if the runner is reduced after each ingate. In contrast, a pressurized gating system delivers metal to the mold in a more turbulent manner but usually flows full and thereby can minimize aspiration, even if a straight sprue is used, once flow has been established. Pressurized gating systems permit somewhat higher yield.

Briggs [9] states that a gating system for high-quality steel castings should be nonpressurized. A 1:4:4 ratio might favor the formation of oxidation defects but a ratio of 1:2:2 or 1:2:1.5 will produce castings nearly free from erosion, will minimize oxidation, and will produce uniform flow. Hans Heine [10] states that a typical gating system for malleable iron using a four-hole strainer and tapered sprue is 1:2:9.5, which is obviously nonpressurized. Wallace and Evans [6] also recommended nonpressurized gating systems for gray iron.

Pouring Time as a Basis for Gating System Design. The optimum pouring time for a casting depends on several factors including fluidity, the thermal gradients, the casting weight, and its thickness. Thin-walled castings must be poured rapidly to avoid laps, whereas thick-walled castings require slower pouring to establish proper temperature gradients to permit adequate feeding. A higher pouring temperature has a considerably greater effect on fluidity than does the alloy composition.

Riser Design. Risers serve a dual function. As reservoirs of liquid metal, they compensate for liquid and solidification contraction or any mold wall movement that may occur (Fig. 7-12). Patternmaker's shrinkage (Tables 7-1 and 7-2) compensates for solid contraction. Total shrinkage varies from negative to zero for certain bismuth and gallium alloys to 12% or more for certain steel and nonferrous alloys. Solidification contraction for various foundry alloys is shown in Table 7-5.

Risers are also reservoirs of heat which aid in establishing proper temperature gradients so that directional solidification can take place (Fig. 7-5). The riser and the gating system are closely interrelated; in fact, in some instances, the riser is a part of the gating system. The riser size for a given application depends primarily on the metal poured, but numerous other factors are also important. For example, gray iron requires little or no risering to compensate for solidification contraction because the precipitation of flake graphite causes compensating expansion. Steel, white iron, and many nonferrous alloys, which have long freezing ranges, may produce sound castings only with the best application of foundry engineering and experience.

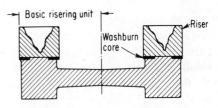

Fig. 7-12. Basic risering unit for feeding castings.[10]

TABLE 7-5. SOLIDIFICATION CONTRACTION FOR SOME CAST METALS

Metal	Contraction, per cent
Low-carbon steel	2.5–3
Carbon steel (1%)	4
White iron	4–5.5
Gray iron	Up to 2.5 expansion
Copper	4.9
70% Cu–30% Sn	5.5
90% Cu–10% Al	4
Aluminum	6.6
Al–4.5% Cu	6.3
Al–12% Si	3.8
Magnesium	4.2
Zinc	6.5

Because risers operate during the phase change from liquid to solid, it is essential to consider how a molten alloy solidifies. When the molten alloy first enters a mold cavity a thin shell of solid metal forms because heat flows rapidly to the mold wall. Dendritic crystals still surrounded by molten metal grow inward from the initially chilled shell. Thus, during solidification, the casting can be divided into three zones:

1. The solid shell of metal at the mold-casting interface.
2. The mushy zone of liquid metal surrounding the crystallized grains.
3. The interior liquid zone.

In a pure metal or alloy of eutectic composition the mushy zone is of little significance because of rapid crystal growth. However, in most cast alloys there is an extended mushy zone within which liquid and solid coexist. Therefore, risers must be able to feed such shrinkage over the entire solidification cycle.

Riser Location. Before the size of a riser can be determined its location must be specified. Any casting, no matter how complex, can be reduced to a series of geometrical shapes consisting of two heavier sections joined by a thinner section. Each heavier section will require a riser. If the connecting thinner sections are not tapered toward a heavier section, centerline shrinkage is likely to occur. The proper use and location of *padding* (the addition of extra metal to create an artificial taper) or chills will eliminate such shrinkage by promoting proper directional solidification toward the riser. Both devices are frequently used in steel and nonferrous castings to increase casting *yield* (the ratio of casting weight to the total weight of metal poured in the mold).

To be effective, a riser must freeze more slowly than the casting that it feeds, and it must supply sufficient metal to compensate contraction from the liquid to the solid state. But some castings will require more than one riser because progressive solidification proceeds with dendritic growth which eventually cuts off sections of the casting from feed metal. In fact, a casting of nonuniform shape such as a wedge block is more easily fed than is a block of uniform cross section.

Any casting, no matter how complex, can be reduced to a series of simple risering systems. The major problem is to reduce a casting in the blueprint stage to the risering system that will give the highest yield and a casting of the required soundness. Improper feeding will result in large internal voids in steel, a depression at the cope surface in iron, or dispersed shrinkage throughout the casting in many nonferrous alloys.

Riser Feeding Distance. For a steel casting it has been shown that an adequate riser will provide complete soundness up to a distance L, equal to 4.5 times the minimum thickness t of the casting.[11] A metal chill of a minimum section equal to the cross section of the casting will, when placed at the end of the casting, merely add 2 in. to the length of the casting that can be fed by a proper riser. However, the same chill placed between two risers will permit them to be 4 in. further apart without impairing soundness. For example, a large annular casting with a mean circumference of 156 in. and a cross section of 4 in. × 12 in. required six risers 10 in. in diam by 7 in. high. When every second riser was removed and replaced by a chill of a

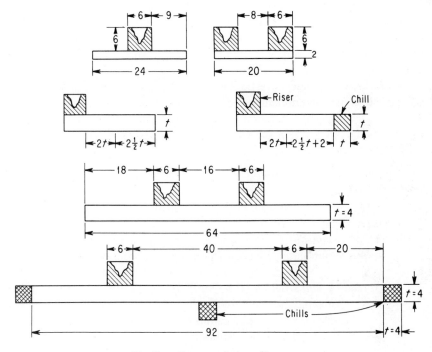

Fig. 7-13. Feeding distance of risers.[11]

section equal to the cross section of the casting, the remaining three risers were adequate when increased to 11 in. diam by 7 in. high. It has also been shown that a riser in the center of a casting has a much greater feeding range than at one end of a casting (Fig. 7-13).

In practical castings ranging from ½-in. plates to castings of 65 tons, the solidification time t was shown to be approximately proportional to the square of the casting volume V, divided by its surface area $S.A.$[12] Thus, as long as the solidification time for the riser is greater than that for the casting, the riser should feed the casting.

SHELL MOLD CASTING

This process derives its name from the use of thin shell-like molds composed of relatively fine sand and a thermosetting plastic binder. The molds are made by dropping the sand-resin mixture on a hot metal pattern. The resin melts and bonds the sand grains together for a depth determined by pattern temperature and the contact or investment time. The removal, by inverting, of the excess sand mix from the sand mix leaves an adhering thin and soft layer which is subsequently cured or hardened in a furnace. After the curing, the shell half is stripped from the pattern. The other shell half is produced in the same way. Any necessary cores are inserted, and the two shell halves are joined together under pressure to form an integral mold.

Patterns, cores, and core boxes for shell molding, in addition to producing the shapes and dimensions of the molds, cores, and castings, perform a precise heat-transfer function. Requirements are reflected in precision tooling such as the pattern shown in Fig. 7-14. Design, therefore, must take into account these factors:

1. *Expansion in the tooling* as it heats to operating temperature, 450–500°F at each investment cycle.
2. *Resin shrinkage* during curing, which may be severe at unconfined dimensions.
3. *Mold expansion* during pouring. This must be controlled by the mold layout, gating, speed of pouring, timing, and volume of metal cost.
4. *Spreading and swelling of the mold,* owing to pressure of the molten metal. This may be countered by increasing mold thickness, increasing the resin content, hard ramming, or a combination of these.
5. *Metal shrinkage* as in conventional sand casting, but complicated by mold spreading and swelling.
6. *Mold collapse,* starting when the mold temperature reaches 730°F approximately. The mold is destroyed in the process. The time overlap between metal cooling and the last remaining mold strength must be such that the metal has enough shape to form into a casting having the planned dimensions. The mold backup and the heat dissipation are important.
7. *Warping and distortion,* not all due to difference in section thicknesses, must be countered by correcting the pattern dimensions after trial runs.
8. *Gating and risering.* The principles of gating and risering previously discussed can be applied to shell mold pattern plates as readily as to match plates or sand molds. Frequently shell molds are designed for bottom pouring and are parted in a vertical plane. As in perma-

nent molding, for some workpiece geometries a horizontal parting is more economical because an increased yield is possible.

The riser size, position and connection are designed using a combination of experience and calculation. The riser must have a greater volume-to-surface area ratio than the casting. The connection must be large enough and close enough to the casting so it will not freeze off prematurely.

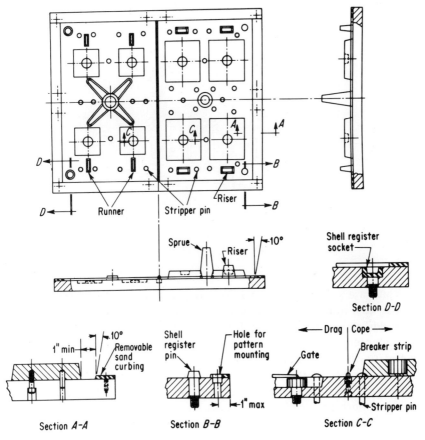

Fig. 7-14. Typical shell-mold pattern.[13]

METAL MOLD CASTING

Casting in metal molds is done in four different ways: (1) permanent mold casting, (2) centrifugal casting, (3) semipermanent mold casting, and (4) die casting. Only the first and the last methods are considered here. Tooling for permanent mold utilizes gravity feeding; tooling for die casting employs pressurized injection.

Tooling for Permanent Mold Casting

Permanent mold castings are produced by pouring molten metal, under pressure of a gravity head or by high-pressure feeding, into a static mold consisting of two (or more) pieces, clamped together. Cores, if used, are also made of metal.

Permanent molds (Fig. 7-15) are made in two or more pieces which, when clamped together, define the shape of the part to be cast, including

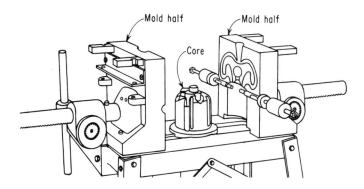

Fig. 7-15. Two-part permanent mold mounted in molding machine, with center core designed to pull down for freeing casting for removal.

gates and risers. Molds, stationary or movable, are usually made of a dense grade of cast iron, as are also most large cores.

Molds should be designed 1 to 2 in. thick. Parting lines, gates, and vents are preferably so located that the molten metal can enter at the bottom by gravity without turbulence or creation of hard spots. To assure progressive solidification of the molten metal from the farthest end of the casting to the point of entry, proper design is required for the gating, risering, external contours, and sectional thickness.

Dimensional standards and tolerances for permanent molding (Table 7-6) must also be included in the production design of the casting.

Thermal Equilibrium. Successful operation of metal molds requires a thermal cycle in which the heat introduced into the mold is virtually equal to the heat that it dissipates. The rate at which the mold temperature rises and the value finally attained are functions of the operating cycle if other conditions are equal. Assuming no external heating or cooling, the heat input to the die per unit time is equal to the heat content of the injected metal plus the heat given up by the solidified alloy as it cools to the ejection temperature and the latent heat of the alloy.

TABLE 7-6. TYPICAL DIMENSIONAL STANDARDS FOR PERMANENT MOLDS AND CORES

(see Fig. 7-16 for identification of dimensional elements)

A. Casting wall thickness t (average):

Casting size, in.	Under 3	3 to 6	Over 26
Wall thickness, in. (min.)	8/64	10/64	12/64

B. Fillet radii:
  Inner radius $B_i = t$ (average wall thickness)
  Outer radius $B_o = 3t$ (blend tangent to walls)

C. Cores (must exceed ¼-in. diam d):
  Unsupported length of cantilever:
   C_a: $10d$ max.
   C_b: $8d$ preferable
   C_c: $4d$ if $d < 5/8$ in.

D. Draft (cores and cavity):

	Da, short cores	Db, other cores	Dc, outside surfaces	Dd, recesses
Minimum	½°	1°	1°	2°
Preferable	2°	3°	3°	5°

E. Typical tolerances, in.:

	Basic tolerance, up to 1 in.	Additional tolerance, over 1 in.
E_a, basic tolerance in one mold section	±1/64	±0.001 in./in.
E_b, between points produced by core or slide and mold	±1/64	±0.002 in./in.
E_c, dimensions across parting lines	±1/64	±0.002 in./in.

F. Machining allowance, in.:

	< 10 in. long	> 10 in. long	Sand cored
Minimum	2/64	3/64	4/64
Preferable	3/64	4/64	5/64

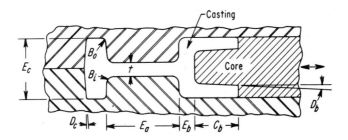

Fig. 7-16. (Lettered designations correspond with those in Table 7-6.)

A necessary feature of the periodic thermal cycle is orderly, progressive cooling from remote sections of the mold toward the ingate(s) or riser. This means that bulky sections of the cavity must be artificially cooled, especially if isolated from the gates by thin sections. As an alternative, or in addition, thin sections or even the riser can be insulated, enclosed in heavier mold walls, or even heated by an external gas jet.

Variations in wall thickness can aid greatly in establishing a uniform heat flow; proper location of the riser and gates near the thinner sections can also help. Heavy casting sections require thin mold walls, or perhaps fins with or without an air blast, or even water jackets. The latter are required for aluminum castings over $\frac{1}{2}$ in. thick. Local chilling of ribs or bosses can be accomplished with steel or copper chills permanently fixed in the mold wall and contoured to the cavity. Such chills may be finned on the outer end. Many of these are corrective measures which are taken after test pourings have disclosed trouble spots.

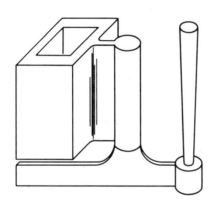

Fig. 7-17. A typical permanent mold gating system.[14]

Gating System Design. Using a gating ratio of 1:2:2 or 1:2:1.5 with a pouring basin, conical sprue, sprue base, and a sprue base well proportioned as in sand casting will give faster flow with least heat loss and turbulence. Most molds are bottom poured. The metal flows from the sprue base well along the runner bottom, feeds the riser, and passes through a slot gate into the casting cavity (Fig. 7-17). Note that a horizontal extension of the base runner receives the initial dross.

Risers. For proper feeding the casting must have directional solidification from thin remote sections toward the riser. The riser in turn is fed last with hot metal and has a surface area-to-volume ratio such that it will freeze more slowly than the casting.

Vents. Vents must discharge the gas in the gating system and mold cavity as fast as the metal enters the mold, but natural venting along sliding members and at the parting line is usually inadequate. Additional venting may be obtained by:

1. cutting slots as deep as 0.010 in. and of suitable width across the parting seal surface.
2. drilling small clusters of holes 0.008 to 0.010-in. diam in the mold wall at a point where venting is needed.

3. drilling one or more ¼-in. diam holes into the area requiring additional venting. Drive into them square pins ¼ in. across the corners (pin vent).
4. drilling holes and installing slotted plugs (plug vent).

Mold Material. The mold material is chosen on the basis of three criteria: material cost, the expected number of pours required, and the casting alloy. Most common and suitable for a permanent mold is high-quality pearlitic gray iron, frequently referred to as Meehanite, inoculated at the ladle to achieve uniform grain size and highly dispersed fine graphite.

DIE CASTING

Design of Die Casting Dies. It has been well established that in a simple die casting die the injected stream travels across the cavity and impinges on the opposite wall. It then spreads out along the cavity walls and finally backfills the entire cavity (Fig. 7-18). In complex dies or those with cores, the flowing mass is more turbulent. This concept of cavity filling is basic to the proper design of these dies.

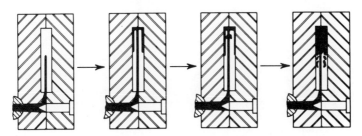

Fig. 7-18. Frommer's concept of die filling.[15]

Gate Location and Size. For a given casting there are many workable gating designs, each more or less approaching the optimum for the prevailing conditions. In general, a gate is usually placed on an edge from which it can be cleanly sheared, or where its trimming will not impair the casting surface. But gate location also affects the character of the metal flow, hence porosity and surface finish.

One authority [16] advocates a gate at the smallest cross section of the die cavity so that the surface of the casting increases with increasing distance from the gate. This technique creates a better heat balance within the die; the hotter metal contacts the die surfaces of smaller area and transfers less heat to the die. As the metal flows farther, it cools and transfers less heat per unit area to the die. Thus, the total heat input to the die block is more uniform throughout the cavity.

The gate area A_g can be determined from the continuity equation

$$Q = \frac{W}{\gamma} = A_g V_g t$$

where Q = casting volume, in.³
 t = cavity filling time, sec
 W = weight of injected metal, lb
 γ = specific weight of injected metal, lb/in.³
 A_g = gate area, in.²
 V_g = velocity at gate, in./sec

In practice, the ingate is made slightly smaller than calculated and recut to a larger dimension if required after a trial run-in. A single large gate is usually preferable to multiple gates because the latter cause increased turbulence when the two streams converge. The gate is usually wide, relatively thin, and so directed that the metal travels across the longer dimension of the die cavity if cores or other features of the geometry do not dictate another arrangement. Die casting gating terminology is defined in (Fig. 7-19). The design of a typical zinc casting is shown in Fig. 7-20.

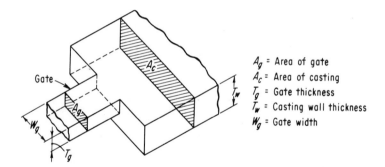

Fig. 7-19. Gating terminology in die casting.

Vents. Porosity is most likely to occur where vents are too small or so poorly located that they become blocked by the inflowing metal. To prevent this, some vents should be located as close to the gate as possible to allow air to escape from that part of the cavity which fills last.

Tests have shown that venting has an important bearing on the injection speed and casting quality.[17] They indicated that the speed of the entering jet is seldom equal to that of a free jet. When the vents were sealed, the jet velocity was reduced by 50%, and increased injection pressure will not increase filling time unless adequate venting is provided.

Vents should not exceed about 0.008 in. thickness if fins are to be avoided, but commercial practice uses vents up to 0.016 in. deep. Ejection pins, slides, insert clearances, and the parting line all add to the total cavity venting area. Some designers use a vent area of at least 50% of the ingate area.

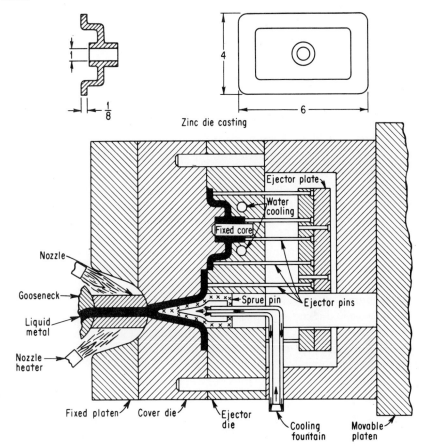

Fig. 7-20. Typical zinc die casting and its die.

Runner Design. Runner designs and locations can be varied to meet the needs of conservation or dissipation of heat to maintain thermal equilibrium and appropriate die temperature gradients. Shapes of runner cross sections vary from round to trapezoidal to elliptical; from wide and flat to narrow and thick. If the heat transfer rate of the runner system is high (high surface area-to-volume ratio), the metal is cooled somewhat by the time it reaches the gate. This is an advantage for a casting of thick cross section but is poor practice for a thin-walled casting because the metal will lose more heat traveling within the cavity, and incomplete fusion and deep striations are likely. In such case, the heat transfer between the runner and cavity can be altered by using a runner more nearly circular in cross section.

In practice most die casting runners are trapezoidal, about two times as wide as they are deep. As they approach a round or square cross section, internal porosity is more likely to appear.

Sprue Pin Design. In edge-gated cavities, which are typical of cold chamber machines, the sprue is merely a small appendage to the main runner and has little influence on metal flow or thermal equilibrium within the die. In center-gated dies for zinc, however, the sprue may have considerable influence on the thermal balance within a die block. Heat transfer from the sprue depends upon the contact area between the sprue pin and the cast metal. A slender sprue pin presents a small heat absorption mass and consequently, because of friction, the sprue pin tip may reach temperatures equal to or greater than the injection temperature of the molten alloy. This impedes solidifications at the gate. Therefore, it is better to use a well-splayed sprue of large diameter in conjunction with a blunt conical sprue pin. Thus, an adequate cross section can be maintained with a relatively thin-walled sprue. This design allows the sprue to be water cooled, so a better die-block temperature gradient can be maintained because a large proportion of heat dissipated from the injected metal is absorbed by the sprue pin and less passes to the adjacent die cavity.

Minimum Section Thickness. Good casting design dictates the use of as uniform a wall thickness as possible, or one that tapers slightly from the thinnest section remote from the gate to the heaviest section at the gate. In addition, alloy fluidity dictates the minimum practical wall thickness for each die-castable metal (Table 7-7). Walls must be thick enough to permit proper filling but sufficiently thin for rapid chilling to obtain maximum physical properties.

TABLE 7-7. MINIMUM WALL THICKNESS FOR DIE CASTINGS [18]

Surface area, in.^{2}	Wall Thickness, in.		
	Tin, lead, zinc	Aluminum, magnesium	Copper
Up to 3.9	0.0236–0.0394	0.0315–0.0471	0.0589–0.0787
4.0–15.5	0.0395–0.0589	0.0472–0.0707	0.0788–0.0982
15.6–77.5	0.0590–0.0787	0.0708–0.0982	0.0983–0.118
Over 77.6	0.0788–0.0982	0.0983–0.118	0.119–0.157

Undercuts and Inserts. Wherever possible a part should be redesigned to eliminate undercuts. In most cases such part modification is more economical than the labor required to handle a loose piece each casting cycle.

Inserts such as bearings, wear plates, bushings, and shafts can be incorporated in die castings, but they must be easily and precisely located in the die. If they are not securely placed, they may slip between the dies during the casting cycle and cause great damage. Inserts must be provided with knurled, crimped, or grooved surfaces so that a good mechanical bond is assured between the insert and casting. If the insert is large relative to the casting better results will be obtained if the insert is preheated (e.g., cylinder liners for automotive engines).

Cored Holes. Holes can be readily cast in all alloys according to the data in Table 7-8.

TABLE 7-8. DEPTHS OF CORED HOLES

Alloy	Min. diam castable, in.	Hole Diameter, in.								
		1/8	5/32	3/16	1/4	3/8	1/2	5/8	3/4	1
Zinc	0.039	3/8	9/16	3/4	1	1½	2	3⅛	4½	6
Aluminum	0.098	5/16	1/2	5/8	1	1½	2	3⅛	4½	6
Magnesium	0.078	5/16	1/2	5/8	1	1½	2	3⅛	4½	6
Copper-base	0.118				1/2	1	1¼	2	3½	5

Data reprinted with the permission of American Die Castings Institute.[19]

Draft Requirements. The amount and location of draft in a die cast part depends upon how it is located in the die and whether it is an external surface or cored hole. Draft on the die surfaces normal to the parting line permits the parts to be ejected without galling or excessive wear on the die impression.

Cores, Slides, and Pins. These moving parts must provide abrasion resistance in addition to heat resistance. Wear can be reduced by:

1. using contacting materials of differing hardness.
2. nitriding one or both surfaces in contact.
3. using a lubricant on the areas of contact (avoid contamination of the molten metal).
4. establishing and maintaining proper clearance between mating parts.
5. polishing the wearing surfaces of the mating parts.

Die Materials. Die casting die materials must be resistant to thermal shock, softening, and erosion at elevated temperatures. Of lesser importance are heat treatability, machinability, weldability, and resistance to heat checking. The performance of die materials is directly related to the injection temperature of the molten alloy, the thermal gradients within the die, and the production cycle. Tool-steels of increasing alloy content are required as the injection temperature of the molten alloy and the thermal gradients within the die increase, and the production cycle becomes shorter. Dies for use with zinc can be prehardened by the manufacturer in a range of R_c 29–34. The higher-melting alloys require hot work tool-steels.

PROBLEMS

1. What is the function of a match plate?
2. What is the disadvantage of bottom gating in sand casting?
3. What pattern material is recommended for the production of 1000 sand molds?
4. What two factors must be considered in the design of core box vents?
5. What detrimental affect does turbulence have in a sand mold gating system?
6. Does Newton's law of inertia apply to the design of a gating system?
7. How does the Reynolds number apply to the design of a gating system?
8. Is the pouring basin design affected by the sprue design?
9. A grey iron part is to be cast in a sand mold employing a single runner having a 2 in. cross sectional area. Determine the diameter of the well at the bottom of the sprue.
10. In the problem above, determine the ingate size and the sprue base size.
11. What factors affect pouring time?

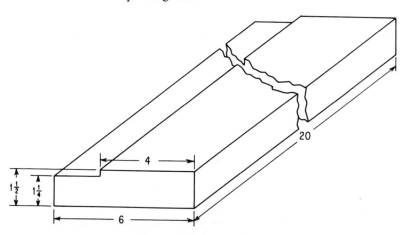

Fig. 7-21. Workpiece for Problem 12.

12. The steel part shown in Fig. 7-21 is to be cast. Determine the number of risers necessary to properly feed the casting. Assume that the risers will feed a distance equal to three times the minimum casting thickness.
13. What is the difference between permanent molding and die casting?
14. Figure 7-22 shows a part which is to be produced by some casting method. It has been determined that permanent mold casting would

provide a satisfactory surface finish and is economically feasible. Does the product design allow for the use of this casting process?
15. What are the three basic criteria in material selection for permanent molds?
16. What is the maximum depth of vents used in commercial practice?
17. Can threads be incorporated in die castings through the use of inserts?

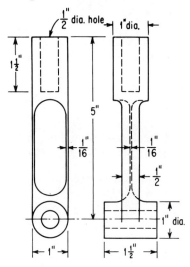

Fig. 7-22. Workpiece for Problem 14.

REFERENCES

1. *Steel Casting Handbook,* Steel Founder's Society of America, 1960.

2. Ebey, D. C., and Winter, W. P., *Introduction to Foundry Technology,* McGraw-Hill Book Co., Inc., New York, 1958.

3. Myskowski, E. T., Bishop, H. F., and Pellini, W. S., "Application of Chills to Increasing the Feeding Range of Risers," *Trans. A.F.S.,* **60** (1952).

4. Campbell, J. S., *Principles of Manufacturing Materials and Processes,* McGraw-Hill Book Co., Inc., New York, 1961.

5. Ruddle, R. W., "Fluid Dynamics of the Flow of Steel," *The Flow of Steel in Molds,* Steel Founder's Society of America, 1958.

6. Wallace, J. F., and Evans, E. B., "Principles of Gating," *Foundry,* **87**:10 (1959).

7. Eastwood, L. W., "Tentative Design of Horizontal Gating Systems for Light Alloys," *Symposium on Principles of Gating,* Chicago, 1951.

8. Kura, J. G., "Toward a Science of Gating," *Battelle Technical Review,* **9**:12 (December, 1960).

9. Briggs, C. W., "Influences of Gating on Steel Casting Quality," *The Flow of Steel in Molds,* Steel Founder's Society of America, 1958.

10. Caine, J. B., "A Theoretical Approach to the Problems of Dimensioning Risers," *Trans. A.F.S.,* **56** (1948).

11. Bishop, H. F., and Pellini, W. S., "The Contribution of Riser and Chill Edge Effects to Soundness of Cast Steel Plates," *Trans. A.F.S.,* **58** (1950).

12. Chvorinov, N., "Theory of the Solidification of Castings," *Giesserei,* **27** (1940).

13. Winter, O. W., "How Precision Toolmaking Affects Shell-molding Success," *American Machinist,* October 21, 1957.

14. *Casting Kaiser Aluminum,* Kaiser Aluminum and Chemical Sales, Inc., 1956.

15. Frommer, L., *Handbuch der Spritzgusstechnik,* J. Springer, Berlin, 1933.

16. Richter, F., "On the Importance of Compression Pressure in High-pressure Casting," *Giesserei,* August 30, 1956.

17. Koester, W., and Goehring, K., "The Study of the Flowing and Filling Process in Die Casting by Means of Motion Pictures," *Giesserei,* **28**:26 (1941).

18. Lieby, G., *Design of Die Castings,* American Foundrymen's Society, 1957.

19. "ACD1-E1, E2, Ed-55T, Linear Dimensional Tolerances," Product Standards for Die Casting, American Die Casting Institute, 1955.

8

GENERAL CONSIDERATIONS IN TOOL DESIGN

This chapter covers related details that are for the most part fundamentally inseparable from prior discussions on tool design. Every tool that is designed must perform a specific function; it must meet certain minimum precision requirements; its cost must be a feasible minimum; it must be available when the production schedule calls for it; and it must meet various auxiliary requirements such as for safety, adaptability to the machine on which it is to be used, and acceptable tool life. The objective of tool design is to help deliver production of the quantity and quality specified, at the lowest cost, and when needed.

The design fundamentals previously presented can be employed to create many alternative designs, for which many different materials are available, and many choices, such as in specifying dimensions, degree of surface smoothness, and safety precautions. Out of this complex of many and conflicting factors, decisions must be made.

Below is a basic pattern to guide the tool designer in making his analyses and to help assure that no primary criterion is overlooked.

BASIC PATTERN FOR TOOL DESIGN ANALYSIS

Create	Analyze in terms of these criteria				
Alternatives	Function	Quality	Cost	Date	Auxiliary
A	x	x	x	x	x
B	x	x	x	x	x
C	x	x	x	x	x
.	.	.	.	.	.
.	.	.	.	.	.
.	.	.	.	.	.
n	x	x	x	x	x

Following this method of analysis will force one to design more creatively and with a greater probability of developing the optimum design. For example, if one is assigned to design a tool, whatever its nature, it is necessary first to "define the problem" or the conditions that will govern every decision regarding geometry of the tool, the choice of materials and their heat treatment, tolerances, degree of complexity, and any auxiliary criteria. The specific criteria are given in the pattern of analysis. It specifies consideration of the questions: (1) will the tool perform the function intended? (2) will the specific quality requirements be met? (3) what are the limitations on money available for design and construction of the tool? (4) when must the tool be completed? and (5) what auxiliary factors as suggested above will affect the design of the tool?

If a tool design effort is to be of optimum value, it must be a creative effort. This suggests that before the details of a tool can or should be developed it is essential to create alternative ways in which the production operation may be tooled up. In other words, creative tool design recognizes the old adage, "There are more ways than one to skin a cat."

Creation of alternatives cannot be avoided if the basic pattern of analysis presented above is followed. The first column heading specifies positively: "Create alternatives."

The extent of creative effort will be suggested by the number of alternatives developed. The process of creating alternatives, however, could go on infinitely. Where does one stop? Here again the pattern of analysis comes to the tool designer's aid. It says, "There is (or must be) a *date* specification to this project." Consequently, there is a definite basis for control of the length of time allowable for the creative phases of each design project.

Since alternatives arise from creative tool design effort, which alterna-

Chap. 8 General Considerations in Tool Design

tive should one select? The answer again must come from an analysis of results summarized in the above pattern of analysis. The choice must be that which best satisfies all five sets of criteria listed. Since seldom does one alternative have the optimum ranking on each of these criteria, the choice generally reflects the best balance between the factors, which often are conflicting. Examples of this type will be given in later subsections of the text.

In summary, it is emphasized that there are five basic criteria that govern every decision in selection of materials, tool geometry, process method, and so on. The degree to which the pattern of analysis presented above is employed in every tool design decision determines the extent to which the tool design effort is contributing to optimum design.

Original Directions for Design

Engineers responsible for manufacturing planning must write down their plans in some form that can be circulated and used by many people to fulfill their individual parts in the total manufacturing operation. Many industries call the document describing all steps in manufacture of a given product a "route sheet." This usually states the raw material, standard components, or supplies required, and every step of procedure to create the finished item to be inspected against a part print. It identifies all machinery required at each step, and any special tooling such as jigs, fixtures, dies, gages, or special items, either existing or that may require new design. An order of some kind is usually originated directing procurement of all such items and conveying some ideas as to the design wanted, or at least the required function of the tooling. All directives of this sort are the first step in design and must be clearly stated and well coordinated in order to get the design started right.

The design of tools and special items requires a routing, a part print, and an order form stating any directives the originator feels necessary. Basically, the planner must visualize the part and be able to understand how the item to be designed will fit into the development of that part during the manufacturing process. The planner generally does not design the tools, but he should direct procurement by words and sketches that will clearly indicate what he wants.

Tooling Layout

The actual work of creating on paper the assembly design of equipment or tools for manufacturing processes should be done within the general framework of the following rules:

1. Lay out the part in an identifying color (red is suggested).

2. Lay out any cutting tools. Possible interference or other confining items should be indicated in another identifying color (blue suggested). Use of the cutting tool should not damage the machine or the fixture.
3. Indicate all locating requirements for the part. There are three locating planes; use three points in one, two points in the second, and only one point in the third plane. Do not locate on parting line of castings or forgings. All locators must be accessible for simple cleaning of chips and dirt.
4. Indicate all clamping requirements for the part. Plan to avoid the marking or deforming of finished or delicate surfaces. Consider the clamping movements of the operator so that injury to the hands or unsafe situations are eliminated. Be sure it is possible to load and unload the part.
5. Lay out the details with due consideration to stock sizes, so as to minimize machining requirements.
6. Use full scale in the layout, if at all possible.
7. Indicate the use of standard (purchasable as shelf items) fixture parts whenever possible.
8. Identify each different item or detail of any design of a tool by the use of balloons with leaders and arrows pointing to the detail in the view that best shows the outline of it. These should *not* go to a line that is common to other details.

Detailing, Stock Lists, and Notes

Many companies advocate rather complete detailing of jigs, fixtures, and gages, but only dimension the assemblies for dies. Certain common practices have evolved for detailing which define the material, dimension all of its shape characteristics, and indicate the finish and hardness of the end item.

1. Draw and dimension with due consideration for someone using the drawing to make the item in the toolroom.
    a. Do not crowd the views or the dimensions.
    b. Analyze each cut to be sure it can be done with standard cutting tools.
2. Use only as many views as necessary to show all required detail.
3. Surface roughness should be specified.
4. Tolerances and fits peculiar to tools need special consideration. It is not economical as a rule to tolerance both details of a pair of mating parts as is required on production part detailing. In cases where a hole and a plug are on different details to be made and

Chap. 8 General Considerations in Tool Design 443

mated, the fit tolerance should be put on the male piece and the hole should carry a nominal size. This allows a toolmaker to ream the hole with a nominal-size tool and grind the plug to fit it, although nominal may vary several thousandths from exact.

5. The stock list of any tool drawings should indicate all sizes required to obtain the right amount for each detail. It is necessary to allow finishing stock in almost all cases, although some finished stock is available today. As far as possible, stock sizes known to be on hand should be used, but in all cases available sizes should be specified. A proper finished detail is dependent upon starting with the right material.
6. Notes are required to convey certain ideas that cannot be communicated by conventional drawing. Heat treatments and finishes are usually identified as specification references rather than being spelled out on each drawing.

SAFETY AS RELATED TO TOOL DESIGN

Safety laws vary greatly from one state to another. All states have some laws requiring protective guards and devices to safeguard workers.

Safety should be designed into the tooling. One of the first and least expensive requirements should be that of breaking all sharp edges and corners. More minor injuries result from this cause than from most other causes. One should never cut against a clamp, because of vibration and tool chatter. The safe way is to avoid machining operations against a clamp. Instead, parts should be nested against pins in order to take the cutter load. Rigidity and foolproofing should always be built into the tooling. Drill jigs should be made large enough to hold without the danger of spinning. Small drill jigs should always be clamped in a vise or against a bar or backstop. Plexiglas guards should be installed around all milling and flycutting operations where chips endanger other workers or work areas. High-speed open cutters on production milling, drilling, turning, and jig borers are almost impossible to safeguard because of the varied size and location of the workpieces to be machined.

In guarding punch presses, no one type of guard is practical for all operations. Ring guards work well on small punching setups where the ring meets an obstruction and the downward motion of the ram is stopped. Larger die sets and presses can be better protected by gate guards which must be positioned after the work is loaded and then interlocked with the clutch mechanism. Barrier-type guards protect by preventing hands and arms from being placed inside the work area. These generally are telescoping perforated sheet metal coverings with an opening to feed the stock. Sweep guards actually sweep clear the danger area as the press ram descends and even provide protection during an accidental descent of the ram. Wire-cage guards are still another way to protect the operator's hands; the operator feeds the work through a small opening. This type of guard is useful for secondary operations. Another method of protection is to provide space between the punch and die too small for the operator's hand to enter; this distance should not exceed $\frac{3}{8}$ inch.

Limit switches can be used extensively to protect worker and product. In punches and dies, limit switches can detect a misfeed or buckling of stock, and check the position of parts in assembling. Photoelectric equipment to protect the operator's hands or body operates when the set beam is interrupted. It is good practice for all punch presses and air or hydraulically operated tooling to be installed with a double-button interlocking protection system, requiring both buttons to be activated before the tool can be used. Other interlocking systems can include two valves operated by foot or a combination foot or knee valve in conjunction with a hand valve. These systems should be so designed that, should one of the buttons be locked in a closed position, the tooling or the press will not operate. The device should be located in such a position or guarded in such a manner that the operator cannot operate the tooling or press while he is in the danger area. Feed mechanisms should be provided for all high-speed punching, machining, or assembling operations. The safety function of a feeding device is to provide a means of moving the part into the nest by gravity or mechanical motion so that there is no necessity for the operator to place his hands in the danger zone. High-speed presses equipped with automatic feeds operate at such speeds that it would be impractical as well as hazardous for an operator to attempt to feed the stock.

In designing tools and additions to machinery involving any electrical equipment, it is mandatory that the system be grounded. Portable electrical equipment should be checked periodically because of rough handling, and all should be grounded to prevent injuries to personnel. Locks provided for electrical switches to hold them in an open position, while tools or punches and dies are being repaired or set, help to prevent accidental damage. Local regulations can be found in each state's electrical code book covering all applications.

Tooling for various industries requires different treatment to insure safe operations, e.g., special electrical controls and motors for industries handling explosive material. The materials used in tooling for the chemical industry should be designed to withstand the corrosive actions, and electrical equipment should be properly sealed. Careful analysis must be given to tooling for each type of application in order to provide maximum protection and long life.

Machining of various plastic materials should employ an exhaust system because of abrasive dust particles and/or some materials that give off poi-

Chap. 8 General Considerations in Tool Design

sonous fumes when hot. Safe procedures should be practiced when handling various metals. Areas around magnesium machining should be kept clean and tools kept sharp to reduce the fire hazard. Metal powders present a hazard of combustion during grinding operations. Data may be found in chemical, electrical, and industrial handbooks to help in tooling for specialized problems involving these industries.

Tooling and additions to machines should be designed so that the operator does not have to lean across a moving cutter or table. All adjustments and clamping should be easily accessible from the front or operator's position. Body geometrics should be considered in designing tooling—not only in terms of safety but also of production.

Tooling involving welding must be guarded to prevent severe burns or eye injury from high-intensity arc welding rays.

All belts, chain drives, gears, sprockets, couplings. keys, and pulleys should be well guarded by sheet metal and panel-type guards. These guards should be strong enough to support and protect in the event that someone or something falls against the guard or in case the belt or chain should break. Always provide an adequate factor of safety in the design of all tools and tooling applications.

Safety glasses should be worn during all machining, grinding, and buffing operations.

Safety standards are available covering all types of industry and applications, and personnel concerned with plant safety should be familiar with them. Some of the agencies handling such information are: American Standards Association; National Bureau of Standards, U.S. Department of Commerce; National Safety Council; National Board of Fire Underwriters; and Association of Casualty and Surety Companies.

Tooling is not restricted solely to machining operations. Equipment of the automatic assembling and inserting type may be classified as either a tool or a machine. One of today's complex tools might contain any combination or all of electric motors, air cylinders, hydraulic equipment, conveyors, and precision indexing tables. Safety is of the utmost concern when a production tool involves such operating mechanisms.

TOOL MATERIALS

Criteria for Material Selection

Tool materials should be selected after a study of the physical and, in some cases, chemical properties desired. In most applications more than one type of material will be satisfactory and final choice will be governed by availability and economies.

The principal tooling materials are tool-steels, but in many applications cast irons, other steels, and nonferrous materials may be successfully used.

In some applications a combination of tool-steels, nonferrous materials, and even composite sections may work out very well.

There are fundamental physical properties that should be understood and taken into account when selecting materials for tooling. Complete data are usually available from manufacturers' literature and handbooks.

Hardness. Hardness is the ability to resist penetration, or the ability to withstand abrasion. It is an important property in selecting tool materials. Hardness alone does not determine the wear resistance or abrasion resistance of a material. In alloy steels, especially tool-steels, the resistance to wear or abrasion varies with the alloy content when the alloy is applied at the same hardness level. (Cf. Tables 8-1, 8-2, and 8-3.)

Rockwell Hardness. This is the most widely used method for measuring the hardness of steel. The test is made by forcing a penetrator into the surface of the metal being tested, by a dead weight acting through a series of levers. A micrometer dial gauge tells the depth to which the penetrator sinks. The softer the metal being tested, the deeper it will penetrate with a given load. The dial gauge does not read directly in depth of penetration, but shows arbitrary scales of "Rockwell numbers." A variety of loadings can be used, each designated by a different letter, and the relative hardness or softness is measured.

Two types of penetrators are used: a diamond cone known as a *brale,* for hard materials such as a hardened tool steel, and a hardened steel ball for testing soft materials.

Brinell Hardness. This method of hardness measurement is much older than the Rockwell. It operates very similarly to the Rockwell ball-test principle. In the Brinell machine a much larger steel ball is used, and this is forced into the material being tested under a load of 3000 kilograms (approximately 6500 pounds). Instead of measuring the penetration, diameter of the impression in the test piece is measured with a small hand microscope with a lens calibrated in millimeters. The measured diameter is converted by means of a table into a Brinell hardness number.

The Brinell hardness measurement is most useful on soft and medium hard materials. On steels of high hardness, the impression is so small it is difficult to read; therefore, the Rockwell tester is more commonly used for such materials.

Tensile Strength. This physical property of materials is the value obtained by dividing the maximum load observed during tensile testing by the specimen's cross-sectional area before testing.

Tensile strength is an important property to take into consideration when designing large fixtures or other tooling. It is of lesser importance in tools and dies except where soft or medium hard ferrous or nonferrous materials are used.

If a steel will stretch slightly before it will break, a reasonably accurate tensile figure can be obtained. However, if the tool material is so hard that it breaks before it stretches, the specimen will rupture in test long before the true strength is obtained.

The tensile tests successfully made on tool-steel involve the use of drawing temperatures much higher than actually used on tools. Tool-steels used for hot work, fatigue, or impact applications are usually used at lower hardness levels and for these types tensile properties can be obtained and are available.

Compressive Strength. Compressive force plays an important part in tool design. It is the maximum stress that a metal, subject to compression, can withstand without fracture.

This test is used on hardened tool-steels, especially at high hardness levels. For all ductile materials the specimens flatten out, under load, and there is no well-marked fracture. For these types of material the compressive strength is usually equal to the tensile strength.

Shear Strength. The shear strength of materials is of significance, especially in designing machines and members subjected to torsion. It may be defined as the value of stress necessary to cause rupture in torsion.

For most steels, except tool and other highly alloyed steels, the shear strength lies between 50 and 60% of the yield strength; hence the yield strength in tension serves fairly well as an index of shear strength.

Hardened tool-steels are considered to be brittle because they deform very little before fracture in tension or bending, yet when subjected to torsion they exhibit considerable ductility. The torsion method of testing offers advantages for the "toughness testing" of hard tool-steels. This is particularly true in those instances where very small amounts of deformation occur. Toughness is the property of absorbing considerable energy before fracture. It involves both ductility and strength.

Yield Strength. This is the property of a material that generally limits its strength in application. It is the stress level at which a material will show a permanent elongation after the load is released.

To compare the elastic properties of steels, both soft and hard, a definite amount of permanent elongation is used as the criterion of yield strength. This is generally 0.2% of the 2-inch gauge length used. Heat treatment is used to improve the yield strength.

Modulus of Elasticity (Bending). This is a measure of the stiffness of a material. It is indicated by the slope of the line generated below the elastic limit during tensile testing. For steel the average value is 30,000,000 psi. Few steels have moduli that deviate widely from this value. The modulus cannot be materially altered by heat treatment.

This is an important property to take into consideration when designing long tools and machine parts. Section must then be taken into consideration.

The moduli of cast irons vary from 10,000,000 to 25,000,000 psi depending on strength and ductility, with nonferrous tooling alloys generally lower in value.

Modulus of Elasticity (Torsion). This modulus corresponds to the modulus of elasticity in the tensile test except that it is measured in a torsion test and is the ratio of the unit shear stress to the displacement caused by it per unit length in the elastic range. It is a usable value when designing shafts, taps, twist drills, or other tools working in torsion. Values should be used as a guide when available.

Impact. Toughness or the ability to resist breaking is measured by the impact test.

There are three general ways of testing the impact strength of a material at a specific hardness level: the Izod, the Charpy, and the torsion impact tests. The first two tests give useful results only on steels that possess some ductility, i.e., those that bend before they break.

The Izod toughness testing machine is built on the pendulum principle. The machine consists of a vise for holding the test piece and a heavy pendulum that acts as a hammer. The pendulum is pulled back a definite distance and allowed to fall of its own weight upon the test piece. The toughness of the test piece is measured by the amount of overswing of the pendulum. The more brittle the steel, the greater the overswing.

The Charpy toughness tester works on the pendulum principle, but the test piece is supported at both ends and the knife edge of the pendulum strikes the middle of the test piece.

These impact machines were designed to test tough materials. None of them can be used to accurately measure the toughness of high-hardness steels or tool-steel. A hardened tool-steel specimen will fracture and show low erratic readings.

The torsional testing method is applicable to steels at high hardness levels. The sample is broken by a torsional blow. This puts the load on the cross section at one time and gives accurate results. This method has advantages for the toughness testing of hard tool-steels. This is true in those instances where very small amounts of deformation occur.

Fatigue, Corrosion. Fatigue is an important physical property to consider in tooling. Fatigue may be defined as a tendency for a metal to break under conditions of repeated cyclic stressing below its ultimate tensile strength.

In the designing of tools, especially punches and other impact tools, fractures start from a poorly designed fillet or from other localized concentration of stress. Under long-continued repetitive stresses a crack starts at this locally overstressed point and proceeds to final failure. Localized stresses should therefore be kept to a minimum.

Corrosion is seldom a problem in tooling design or performance. In some cases frictional heat causes the lubricant to break down, resulting in chemical or electrochemical attack on the tooling material.

Ferrous Tool Materials

Many ferrous materials may be used in tool construction. The supporting components of fixtures usually are made of low-carbon steels while the wear pads are made of hardened tool-steel. Dies will be made of a tool-steel selected because of performance requirements. The selected steel may be only one of many types equally applicable.

Tool and Die Steels. Proper selection of tool and die steels is complicated by their many special properties. The five principal ones are: heat resistance, abrasion resistance, shock resistance, resistance to movement or distortion in hardening, and cutting ability.

Because no one steel can possess all of these properties to the optimum degree, hundreds of different tool steels have been developed to meet the total range of service demands.

The steels listed in Table 8-1 will adequately serve 95 per cent of all metal stamping operations. The list contains thirty-one steels, nine of which are widely applied and readily available. The other steels included represent slight variations for improved performance in certain instances, and their use is sometimes justified because of special considerations.

The tool-steels are identified by letter and number symbols. All the steels in the list except those in the S and H groups can be heat treated to a hardness greater than Rockwell C62 and, accordingly, are hard, strong, wear-resistant materials. Frequently hardness is proportional to wear resistance, but this is not always the case, because the wear resistance usually increases as the alloy content, and particularly the carbon content, increases.

The toughness of the steels, on the other hand, is inversely proportional to the hardness and increases markedly as the alloy content or the carbon content is lowered.

Table 8-2 lists the basic characteristics, and Table 8-3 the hardening and tempering treatments, of the various steels listed.

The general nature and application of the various standard tool steel classes are as follows:

W, Water-Hardening Tool-Steels. This group includes the plain carbon (W1) and the carbon vanadium (W2) types. The carbon steels were the original tool-steels. Because of their low cost, abrasion-resisting and shock-resisting qualities, ease of machinability, and ability to take a keen cutting edge, the carbon grades are widely applied. Both types are shallow-hardening and are readily available.

O, Oil-Hardening Tool-Steels. Types O1 and O2 are manganese oil-hardening tool-steels, readily available and of low cost. These steels have

TABLE 8-1. AISI IDENTIFICATION AND CLASSIFICATION OF TOOL STEELS
(partial listing)

Steel types *	Average Composition, per cent							
	C	Mn	Si	Cr	W	Mo	V	Other
W1	1.00							
W2	1.00						0.25	
O1	0.90	1.00		0.50	0.50			
O2	0.90	1.60						
O7	1.20			0.75	1.75	0.25		
A2	1.00			5.00		1.00		
A4	1.00	2.00		1.00		1.00		
A5	1.00	3.00		1.00		1.00		
A6	0.70	2.00		1.00		1.00		
D2	1.50			12.00		1.00		
D3	2.25			12.00				
D4	2.25			12.00		1.00		
D6	2.25		1.00	12.00	1.00			
S1	0.50			1.50	2.50			
S2	0.50		1.00			0.50		
S4	0.50	0.80	2.00					
S5	0.50	0.80	2.00			0.40		
H11	0.35			5.00		1.50		
H12	0.35			5.00	1.50	1.50	0.40	
H13	0.35			5.00		1.50	1.00	
H21	0.35			3.50	9.00			
H26	0.50			4.00	18.00		1.00	
T1	0.70			4.00	18.00		1.00	
T15	1.50			4.00	12.00		5.00	5.00 Co
M2	0.85			4.00	6.25	5.00	2.00	
M3	1.00			4.00	6.00	5.00	2.40	
M4	1.30			4.00	5.50	4.50	4.00	
L2	0.50			1.00			0.20	
L3	1.00			1.50			0.20	
L6	0.70			0.75				1.50 Ni
F2	1.25				3.50			

* W, water-hardening; O, oil-hardening, cold-work; A, air-hardening, medium-alloy; D, high-carbon high-chromium, cold-work; S, shock-resisting; H, hot-work; T, tungsten-base high-speed; M, molybdenum-base high-speed; L, special-purpose, low-alloy; F, carbon-tungsten, special-purpose.

Chap. 8   General Considerations in Tool Design

TABLE 8-2. COMPARISON OF BASIC CHARACTERISTICS OF STEELS USED FOR PRESS TOOLS

AISI Steel No.	Non-deforming properties	Safety in hardening	Toughness	Resistance to softening effect of heat	Wear resistance	Machin-ability
W1	Poor	Fair	Good	Poor	Fair	Best
W2	Poor	Fair	Good	Poor	Fair	Best
O1	Good	Good	Fair	Poor	Fair	Good
O2	Good	Good	Fair	Poor	Fair	Good
O7	Good	Good	Fair	Poor	Fair	Good
A2	Best	Best	Fair	Fair	Good	Fair
A4	Best	Best	Fair	Poor	Fair	Fair
A5	Best	Best	Fair	Poor	Fair	Fair
A6	Best	Best	Fair	Poor	Fair	Fair
D2	Best	Best	Fair	Fair	Good	Poor
D3	Good	Good	Poor	Fair	Best	Poor
D4	Best	Best	Poor	Fair	Best	Poor
D6	Good	Good	Poor	Fair	Best	Poor
S1	Fair	Good	Good	Fair	Fair	Fair
S2	Poor	Fair	Best	Fair	Fair	Fair
S4	Poor	Fair	Best	Fair	Fair	Fair
S5	Fair	Good	Best	Fair	Fair	Fair
H11	Best	Best	Best	Good	Fair	Fair
H12	Best	Best	Best	Good	Fair	Fair
H13	Best	Best	Best	Good	Fair	Fair
H21	Good	Good	Good	Good	Fair	Fair
H26	Good	Good	Good	Best	Good	Fair
T1	Good	Good	Fair	Best	Good	Fair
T15	Good	Fair	Poor	Best	Best	Poor
M2	Good	Fair	Fair	Best	Good	Fair
M3	Good	Fair	Fair	Best	Good	Fair
M4	Good	Fair	Fair	Best	Best	Poor
L2	Fair	Fair	Good	Poor	Fair	Fair
L3	Fair	Poor	Fair	Poor	Fair	Good
L6	Good	Good	Good	Poor	Fair	Fair
F2	Poor	Poor	Poor	Fair	Best	Fair

TABLE 8-3. HARDENING AND TEMPERING TREATMENTS FOR PRESS TOOLS

AISI Tool-steel	Preheat temp., °F	Rate of heating for hardening	Hardening temp., °F	Time at temp., min.	Quenching medium	Tempering temp., °F	Depth of hardening	Resistance to decarburizing
W1		Slow	1425–1500	10–30	Brine or water	325–550	Shallow	Best
W2		Slow	1425–1550	10–30	Brine or water	325–550	Shallow	Best
O1	1200	Very slow	1450–1500	10–30	Oil	325–500	Medium	Good
O2	1200	Very slow	1400–1475	Do not soak	Oil	325–600	Medium	Good
O7	1200	Slow	1575–1625	10–30	Oil	350–550	Medium	Good
A2	1450	Very slow	1700–1800	30	Air	350–700	Deep	Fair
A4	1250	Slow	1450–1550	15–30	Air	300–500	Deep	Very good
A5	1250	Slow	1450–1550	15–30	Air	300–500	Deep	Very good
A6	1250	Slow	1500–1600	15–30	Air	300–500	Deep	Very good
S1		Slow to 1400	1650–1750	10–30	Oil	500–600	Medium	Fair
S2		Slow	1525–1575	10–30	Brine or water	350–700		Fair
S4		Slow	1550–1650	10–30	Brine or water	350–700		Poor
S5			1600–1700		Oil	350–700	Medium	Poor
H11	1400	Slow	1800–1850	15–60	Air	900–1200	Deep	Good
H12	1400	Slow	1800–1850	15–60	Air	900–1200	Deep	Good
H13	1400	Slow	1800–1850	15–60	Air	900–1200	Deep	Good
H21	1550	Medium	2000–2200	5–15	Air, oil	1000–1200	Deep	Good
H26	1550	Medium	2000–2200	5–15	Air, oil	1000–1200	Deep	Good
T1	1500–1600	Rapid from preheat	2150–2300	Do not soak	Air, oil or salt	1025–1200	Deep	Good
T15	1500–1600		2125–2270	Do not soak	Air, oil or salt	1000–1200	Deep	Fair
M2	1500	Rapid from preheat	2125–2225	Do not soak	Air, oil or salt	1025–1200	Deep	Poor
M3	1450–1550		2125–2225	Do not soak	Air, oil or salt	1025–1200	Deep	Poor
M4	1450–1550		2125–2225	Do not soak	Air, oil or salt	1025–1200	Deep	Poor
L2		Slow	1550–1700	15–30	Oil	350–600	Medium	Good
L3		Slow	1425–1500	10–30	Brine or water	300–800	Medium	Good
L3		Slow	1500–1600	10–30	Oil	300–800	Medium	Good
L6		Slow	1450–1550	10–30	Oil	300–1000	Medium	Fair
F2	1200	Slow	1525–1625	15–30	Brine or water	300–500	Shallow	Good

less movement than the water-hardening steels, and are of equal toughness with the water-hardening steels when the latter are hardened throughout. Wear resistance is slightly better than that of water-hardening steels of equal carbon content. Steel 07 has greater wear resistance because of its increased carbon and tungsten content.

A, Air-Hardening Die Steels. Type A2 is the principal air-hardening tool-steel. It has minimum movement in hardening and has higher toughness than the oil-hardening die steels, with equal and greater wear resistance. Steels A4, A5, and A6 can be hardened from lower temperatures, but have lower wear resistance and better distortional properties.

D, High-Carbon High-Chromium Die Steels. Type D2 is the principal steel in this class. It finds wide application for long-run dies. It is deep-hardening, fairly tough, and has good resistance to wear. Steels D3, D4, and D6, containing additional carbon, have very high wear resistance and lower toughness. Steels D2 and D4 are air-hardened.

S, Shock-Resisting Tool-Steels. These steels contain less carbon and have higher toughness. They are applied where heavy cutting or forming operations are required, and where breakage is a serious problem. Steels S1, S4, and S5 are readily available. Steels S4 and S5 are more economical than S1.

H, Hot-Work Die Steels. These steels must combine red hardness with good wear resistance and shock resistance. They are air-hardening and on occasion are used for cold-work applications. They have relatively low carbon content and intermediate to high alloy content.

T and M, Tungsten and Molybdenum High-Speed Steels. Steels T1 and M2 are equivalent in performance and have good red hardness and abrasion resistance. They have higher toughness than many of the other die steels. They may be hardened by conventional methods or carburized for cold-work applications. Steels M3, M4, and T15 have greater cutting ability and resistance to wear. They are more difficult to machine and grind because of their increased carbon and alloy contents.

L, Low-Alloy Tool-Steels. Steels L3 and L6 are used for special die applications. Other L steels find application where fatigue and toughness are important considerations, such as in coining or impression dies.

F, Finishing Steels. Steel F2 is of limited use but occasionally applied where extremely high wear resistance in a shallow-hardening steel is desired.

Cast Iron. Cast iron is essentially an alloy of iron and carbon, containing from 2 to 4 per cent carbon, 0.5 to about 3.00 per cent silicon, 0.4 to approximately 1 per cent manganese, plus phosphorus and sulphur. Other alloys may be added depending on the properties desired.

The high compressive strength and ease of casting of the gray irons are utilized in large forming and drawing dies to produce such items as

TABLE 8-4. APPLICATIONS OF TOOL STEELS

Application	Suggested AISI tool-steels	Rockwell C hardness range
Arbors	L6, L2	47–54
Axle burnishing tools	M2, M3	63–67
Boring bars	L6, L2	47–54
Broaches	M2, M3	63–67
Bushings (drill jig)	M2, D2	62–64
Cams	A4, O1	59–62
Centers, lathe	D2, M2	60–63
Chasers	M2	62–65
Cutting tools	M2	62–65
Dies, blanking	O1, A2, D2	58–62
Dies, bending	S1, A2, D2	52–62
Dies, coining	S1, A2, D2	52–62
Dies, cold heading:		
Solid	W1, W2	56–62
Insert	D2, M2	57–62
Dies, hot heading	H12, H13	42–48
Dies, lamination	D2, D3	60–63
Dies, shaving	D2, M2	62–64
Dies, thread rolling	D2, A2	58–62
Die casting:		
Aluminum	H13	42–48
Form tools	M2, M3	63–67
Lathe tools	M2, T1	63–65
Reamers	M2	63–65
Shear blades:		
Light stock	D2, A2	58–61
Heavy stock	S1, S4	52–56
Rolls	A2, D2	58–62
Taps	M2	62–65
Vise jaws	L2, S4	48–54
Wrenches	L2, S1	40–50

automobile panels, refrigerator cabinets, bath tubs, and other large articles. Conventional methods of hardening result in little distortion.

Alloying elements are added to contract graphitization, to improve mechanical properties, or to develop a special characteristic.

Stainless Steel. Stainless steels are corrosion-resistant alloys that are used where other steels would be attacked by oxidation. They derive their high resistance to corrosion from the presence of chromium.

The hardenable types are used for bearings and cutlery. Nonhardenable types are used for chemical structural purposes or in tooling where a nonmagnetic material may be required and corrosion is a problem.

Chap. 8   General Considerations in Tool Design   455

Nonferrous Tool Materials

Nonferrous alloys are used to some degree as die materials in specialized applications, and generally for limited-production requirements. On the other hand, in jig and fixture design some of them find extensive use where lightness of weight of tools may be an important factor.

Aluminum. Dural (Duralumin) aluminum 17S sheet is used as a facing over form blocks. Its composition is copper, 4.0 per cent, manganese, 0.5 per cent, magnesium 0.5 per cent, with aluminum and normal impurities constituting the remainder.

Aluminum bronze cast to the specified die shape is used for forming and drawing stainless steel without scratching or galling. These alloys are characterized by compressive strengths of 160,000 to 170,000 psi, and hardnesses of Rockwell C27 to 35.

Magnesium. Sheet magnesium and magnesium alloys are used as a facing material over blocks. Their application is limited to very short production runs.

Zinc-base Alloys. Zinc-base alloys are quickly cast into punch-and-die shape at low cost. The production of experimental parts with such dies can prove out the design of both the part and the die before permanent tooling is started.

Lead-base Alloys. Lead punches, composed of 6 to 7 per cent antimony and the remainder lead, have been used with Kirksite dies. These, again, are used on limited-production runs.

Cast Beryllium Copper Alloys. Cast alloys of beryllium, cobalt, and copper have characteristics comparable with the aluminum bronze.

Bismuth Alloys. The alloys of bismuth are used chiefly as a matrix material for securing punch-and-die parts in a die assembly, and as cast punches and dies for short-run forming and drawing operations.

Nonmetallic Die Materials

Nonmetallic die materials are chiefly used where production parts requirements are limited and tool-steel tooling would not be economically practical.

Hardboard. Sheets composed of compressed wood fiber are used as punches and die material in drawing and forming operations, as form blocks in rubber forming, and in stretch dies. Die stock processed for higher tensile strength and thickness is available.

Densified Wood. Various woods are impregnated with a phenolic resin after which the laminated assembly is compressed to about 50 per cent of the original thickness of the wood layers.

Densified wood punches and dies are used in forming and drawing dies;

in the latter, scoring of the part is infrequent because of the low coefficient of friction of densified wood when properly finished.

Rubber. Molded-rubber female dies and rubber-covered punches are used in difficult forming operations, such as the forming of deeply fluted lighting reflectors.

Plastics. Molded or machined form, draw, and stretch dies of thermosetting plastic are being used. Production life of the newer plastics used as die materials is showing much improved performance. On thin-gage steel drawn into simple shapes production runs of 50,000 pieces are normal. For more extended runs (100,000 or more pieces), metal wearplates can be inserted. Plastics can be impregnated with metallic powders and other wear-resisting materials.

HEAT TREATING

The purpose of heat treatment is to control the properties of a metal or alloy through the alteration of the structure of the metal or alloy by heating it to definite temperatures and cooling at various rates. This combination of heating and controlled cooling determines not only the nature and distribution of the microconstituents, which in turn determine the properties, but also the grain size.

Heat treating should improve the alloy or metal for the service intended. Some of the various purposes of heat treating are:

1. to remove strains after cold working.
2. to remove internal stresses such as those produced by drawing, bending, or welding.
3. to increase the hardness of the material.
4. to improve machinability.
5. to improve the cutting properties of tools.
6. to increase wear-resisting properties.
7. to soften the material, as in annealing.
8. to improve or change the physical properties of a material such as corrosion resistance, heat resistance, magnetic properties, or others as required.

Treatment of Ferrous Materials

Iron is the major constituent in the steels used in tooling, to which carbon is added in order that the steel may harden. Alloys are put into steel to enable it to develop properties not possessed by plain carbon steel, such as ability to harden in oil or air, increased wear resistance, higher toughness, and greater safety in hardening.

Heat treatment of ferrous materials involves several important operations which are customarily referred to under various headings, such as normalizing, spheroidizing, stress relieving, annealing, hardening, tempering, and case hardening.

Normalizing. This is the operation of heating to a temperature about 100° to 200°F above the critical range and cooling in still air. This is about 100°F over the regular hardening temperature.

The purpose of normalizing is usually to refine grain structures that have been coarsened in forging. With most of the medium-carbon forging steels, alloyed and unalloyed, normalizing is highly recommended after forging and before machining to produce more homogeneous structures and in most cases improved machinability.

High-alloy air-hardening steels are never normalized, since to do so would cause them to harden and defeat the primary purpose.

Spheroidizing. This is a form of annealing which, in the process of heating and cooling steel, produces a rounded or globular form of carbide—the hard constituent in steel.

Tool-steels are normally spherodized for best machinability. This is accomplished by heating to a temperature of 1380–1400°F for carbon steels and higher for many alloy tool-steels, holding at heat one to four hours, and cooling slowly in the furnace.

Stress Relieving. This is a method of relieving the internal stresses set up in steel during forming, cold working, and cooling after welding or machining. It is the most simple heat treatment and is accomplished by merely heating to 1200–1350°F followed by air or furnace cooling.

Large dies are usually roughed out, then stress-relieved and finish-machined. This will minimize change of shape not only during machining but during subsequent heat treating as well. Welded sections will also have locked-in stresses owing to a combination of differential heating and cooling cycles as well as to changes in cross section. Such stresses will cause considerable movement in machining operations.

Annealing. The process of annealing consists of heating the steel to an elevated temperature for a definite period of time and, usually, cooling it slowly. Annealing is done to produce homogenization and to establish normal equilibrium conditions, with corresponding characteristic properties.

Tool-steel as purchased is generally in the annealed condition. Sometimes it is necessary to rework a tool that has been hardened, and the tool must then be annealed. For this type of anneal, the steel is heated slightly above its critical range and then cooled very slowly.

Finished parts may be annealed without surface deterioration by placing them in a closed pot and covering with compounds that will combine with the air present to form a reducing atmosphere. Partially spent carburizing compound is widely used, as well as cast iron chips, charcoal, and commercial neutral compounds.

Hardening. This is the process of heating to a temperature above the critical range, and cooling rapidly enough through the critical range to appreciably harden the steel. (See Table 8-3 for specific treatment.)

A simplified theory of hardening steel is that iron has two distinct and different atomic arrangements, one existing at room temperature (or again near the melting point), and one above the critical temperature. Without this phenomenon it would be impossible to harden iron-base alloys by heat treatment.

What happens in the heat treatment of die steels is represented graphically in Fig. 8-1. Starting in the annealed condition at A, the steel is soft, consisting of an aggregate of ferrite and carbide. Upon heating above the critical temperature to B, the crystal structure of ferrite changes, becomes austenite, and dissolves a large portion of the carbide. The new structure,

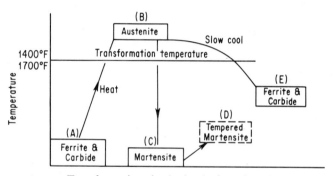

Fig. 8-1. Transformations in the hardening of steel.

austenite, is always a prerequisite for hardening. By quenching it (cooling rapidly to room temperature), the carbon is retained in solution, and the structure known as martensite (C in diagram) results. This is the hard matrix structure in steels. It is initially highly stressed, for the change from austenite involves some volumetric expansion against natural stiffness of the steel, so it must be reheated to an intermediate temperature (D) to soften it slightly and relieve those internal stresses which unduly embrittle the steel.

If quenching is not rapid enough, the austenite reverts to ferrite and carbide (E) and high hardness is not obtained. The rate at which quenching is required to produce martensite depends primarily on the alloy content. Low-alloy die steel is water or oil hardening, while highly alloyed steel usually can be hardened in air, i.e., quenched at a much slower rate. The high alloys make the reaction more sluggish.

Tempering. This is the process of heating quenched and hardened steels and alloys to some temperature below the lower critical temperature to reduce internal stresses set up in hardening. Thus the hard martensite resulting from the quenching operation is changed in tempering in the direction of the equilibrium properties, the degree being dependent on the tempering temperature and rate of cooling. (See Table 8-3 for specific treatment.)

Case Hardening. The addition of carbon to the surface of steel parts and the subsequent hardening operations are important phases in heat treating. The process may involve the use of molten sodium cyanide mixtures, pack carburizing with activated solid material, such as charcoal, or coke, gas, or oil carburizing, and dry cyaniding.

Whether a solid carbonaceous packing material is used, or a liquid gas, the objective is to produce a hard, wear-resistant surface with a core of such hardness or toughness as is best suited for the purpose. The carbon content of the surface is raised to 0.80–1.20% and the case depth can be closely controlled by the time, the temperature, and the carburizing medium used. Pack carburizing is generally done at 1700°F for eight hours to produce a case depth of $\frac{1}{16}$ in. Light cases up to 0.005 in. can be obtained in liquid cyanide baths and case depths to $\frac{1}{32}$ in. are economically practical in liquid carburizing baths.

Usually low-carbon and low-carbon alloy steels are carburized. The usual carbon range is 0.10 to 0.30% carbon, though higher-carbon-content steels may be carburized as well.

Treatment of Nonferrous Materials

The heat treatment of nonferrous metals and alloys closely approximates that of steel except that the temperature ranges used are lower, and hardening is accomplished by the precipitation of hard metallic compounds or particles.

Nonferrous metals and alloys that are not heat-treatable harden by cold work only.

For the heat-treatable alloys of aluminum, hardening is accomplished by precipitation. When an alloy is water-quenched from the "hardening heat" it is very soft; this is known as the *solution treatment*. Hardness is accomplished by aging, which follows the quenching operation. The aging temperature for some aluminum alloys is room temperature; others may require an elevated temperature such as 290–360°F, depending on the alloy. As a rule, the lower the aging temperature, the longer is the time required for the alloy to reach full hardness.

Beryllium copper is a precipitation-hardening alloy and is usually furnished by the manufacturer in the very soft solution-treated condition. It has excellent forming properties in this condition. Formed parts are hardened by aging at 560–620°F for two hours at heat. A hardness of 38–42 Rockwell C can be expected.

All other brass and bronze alloys are hardenable only by cold working and may be softened to various degrees by stress relieving or annealing.

SURFACE ROUGHNESS

Definition

By definition surface roughness is relatively finely spaced surface irregularities, which have height, width, direction, and shape that establish the predominant surface pattern.* In order to better understand surface roughness, let us look at an enlarged drawing of the surface of a machined part and see that the main factors we must consider are roughness, waviness, and lay (see Fig. 8-2).

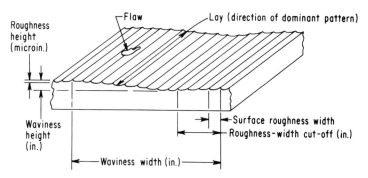

Fig. 8-2. Basic surface characteristics of roughness, waviness, and lay.

The symbol used to designate surface irregularities is the check mark with horizontal extension. A completed roughness, waviness, and lay specification is illustrated in Fig. 8-3.

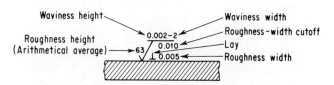

Fig. 8-3. A typical roughness, waviness, and lay specification.

* Adapted from MIL-STD-10A.

Roughness height is measured in microinches (0.000001 in.). Width may be measured by several methods to be discussed later; however, standard roughness-width cutoff has been specified as 0.030 in. by the U.S. Government, and this standard roughness width is used unless specified otherwise. Product specifications covering roughness, waviness, and lay must be specified by the product designer, and for the tool by the tool designer. Surface characteristics should not be controlled on a drawing or specification unless such control is essential to performance of the tool or product. A thorough treatment of the subject will be found in *Tool Engineers Handbook*.[3] *

Method of Manufacture and Cost Relationships

After the surface roughness and tolerance requirements have been specified, the method of manufacture is dictated to a certain extent. The surface roughness requirement and tolerances do not dictate a specific method of manufacture, but they do narrow the breadth of selection since each method of manufacture is capable of producing surfaces having a roughness value within a certain range. Figures 8-4 and 8-5 show that there is considerable overlapping of machine capabilities, but also suggest a relatively limited choice of available machine methods.

The relationships of cost, tolerance, and surface roughness are shown in Fig. 8-6. Machining to close tolerances and low surface roughness consumes more time than does rougher work; hence, the cost is higher.

Range of total tolerance, in.	Symbol *	Typical machining operations
0.062 to 0.125	1000	Flame cutting
0.015 to 0.062	500	Snag grind, sandcasting
0.010 to 0.015	250	Saw, forging, permanent mold casting
0.005 to 0.010	125	Rough turn, drill, shape, mill, bore
0.002 to 0.005	63	Smooth turn, shape, mill, bore, ream
0.0005 to 0.002	32	Grind, smooth turn and polish
0.0002 to 0.0005	16	Grind, hone, burnish
0.0001 to 0.0002	8	Grind, hone, burnish
0.0004 to 0.0001	4	Lap, polish, superfinish

*/ Values are in microinches

Fig. 8-4. A company standard for process finish capabilities. (*Perfecting Service Co.*)

* Superior numbers indicate specific references listed at the end of this chapter.

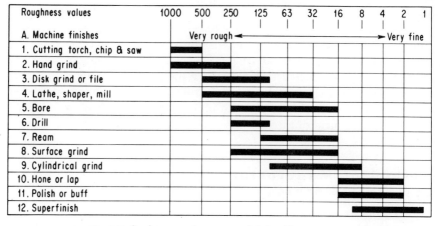

Fig. 8-5. Surface roughness associated with common production operations.

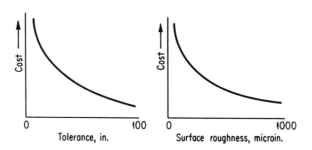

Fig. 8-6. Relationships between tolerance, roughness, and cost.

Measuring Surface Roughness

Once the tolerances and surface requirements have been determined and specified, these requirements can be accurately measured on the part after it has been manufactured.

Available measuring instruments employ an electronic circuit to amplify and measure the vertical movement of a diamond stylus. A standard specimen is available for calibration of the instrument. The instrument stylus is moved over the part being measured and the surface roughness read accurately from the instrument in the range of one to one thousand microinches.

The tool designer must be ever watchful for overpreciseness in his own work as well as that of the product designer. This is particularly true where the tool designer is not consulted during the product design stage. The product designer who specifies the surface roughness, the tool designer who designs the tools to make the part, the production personnel who make the part, and the quality control personnel who inspect the part should all work together for realistic surface roughness specification and realization.

FITS AND TOLERANCES

Definitions

Tolerance is the allowable variation in size from the nominal or specified size. The metalworker constantly deals with the tolerance on a part and the allowances or fits between mating parts. *Allowance* is an intentional difference between maximum material limits of the mating parts. The use of the parts to be manufactured determines the tolerance and allowances and is specified by the product designer. Similarly, tolerances and allowances must also be specified by the tool designer when designing tooling for the parts.

When the tool designer begins work on a die, jig, or fixture, he must be quite familiar with tolerance as well as with clearance and interference fits and when and how to use each.

In Chap. 5, tolerances and allowances are discussed with primary references to gages and gaging practices. Data in the section are therefore chiefly confined to fits and limits as set forth in American Standard ASA B4.1-1955.[2]

Also treated in Chap. 5 are the concepts of geometrical tolerances or tolerances of *form*.

Unilateral Tolerance System

In this system a tolerance is applied in one direction from basic size, and may be plus or minus as shown in Fig. 8-7. This figure shows in graphic form how tolerances are assigned to a mating shaft and hole, and the resulting fit. Two sets of tolerances are shown for a 1-in. shaft and hole and the resulting fits. Fits are given in order of grade to fit, H_1/S_1 being the finest (smallest clearance, see Table 8-5). The difference is in grade of fit only and not in class of fit. The chief advantage of this system is that it allows changing of tolerances without improving the essential class of fit.

In tooling a product under the unilateral tolerance system, it is possible to first tool for a close tolerance and H_1/S_1 fit for the mating parts, and after experience has been gained in the field with the product in use, it may be found possible to increase the tolerance and use H_2/S_2 fit. The change would in most cases lower the manufacturing costs.

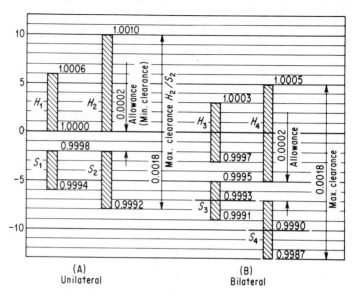

Fig. 8-7. Allocation of unilateral and bilateral tolerances.

TABLE 8-5. VALUES OF UNILATERAL CLEARANCES
(all dimensions in inches)

	Combination			
	H_1/S_1	H_1/S_2	H_2/S_1	H_2/S_2
Allowance (minimum clearance)	0.0002	0.0002	0.0002	0.0002
Hole tolerance	0.0006	0.0006	0.0010	0.0010
Shaft tolerance	0.0004	0.0006	0.0004	0.0006
Maximum clearance	0.0012	0.0014	0.0016	0.0018

Bilateral Tolerance System

This system consists of applying the tolerance in both directions from nominal size of the hole and also the shaft. Tolerances applied in this manner make it impossible to change tolerances on holes and shafts and still retain the same class of fit.

Figure 8-7 shows in graphic form how the tolerances are assigned to a mating shaft and hole, and the resulting fit. The fits are given in order of increasing clearance, H_3/S_3 being the highest grade and class of fit. The difference is in both grade and class of fit as shown in Table 8-6.

From the table the combinations H_3/S_4 and H_4/S_3 result in an allowance of 0.0004 and 0.0000 in. respectively, which constitutes two classes of fits. Combinations H_3/S_3 and H_4/S_4 have the same allowance and are the

TABLE 8-6. VALUES OF BILATERAL CLEARANCES
(all dimensions in inches)

	Combination			
	H_3/S_3	H_3/S_4	H_4/S_3	H_4/S_4
Allowance (minimum clearance) . .	0.0002	0.0004	0.0000	0.0002
Hole tolerance	0.0006 *	0.0006 *	0.0010 †	0.0010 †
Shaft tolerance	0.0004	0.0006	0.0004	0.0006
Maximum clearance	0.0012	0.0016	0.0014	0.0018

* ±0.0003.   † ±0.0005.

same classes of fit, but it is still impossible to change smoothly from one grade of work to another when using the bilateral system. This system is not as widely used as the unilateral system and, therefore, most of the standards in use today are based on the unilateral system.

Basic-hole and Basic-shaft Systems

The designer must choose either of the two systems in order to secure interchangeability. The decision usually depends upon the method of manufacture, the product, or the condition of readily available raw stock. In the United States, common practice is to use the basic-hole system because of the ease of manufacturing and measuring shafts to size as opposed to the greater difficulty of manufacturing and measuring holes to size. Figure 8-8 shows graphically the difference in the two systems.

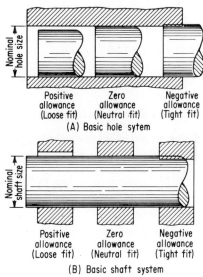

Fig. 8-8. Basic-hole and basic-shaft systems.

Standard Tolerances

Standard tolerances have been determined (Table 8-7) for use in many grades of work. The given tolerances for any one grade represent approximately similar production difficulties throughout the range of sizes for that grade. By using the table, it is possible to select appropriate tolerances for holes and shafts. Use of the table also insures that the parts can be gaged by the use of standard gages during the production process.

TABLE 8-7. STANDARD TOLERANCES ARRANGED BY GRADE
(tolerance values are in thousandths of an inch)

Nominal Size Range, in.		Grade 4	Grade 5	Grade 6	Grade 7	Grade 8	Grade 9	Grade 10	Grade 11	Grade 12	Grade 13
Over	To										
0.04	0.12	0.15	0.20	0.25	0.4	0.6	1.0	1.6	2.5	4	6
0.12	0.24	0.15	0.20	0.3	0.5	0.7	1.2	1.8	3.0	5	7
0.24	0.40	0.15	0.25	0.4	0.6	0.9	1.4	2.2	3.5	6	9
0.40	0.71	0.2	0.3	0.4	0.7	1.0	1.6	2.8	4.0	7	10
0.71	1.19	0.25	0.4	0.5	0.8	1.2	2.0	3.5	5.0	8	12
1.19	1.97	0.3	0.4	0.6	1.0	1.6	2.5	4.0	6	10	16
1.97	3.15	0.3	0.5	0.7	1.2	1.8	3.0	4.5	7	12	18
3.15	4.73	0.4	0.6	0.9	1.4	2.2	3.5	5	9	14	22
4.73	7.09	0.5	0.7	1.0	1.6	2.5	4.0	6	10	16	25
7.09	9.85	0.6	0.8	1.2	1.8	2.8	4.5	7	12	18	28
9.85	12.41	0.6	0.9	1.2	2.0	3.0	5.0	8	12	20	30
12.41	15.75	0.7	1.0	1.4	2.2	3.5	6	9	14	22	35
15.75	19.69	0.8	1.0	1.6	2.5	4	6	10	16	25	40

Classes of Standard Fits

As an aid in learning the classes of fit we will use symbols, but in actual practice dimensions, not symbols, are shown on manufacturing drawings.

    RC—running or sliding fit
    LC—locational clearance fit
    LT—transition fit
    LN—locational interference fit
    FN—force or shrink fit

Use of the symbols above, together with a number, represents a complete fit and can, by the use of the standard tolerance tables, be converted into actual dimensions of the mating parts. An example of a fit using the symbol and number system would be FN4, which means a Class 4 force fit.

Selection of fits is made by an analysis of the service required of the mating parts and then converted into actual dimensions of the parts.

It is beyond the scope of this text to present a complete set of tables, but they are included in American Standard ASA B4.1-1955. Instead, some graphs of the various fits will be shown that will help to visualize the difference in fits, and to give a general knowledge of the relative difference in fits. (See Figs. 8-10 through 8-12.)

Chap. 8 *General Considerations in Tool Design* 467

Figure 8-9 is a graph which gives a comparison of clearance or interference for all classes of fits for a one-inch shaft and hole. On the graph, note particularly the LT classes which show both interference and clearance within this group of fits.

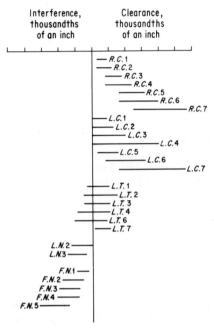

Fig. 8-9. Clearance or interference for each ASA class of fit, for 0.71 to 1.19-in. nominal diameter range.

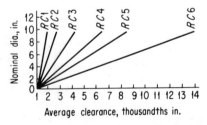

Fig. 8-10. Ranges of clearances for running and sliding fits. RC1—Close sliding fits are intended for accurate location of parts that must assemble without perceptible play. RC2—Sliding fits are intended for accurate location but with greater maximum clearance than RC1. RC3—Precision running fits are about the closest fits that can be expected to run freely at slow speeds and light loads. RC4—Close running fits are intended for running fits on accurate machining and at moderate speeds. RC5 and RC6—Medium running fits are intended for higher running speeds or heavier loads or both. RC7—Free running fits are intended for use when accuracy is not essential or where larger temperature variations are likely to be encountered or both.

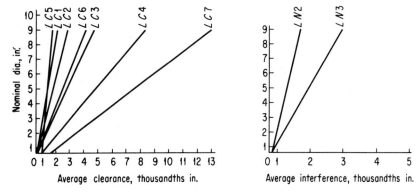

Fig. 8-11. Ranges of clearances and interferences for ASA locational and transition fits. LC—Locational clearance fits are intended for parts that are normally stationary but that can be freely assembled and disassembled. LT—Transition fits are a compromise between clearance and interference fits. They are intended for use when location is important but a small amount of clearance or interference is permissible. LN—Locational interference fits are used when accuracy of location is of prime importance.

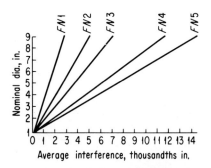

Fig. 8-12. Ranges of interference for ASA force or shrink fits. FN1—Light drive fits require light assembling pressures and produce more or less permanent assemblies. FN2—Medium drive fits are suitable for ordinary steel parts or for shrink fits on thin sections. They are about the tightest fits that can be used with high-grade cast iron outer members. FN3—Heavy drive fits are suitable for heavier steel parts or for shrink fits in medium sections. FN4 and FN5—Force fits are suitable for parts that can be highly stressed or for shrink fits where a heavy assembly press force is impractical.

Chap. 8 *General Considerations in Tool Design* 469

TOOLING ECONOMICS*

The over-all functions of tool and manufacturing engineers embrace the entire span from participation in product development committees, through all process planning, methods engineering, and production, through to final inspection.

At every stage in total manufacturing, these engineers are confronted with the dollar sign. Ever more and better production is demanded, at higher speeds and lower costs.

Tool design is but one somewhat narrow function in the manufacturing complex, but a function that can in many instances make or break the economic success of the entire operation. It is far from adequate merely to accept the available tooling budget and make a good tool within the financial limits set.

The tool designer should know enough economics to determine, for example, whether temporary tooling would suffice even though funds are provided for more expensive permanent tooling. He should be able to check his design plans sufficiently well to initiate or defend a planning decision on writing off the tooling on a single run as opposed to write-off distributed against probable future reruns. He should have an opinion, in some instances and backed by economic proof, of certain process changes that would make optimum use of the tools.

Many aspects of manufacturing economics are beyond the scope of this book, but excellent literature is available.

Analysis of Small-tool Costs

The following analyses are particularly applicable where production rates are small and fixed.

Let N = number of pieces manufactured per year
C = first cost of fixture
I = annual allowance for interest on investment, per cent
M = annual allowance for repairs, per cent
T = annual allowance for taxes, per cent
D = annual allowance for depreciation, per cent
S = yearly cost of setup
a = saving in labor cost per unit
t = percentage of overhead applied on labor saved
V = yearly operating profit over fixed charges
H = number of years required for amortization of investment out of earnings

* Adapted from *Tool Engineers Handbook*.[1]

For simplification, the following values are used in the following examples:

$$a = .03, \quad t = 50\%, \quad \text{cost of each setup} = \$10$$
$$I = 6\%, \quad T = 4\%, \quad M = 10\%, \quad D = 50\%$$
$$I + T + D + M = 70\%$$
$$H = \frac{1}{D} = 2 \text{ years}$$

Number of pieces required to pay for fixture:

$$N = \frac{C(I + T + D + M) + S}{a(1 + t)} \tag{8-1}$$

Example. If a fixture costs $400 and one run is made per year

$$N = \frac{(400 \times 0.70) + 10}{0.03 \times 1.5} = 6450 \text{ pieces}$$

Therefore, on 2 yearly runs of 6450 pieces, the fixture will pay for itself. If 6 runs per year are made,

$$N = \frac{(400 \times 0.70) + (6 \times 10)}{0.03 \times 1.5} = 7550 \text{ pieces}$$

If a fixture must pay for itself in a single run, $H = 1$ and $D = 100$ per cent and $(I + D + T + M) = 120$ per cent,

$$N = \frac{(400 \times 1.20) + 10}{0.045} = 10{,}900 \text{ pieces}$$

Economic investment in fixtures for given production:

$$C = \frac{Na(1 + t) - S}{I + T + D + M} \tag{8-2}$$

Example. For a single run of 10,900 pieces with an estimated saving of 3 cents per piece,

$$C = \frac{(10{,}900 \times 0.045) - 10}{1.20} = \$400$$

If 7550 pieces are made in 6 runs per year and the fixture is to pay for itself in 2 years and $(I + D + T + M) = 70$ per cent,

$$C = \frac{(7550 \times 0.045) - 60}{0.70} = \$400$$

Number of years required for a fixture to pay for itself:

$$H = \frac{C}{Na(1 + t) - C(I + T + M) - S} \tag{8-3}$$

Example. In the preceding example, $C = \$400$; then,

$$H = \frac{400}{(7550 \times 0.045) - (400 \times 0.20) - 60} = 2 \text{ years}$$

Profit from improved fixture designs:

$$V = Na(1 + t) - C(I + T + D + M) - S \tag{8-4}$$

The previous equations have assumed a break-even cost.

Example. Assume that $C = \$250$ instead of $\$400$. Then, with 7550 pieces per year and 6 runs per year,

$$V = (7550 \times 0.045) - (250 \times 0.07) - 60 = \$105 \text{ per year}$$

Equation (8-2) can be used for comparing costs of alternative fixtures having different setup and labor costs.

Example. In a case of a 2000-piece run, and $D = 100$ per cent, assume (1) a fixture with $S = \$10$ and $a = \$0.03$ and (2) a fixture with $S = \$15$ and $a = \$0.05$.

$$C_1 = \frac{(2000 \times 0.03 \times 1.5) - 10}{1.20} = \$66.66$$

$$C_2 = \frac{(2000 \times 0.05 \times 1.5) - 15}{1.20} = \$112.50$$

Tooling Economies in Combined Operations. Analysis may sometimes show that operations can be advantageously combined. The total cost of tooling may thus be reduced, or the production costs, or both. Table 8-8 illustrates a case where the cost of combined tools was less than the total cost of the separate tools otherwise required. The combined operation was done at the speed of the blanking operation above.

TABLE 8-8. COSTS OF COMBINED VS. SEPARATE OPERATIONS [3]

Costs	Blanking operation alone	Forming operation alone	Total blank and form	Combined operations
Tools	$40.00	$30.00	$ 70.00	$50.00
Setup	2.00	2.00	4.00	3.00
Maintenance	2.00	..	2.00	2.00
Processing	4.00	30.00	34.00	4.00
Total cost	$48.00	$62.00	$110.00	$59.00

Process-cost Comparisons. During process planning, many possible methods of manufacturing may be reduced to a few based upon alternate process steps, use of available equipment, or combined operations. Under these conditions, a comparison of costs for the different tools and process steps may quickly reveal the combination that will result in the lowest total cost per part.

Let N_t = total number of parts to be produced in a single run
 N_b = number of parts for which the unit costs will be equal for each of two compared methods Y and Z ("break-even point")
 T_y = total tool cost for method Y
 T_z = total tool cost for method Z
 P_y = unit tool process cost for method Y
 P_z = unit tool process cost for method Z
 C_y, C_z = total unit cost for methods Y and Z, respectively.

Then

$$N_b = \frac{T_y - T_z}{P_z - P_y} \quad (8\text{-}5)$$

$$C_y = \frac{P_y N_t + T_y}{N_t} \quad (8\text{-}6)$$

$$C_z = \frac{P_z N_t + T_z}{N_t} \quad (8\text{-}6a)$$

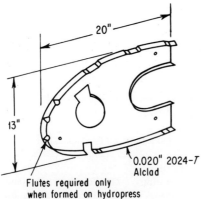

Fig. 8-13. General specifications for aircraft-flap nose rib.

Example. The aircraft-flap nose rib shown in Fig. 8-13 of 0.02-in. 2024-T Alclad was separately calculated to be formed by Hydropress, drop hammer, Marform, steel draw die, and hand forming. For such reasons as die life, equipment available, and handwork required, the choice narrowed down to Hydropress vs. steel draw die. With Hydropress, the flanges had to be fluted, and for piece quantities over 100 a more expensive steel die costing $202 had to be used.

Actual die and processing costs for both methods are listed in Table 8-9. P_y and P_z are processing costs, and T_y and T_z are die costs, for the steel draw die and Hydropress methods, respectively.

Figures in the last column of Table 8-9 were not stated in the original report but can properly be extrapolated on the basis of apparent stability of P_y and P_z at $N_t = 500$, and assuming their stability at higher production.

Chap. 8   General Considerations in Tool Design

TABLE 8-9. COST COMPARISON OF METHODS FOR PRODUCING
AIRCRAFT-FLAP NOSE RIB

N_t*	5	25	50	100	500	780†
P_y	$ 3.00	$ 1.18	$ 1.11	$ 1.05	$ 1.05	$ 1.05
P_z	4.40	2.05	1.96	1.85	1.85	1.85
T_y	810.00	810.00	810.00	810.00	810.00	810.00
T_z	103.00	103.00	103.00	103.00	202.00	202.00
C_y	165.00	33.60	17.30	9.15	2.67	2.12
C_z	25.00	6.17	4.02	2.88	2.25	2.12

* Symbols in this column are the same as in Eqs. (8-6) and (8-6a).
† Extrapolated.

On the basis of listed figures at $N_t = 500$, and from Eq. (8-5), the production at which total unit costs C_y and C_z will be the same for both methods is

$$N_b = \frac{810 - 202}{1.85 - 105} = 760 \text{ pieces}$$

As an alternate method for calculating the break-even point between two machines, e.g., a turret lathe and an automatic, the Warner & Swasey Co. uses a formula based on known or estimated elements that make up production costs. The break-even-point formula is

$$Q = \frac{pP(SL + SD - sl - sd)}{P(l + d) - p(L + D)} \qquad (8-7)$$

where $Q =$ quantity of pieces at break-even point
$p =$ number of pieces produced per hour by the first machine
$P =$ number of pieces produced per hour by the second machine
$S =$ setup hours required on the second machine
$s =$ setup hours required on the first machine
$L =$ labor rate for the second machine, in dollars
$l =$ labor rate for the first machine, in dollars
$D =$ hourly depreciation rate for the second machine (based on machine-hours for the base years period)
$d =$ hourly depreciation rate for the first machine (based on machine-hours for the base years period)

Example. Assume that it is desired to find value Q when the various factors are as follows: $p = 10$ pieces per hour; $P = 30$ pieces per hour; $S = 6$ hr; $s = 2$ hr; $L = \$1.50$ per hour; $l = \$1.50$ per hour; $D = \$1.175$ per machine-hour; and $d = \$0.40$ per machine-hour (10-year period).

Then, substituting these values in the formula,

$$Q = \frac{10 \times 30(6 \times 1.5 + 6 \times 1.175 - 2 \times 1.5 - 2 \times 0.4)}{30(1.5 + 0.4) - 10(1.5 + 1.175)}$$

Answer. $Q = 121$ pieces, or the quantity of parts on which the cost is the same for either machine.

Effect of Tool Material on Minimum-cost Tool Life. The tool designer can often influence the choice of cutting-tool materials. Certainly he should know the tool-life economics involved.

Typical tool-life curves are shown in Fig. 8-14 for high-speed-steel, sintered carbide, and oxide tools. The wear characteristics, as denoted by the n values, show that economic life T_c for carbide tools is shorter than that for HSS tools (T_c equals 15 min for carbide vs. 35 min for HSS). It is even more important to use higher speeds and shorter tool life for oxide tools, since T_c equals 5 min.

From Table 8-10, it is evident that most economic metal removal demands high horsepower and speeds when cutting with carbide and oxide tools. Data in the table assume equal cost per tool of $0.50, a labor overhead rate of $6.00 per hour, and a tool change time (TCT) of 2 min.

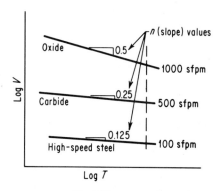

Fig. 8-14. Tool-life comparisons for various tool materials.

TABLE 8-10. PERFORMANCE DATA FOR VARIOUS TOOL MATERIALS

Tool material	$1/n - 1$ * values	Tool Life, min T_c	T_p	Minimum-cost cutting speed, V_c, sfpm
High-speed-steel	7	35	14	107
Tungsten carbide	3	15	6	701
Oxide	1	5	2	3500

* The term $1/n - 1$ is derived from differentiation of the tool-life equation $VT^n = C_t$, where $V =$ cutting speed, $T =$ actual cutting time between sharpenings, and C_t is a constant numerically equal to the cutting speed that gives a 1-minute tool life under the actual cutting conditions.

Economic Lot Sizes

Economic lot sizes are calculated and employed to obtain the minimum unit cost of a given part or material. This minimum is reached when the costs of planning, ordering, setting up, handling, and tooling equal the costs of storage of finished parts. These costs may be equated and the lot size determined by mathematical calculation. Depending on the number of variables to consider, the formula can range from one of relative simplicity to one that is relatively complex.

By assuming the number of pieces required per month as constant, the inventory increasing until the lot ordered is completely sent to stock, and decreasing uniformly with use, a relatively simple formula for calculation of economic lot size can be devised.

$$L = \sqrt{\frac{24mS}{kc(1 + mv)}} \qquad (8\text{-}8)$$

where c = value of each piece, dollars
  k = annual carrying charge per dollar of inventory, dollars
  L = lot size, pieces
  m = monthly consumption, pieces
  S = setup cost per lot, dollars
  v = ratio of machining time to lot sizes, months per piece

Labor, material, and other costs not related to lot size are also omitted because only costs pertinent to lot size have any influence on lot size. In actual practice, reasonable values for S, m, and c are obtained only with difficulty unless the groundwork has been done by the standards, sales, or methods departments.

Formulas for calculating economic lot sizes can never prove out exactly because of the assumptions upon which many of the factors may or must be based. Such a formula should be regarded as just a useful guide, to be applied with mature judgment.

Break-even Charts

Break-even charts are perhaps most widely used to determine profits based on anticipated sales. Other uses, however, can be made of them, such as for selecting equipment or for measuring the advisability of increased automation.

In determining which of two machines is most economical, the fixed cost of the equipment is plotted (Fig. 8-15) with the related variable costs (found by multiplying the number of pieces by the unit-piece cost). The total cost is composed of the sum of the fixed and variable costs.

For example, assume the initial cost for machine A is $1500 and the unit production cost on the machine is $0.75 each. For the other machine B the initial cost is $6000 and the unit production cost $0.15 each.

The chart shows that it is more economical to purchase machine A if production never exceeds 7500 pieces. For higher production quantities the economy lies with machine B.

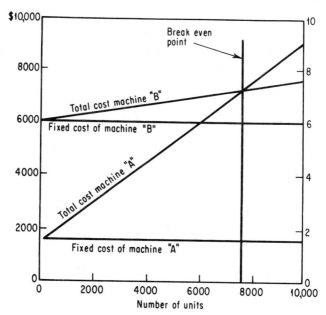

Fig. 8-15. Break-even chart for machine selection; choice is based upon volume of production.

MATERIAL HANDLING AT THE WORKPLACE

It is beyond the scope of this text to go into detail regarding the bulk handling of materials and parts through the factory, or the many principles that underly the design of a workplace. Full texts are devoted to this subject-matter area. It is the objective of this brief discussion to emphasize the important part that the tool design plays in the total cost of an operation from the standpoint of materials handling.

To pick up a workpiece, place it into a tool, clamp the part, unclamp, and remove and set aside after machining may consume more time than the actual machining. Such operations, however essential, contribute nothing of value to the product, but their performance is paid for at the same rate as the productive effort. Consequently, tooling should be designed so far as possible to reduce the nonproductive time and costs.

Once a tool is designed and built, the methods of handling materials into and out of it are fixed. It is essential then that the original methods planning be carried out with care. The following list of principles * should be applied to help ensure maximum motion economies. It will be noted that the principles are arranged under three general headings: (1) use of the

* Adapted from *Motion and Time Study*.[4]

human body; (2) arrangement of the workplace, and (3) design of tools and equipment.

1. *Use of the human body*
    a. Both hands should begin and end their basic divisions of accomplishment simultaneously and should not be idle at the same instant, except during rest periods.
    b. The motions made by the hands should be made symmetrically and simultaneously away from and toward the center of the body.
    c. Momentum should be employed to assist the worker wherever possible, and should be reduced to a minimum if it must be overcome by muscular effort.
    d. Continuous curved motions are preferable to straight-line motions involving sudden and sharp changes in direction.
    e. The least number of basic divisions should be used, and these should be confined to the lowest possible classifications. These classifications, summarized in ascending order of time and fatigue expended in their performance, are:
        (1) Finger motions
        (2) Finger and wrist motions
        (3) Finger, wrist, and lower arm motions
        (4) Finger, wrist, lower arm, and upper arm motions
        (5) Finger, wrist, lower arm, upper arm, and body motions
    f. Work that can be done by the feet should be arranged so that it is done simultaneously with work being done by the hands.

2. *Arrangement and conditions of the workplace*
    a. Fixed locations should be provided for all tools and materials so as to permit the best sequence and to eliminate or reduce the therbligs *search, select,* and *find*.
    b. Gravity bins and drop delivery should be used to reduce *reach* and *move* times; also ejectors should be provided wherever possible to remove finished parts automatically.
    c. All materials and tools should be located within the normal area in both the vertical and horizontal planes.
    d. A comfortable chair should be provided for the operator, and the height should be so arranged that the work can be efficiently performed by the operator alternately standing and sitting.
    e. Proper illumination, ventilation, and temperature should be provided.
    f. Visual requirements of the workplace should be considered so that eye fixation demands are minimized.

g. Rhythm is essential to the smooth and automatic performance of an operation, and the work should be arranged to permit an easy and natural rhythm wherever possible.

3. *Design of tools and equipment*
   a. Multiple cuts should be taken whenever possible by combining two or more tools in one, or by arranging simultaneous cuts from both feeding devices if available (cross slide and hex turret).
   b. All levers, handles, wheels, and other control devices should be readily accessible to the operator and should be designed so as to give the best possible mechanical advantage.
   c. Holding parts in position should be done by fixtures.
   d. Always investigate the possibility of powered or semiautomatic tools, such as power nut and screw drivers, and speed wrenches.

For the manipulation of materials in and out of presses there are many types of mechanical feeding and ejection devices.* These may be used alone or in combination with manual feeding and ejecting methods. Good design practice calls for full consideration of available types and for selection of the optimum design through critical evaluation of the functional, quality, cost, time, and auxiliary requirements to be met.

RULES FOR GOOD DESIGN

Many principles and specific recommendations for good tool design are stated in previous chapters of this book. However, for readier checking of any proposed tool design it is handy to have the commoner check points listed together in one place.

Fixture Design Questions[5]

Examine the following, as affecting *setup*:

1. Will the fixture fit into the machine for which it is intended?
2. Will the clamping slots or holes in the fixture line up with the T slots in the table?
3. Will the fixture, when in place, overhang the end of the table?
4. On a multispindle machine, will the jig interfere with any other jig or fixture next to it?

---

* An informative discussion of feeding and ejecting devices is contained in *Tool Engineers Handbook*,[1] pp. 56-11 through 56-14.

Chap. 8 General Considerations in Tool Design   479

5. Can the setup man see whether the cutter or drill is correctly set?
6. Can cutting tools be adjusted when the fixture is in place, or easily removed for sharpening?
7. Can setting blocks, bushings, stops, or collars be used in setting up the cutting tools?
8. Have suitable locating plugs been provided for setting up?
9. Does the setup man need more than one size of wrench (undesirable)?
10. Are holddown bolts so placed as to make tightening easy?

As to *locating and holding the workpiece*:

11. Is there any obstruction in the shop layout that would hinder loading of the workpiece into the fixture?
12. Is the fixture design correctly related to the material-handling equipment?
13. How will allowable variations in the shape of the workpiece affect its location in the jig or fixture?
14. If no previous operations have been done on the workpiece, are there suitable datum faces or points from which to locate?
15. Are the locating points as widely spaced as possible?
16. Are centralizing means required to compensate for variations in the workpiece?
17. Is the tolerance on locating points sufficiently close to obtain the specified operational accuracy?
18. Can all subsequent operations be located from the same points?
19. Are the locating points as small in area as practicable?
20. Are locators safe from damage by cutters' overrunning or being set too deep?

As to *loading and unloading the workpiece*:

21. Will cutters such as long drills interfere with workpiece loading or locating?
22. Will the clamps interfere with loading or unloading?
23. Is clearance sufficient to permit the workpiece to be easily lifted over or into the locating and centering devices?
24. Have any sliding pins or other hand-operated locators been provided with comfortable handles?
25. Are all movable locators and adjustments on the side of the fixture nearest the operator?
26. Can the fixture be loaded with one hand while the other hand is unloading the completed workpiece?
27. Are any burrs likely to interfere with unloading?
28. Should an ejector be provided?

As to *the forces involved*:

29. Is the workpiece supported as closely as possible to the point of load application?
30. Is the cutting force resisted by a solid support, and not by the clamp?
31. Can the cutting force be used to help locate and secure the workpiece?
32. Has the clamp sufficient range to accommodate allowable workpiece variations?
33. Is the workpiece supported directly under the clamping points?
34. Will the clamping force unduly distort the workpiece?
35. Will the clamp tend to loosen under cutter chatter or vibration?
36. Can it be planned to have a single standard wrench tighten all the clamps?

As to *sundry requirements*:

37. Does the workpiece size, the required clamping force, or the required speed of action warrant use of pneumatic or hydraulic clamping?
38. Are the jig feet large enough to span the T slots in the machine table?
39. Will the fixture design keep the length of cutter travel to a minimum?
40. Will the operator, when positioning the jig, be able clearly to see all bushings or cutter guides?
41. If the cutters need guiding, are the bushings long enough to provide adequate tool support?
42. Do the tools need guiding for a second operation?
43. Can the use of slip bushings be avoided by the use of stepped drills?
44. If slip bushings must be used, are the heads large enough, fluted for easy gripping, and provided with locking means?
45. Do all supporting pads and pins stand well clear of chip-collecting surfaces?
46. Can channels be provided to allow the coolant to wash the chips away?
47. Is it possible for chips to foul the clamp-lifting springs?
48. Can the cutter flutes discharge chips even when covered up?
49. If the workpiece is to be measured while still in the fixture, can it be easily cleaned?
50. Is there sufficient clearance between tools and the workpiece for easy gaging?

Chap. 8 General Considerations in Tool Design

51. Can the tools be damaged or made inaccurate through incorrect insertion of the workpiece?
52. Does the fixture design protect the operator from coolant spray or flying chips?
53. Is the fixture so heavy as to require lifting lugs, eyebolts, or chain slots?
54. If loose parts are unavoidable, can they be attached to the fixture with keeper screws or light chains?
55. Have standard commercial components been specified wherever practicable?
56. Have blind holes been avoided wherever possible?
57. Are the dowel pins spaced as far apart as practicable?
58. Have breather holes been provided to allow escape of air from close-fitting plunger holes?
59. Provided the production and economics warrant, have all wearing parts been specified to be hardened?

Die Design Questions[3]

Examine the following, as to *preliminary planning*:

1. Has the blank been developed with due regard to best grain direction, to stresses and strains involved, and to the pressworking equipment to be used?
2. Are idle stations needed in a planned progressive die?
3. Can required dimensional accuracy be realized from the planned stock strip layout?
4. Can the burr be so placed as to require no removal?
5. Is the correct side of the blank up with respect to any shaved portions?
6. Will any forming be done across the grain (optimum), or not to exceed 45°?
7. Is material utilization maximum?
8. Have proper provisions been made for clamping the die set to the press?
9. Have the design features been checked against the shut height of the closed die?
10. Have unavoidable delicate projections been designed as inserts, for easy replacement?
11. Has the centerline of pressure been properly established?
12. Have any pilot-hole punches been suitably located?
13. Has the *final* sequence of operations been thoroughly checked and established?

As to *punch planning*:

14. Have any notching punches been located and, if needed, provided with heel blocks or other backup support?
15. Has it been determined whether shedder provision is needed on any forming punches?
16. Where small pierce or blank punches are to be grouped closely together, are they stepped to reduce total shearing pressure?
17. If punches must be used having more than about 4-in. unguided length, have spacers or filler plates been considered?
18. Have heel punch fillets been made as large as possible?
19. Have spanking punches, if any, been located at next-to-last station and, preferably, combined with bending or forming?

As to *die plates and punch plates*:

20. Provided the intended service requires it, has the die block been specified to be finished square on all sides?
21. Have edges of die openings been designed a minimum distance of 1 to 1½ times block thickness from outside edge of block?
22. Have the punches and dies been designed *sectional*, where feasible, for easy construction, hardening, sharpening, and replacement?
23. Are any finger stops so located as to avoid cutting on only one edge of the die?
24. Will inserts and bushings be planned wherever needed to facilitate diemaking, heat treatment, or easy replacement of worn or broken sections?
25. Has a selected die set been checked for parallelism of mounting surfaces? for fit of guide posts in their bushings?
26. Have needed scrap cutters been suitably located?
27. Have adequate provisions been made for scrap disposal?
28. Has doweling been checked for sufficient size to withstand shearing action; for spacing far enough apart; for means of removal from blind holes; for advisable staggering to prevent misassembly?
29. Is the punch plate sufficiently thick to support all punches adequately?
30. Have any necessary clearance holes in die block or stripper been checked for transport of blanks or slugs?
31. Have any hardened punches been designed to be mounted in a soft plug, rather than pressed directly into a hardened punch plate?

As to *general design details*:

32. If die setup pins are to be used, are they large enough for needed rigidity, and far enough apart?

33. Have any needed release or vacuum pins been checked as to location and action?
34. Have blank-hole and scrap-hole clearances been checked?
35. Have the sizes of all springs been calculated?
36. Have bushing decisions been checked as to need, location, and optimum length?
37. Has the planned piloting practice been checked as to removability to facilitate punch grinding, for adjustability, and to avoid misfeeds?
38. Have boltheads in die plates been set sufficiently below the top surface to permit maximum die sharpening?
39. Have any necessary air vent holes been located?
40. Are stop or bumper blocks needed anywhere?
41. Has a thorough check been made to ensure safety to the operator, the die, and the press?

As to *heat treatment of tooling*: Design is the sum total of many variables among which are geometry, mass, surface area, surface finish, material used, method of fabrication, and heat treatment.

Since heat treatment is probably the most severe operation any tool or die must go through, it is necessary that ease or safety in heat treatment be given every possible consideration when designing tools and dies. Some of the basic rules for design, directly related to heat treatment, are:

1. Use sufficiently oversize stock to insure freedom from surface defects and decarburization after grade selection is made (Table 4).
2. Generous fillets should be used whenever possible to minimize stress concentration during heat treatment.
3. Avoid sharp re-entrant angles; also square inside corners.
4. Avoid thin-walled areas. Increase cross section in such areas if possible.
5. Avoid drastic changes in cross section. Use steps or taper whenever possible.
6. Use sectional dies if the design is considered to be hazardous.
7. Avoid the use of blind holes when possible, because they tend to alter uniformity of cooling.
8. Use fillets at base of keyways to minimize stress concentration.
9. Avoid use of large masses. If design permits, incorporate a hole to facilitate cooling.

(See also "Heat Treating," above.)

PROBLEMS

1. What five questions should be asked by the tool designer before starting the basic design?
2. Why is it important to check machine size and capacity before design?
3. Why is oversize stock specified on parts to be heat treated?
4. What is a routing sheet and what is its function?
5. What type of steel should be specified for welded construction?
6. Why should the cutting tool force be in the direction of the nest instead of the clamp?
7. Name two means of testing the hardness of material.
8. What are the three methods of testing impact strength of a material of a specified hardness?
9. When would cast iron be used in designing a fixture?
10. What are some of the purposes of heat treating?
11. How many years would it take and how many pieces would have to be made to reach the break-even point if the fixture cost $500.00?

    Given: $a = 0.02$ $M = 20\%$
    $t = 50\%$ $D = 40\%$
12. Using the values given in question 11, how many pieces must be run to break-even in one run?

REFERENCES

1. American Society of Tool and Manufacturing Engineers, *Tool Engineers Handbook*, 2nd. ed., McGraw-Hill Book Co., Inc., New York, 1959.
2. *Preferred Limits and Fits for Cylindrical Parts*, ASA B4. 1-1955, American Standards Association.
3. American Society of Tool and Manufacturing Engineers, *Die Design Handbook*, McGraw-Hill Book Co., Inc., New York, 1955.
4. Niebel, B. W., *Motion and Time Study*, Richard D. Irwin, Inc., Homewood, Ill., 1958.
5. Checklist of "Jig and Fixture Design," *American Machinist*, Sept. 12, 1955, and Sept. 26, 1955.

INDEX

A

Alignment, 328
Allowances
 bend, 222
 clinch, 395
 die-shift, 301
 gage-wear, 335
 machining, for castings, 413
 scrap-strip, 206
 shrinkage for castings, 413
Amplification of dimensional variations, 345
Angle plates, 126
Angles
 basic tool, 2
 cutting, 4
 rake, 3
 relief, 3
 for single-point tools, 7
Angularity, 330
Annealing, 457
Arbors, 130
ASA Standards for die sets, 208
Asperities, 27

B

Beading dies, 236
Bend allowances, 222
Bending dies
 blank development, 225
 evolution of, 224

Bending methods
 edge bending, 223
 V-bending, 223
Bending pressures, 223
Bernoulli's theorem, 419
Blankholder pressure, 255
Blanks
 development of, 253
 diameters for shells, 254
 reduction, 255
Blanking dies
 evolution of, 211
Blocks, setting, 142
Boring, 151
Brazing
 fixtures, 383
 tools, 382
Breaker, chip, 6
Break-even charts, 475
Brinell hardness, 446
Broaches, 66
Built-up edge, 13
Bulging dies, 237
Bulging operations, 237
Bushings, guide, 147

C

Casting, definition, 406
 tooling for, 406-436
Castings
 machining allowance, 413

Castings (*cont.*)
  minimum thickness, 413
  pouring time, 422
  sand, 407
  shrinkage allowance, 413
Charpy (*see* Tests, impact)
Charts, break-even, 475
Chatter, 86
Chip breakers, 6
Chips
  disposal of, 85
  formation, 9, 11
  types, 11
Chucks
  electrostatic, 128
  lathe, 135
  magnetic, 126
  vacuum, 127
Clamping
  elements, 113
  forces, 111
Clamps
  for welding fixtures, 381
Clearances
  angular, 176
  die, 172
  draw, 258
Coining dies, 241, 281
Coining presses, 291
Collets, 132
Concentricity, 330
Cores
  blowing of, 415
  boxes, 415
  support, 417
Coremaking equipment, 414
Corrosion of metals, 448
Costs (*see* Economics)
Counterbores, 72
Countersinks, 72
Curling dies, 236
Cut size, 16
Cutters
  end-milling, 77
  gear-shaper, 68
  milling, 68, 77
  setting of, 142
Cutting dies
  blanking, 199, 202
  compound, 204
  cut-off, 199
  lancing, 199
  notching, 199
  piercing, 199
  shaving, 199
  steel-rule, 202
Cutting fluids, 21, 23

Cutting forces
  determination of, 176
  direction of, 144
  reduction of, 178
Cutting process (*see* Shearing process)
Cutting speeds, 15
Cutting tools, 1-87 (*see also* Single and multiple-point tools, Milling cutters, Drills, etc.)

D

Degrees of freedom, 94
Dial indicators, 346
Diamond-pin locators, 99
Die blocks
  critical area (Table), 181
  design, 178, 211
  dimensioning of, 178, 211
  thickness of, 179, 211
Die castings
  cored holes in, 435
  draft, 435
  inserts, 435
  runners, 433
  section thickness, 434
  undercuts, 434
Die-casting dies
  cores for, 435
  gates, 431
  materials, 435
  slides, 435
  sprue pins, 434
  vents, 432
  wear reduction, 435
Die-cutting operations, 199
  (*see also* Cutting dies)
Die materials, metallic, 449, 452, 454
Die materials, nonmetallic
  densified wood, 455
  hardboard, 455
  plastic, 456
  rubber, 456
Die sets
  commercial, 208
  selection of, 211
Die sinking, 78
Die springs, 195
Die steels, 449
Dies
  clearances in, 172, 176
  design questions, 481
  die-casting (*see* Die-casting dies)
  drawing (*see* Draw dies)
  drop-forging, 303
  heat treatment, 452

Index

progressive (*see* Progressive dies)
   shear action in, 169
   springs for, 195
   upsetting-machine, 316
Dimensional variations, 345
Distance, linear, 329
Draw dies
   double-action, 252
   draw clearance, 258
   draw radius, 257
   evolution of, 256
   lubrication of, 259
   materials, 258
   metal flow in, 246
   single-action, 249
Draw radii, 257
Draw rings, 258
Drawing, definition, 246
   lubricants for, 259
   pressures in, 255
Drill fixtures, 147
Drill jigs
   box-type, 149
   evolution of, 146
   leaf-type, 149
   pump-type, 121
Drilling
   horse power for, 62
   process of, 146
   thrust in, 62, 146
   torque in, 62, 146
Drills
   core-type, 70
   step-type, 74
   subland, 74
   twist, 58, 68
Drop-forging dies, 303

E

Economics
   lot sizes, 474
   of machining, 43
   of tooling, 469
Edge bending, 223
Ejectors, 194
Elasticity, moduli of, 447
Embossing dies, 233
Embossing operations, 233
End mills, 77
Extrusion dies
   for backward extrusion, 280
   combination, 280
   for diepots, 284
   for forward extrusion, 280

lubrication of, 279
pressure anvils, 84
pressures, 277
punches for, 283
Eyeleting, 394

F

Fatigue of metals, 448
Fillets, 299
Fits, standard, 466
Fixtures
   arc-welding, 375
   design procedure, 160
   design questions, 478
   drill, 147
   elements of, 115
   gas-welding, 374
   indexing milling, 157
   for joining processes, 373
   milling, evolution of, 152
   resistance-welding, 376
   riveting, 393, 399
   spot-welding, 376
   for threaded fasteners, 392
   types of, 115
   for welding, 373, 381
Flatness, 327
Forces
   in clamping, 111
   distribution of, 273
   cutting
      determination of, 176
      direction of, 144
      reduction of, 178
   drilling, 64
   shearing, 176, 178
   in stripping, 195
   tool, 107
Forging
   auxiliary tools for, 303-311
   design for, 296, 317
   draft in, 296
   finish allowances, 302
   flashing in, 309
   machines for, 285, 286, 288, 292
   presses for, 287, 289
   process of, 293
   shrinkage in, 300
   stock gathering, 317
   thickness tolerances, 301
Forging dies
   for forging machines, 319, 321
   gutter dimensioning, 309
   parting lines, 297
   wear, 300

Forming dies
  definition, 225
  solid-type, 226
  pressure-pad-type, 231
Freedom, degrees of, 94

G

Gagemaker's tolerances, 333
Gages
  dial-indicator, 346
  electrical, 353
  flush-pin, 343
  materials for, 336
  optical comparators, 355
  plug-type, 340
  pneumatic, 351
  for positionally toleranced parts, 361
  ring-type, 341
  snap-type, 342
  tolerances, 332
  types of, 338
  wear allowances, 335
Gaging
  applications of, 362
  of concentricity, 366
  of contours, 366
  hole patterns, 364
  by optical methods, 142, 355, 367
  policy, 336
  of positionally toleranced parts, 357, 362
  of squareness, 364
  of straightness, 365
Gating systems
  elements of, 421
  friction losses, 420
  for permanent molds, 430
  pouring basins, 421
  ratio gating, 422
  sprue base wells, 421
  turbulences in, 418
  velocities in, 419
Gear shaper cutters, 68
Geometry, cutting-tool
  effects upon chips, 19
  effects upon tool life, 41
Guide bushings, 147

H

Hardening (heat treatment), 458
Hardness
  Brinell, 446
  Rockwell, 446

Heat treatment
  of die steels, 452
  of ferrous metals, 456
  of nonferrous metals, 459
Helve hammers, 288
Hobs, 81
Hole-extruding dies, 244
Hole flange widths, 245
Hole-flanging dies, 244
Horsepower
  for drilling, 62
  for machining, 51
  for machining different materials (Table), 53
  for milling, 64
  for turning, 52

I

Impact extrusion, 277
Impact tests (*see* Tests, impact)
Indicators, dial, 346
Induction heating
  inductors for, 385
  internal, 389
Inspection tools, 325-370
Izod (*see* Tests, impact)

J

Jaws, vise, 90, 124
Jigs, drill
  box-type, 149
  leaf-type, 149
  pump-type, 121
Joining processes, 372

K

Keys, 142
Knockouts, 197

L

Lathe chucks, 135
Lettering dies, 234
Location
  accuracy of, 120
  axial, 133
  center of pressure, 170
  methods, 94
  radial, 99
  3-2-1 method, 94

Index 489

Locators
  cavity type, 104
  diamond-pin, 99
  nest-type, 104
  pin-type, 94, 97
  V-type, 100
Lubricants, 258

M

Machinability, tool-wear, effects on, 29
Machining
  allowances for castings, 413
  economics, 43
  elements of, 1
  horsepower for, 12, 51
Magnification of dimensional variations, 345
Mandrels
  combination, 131
  roll-lock, 131
  solid, 130
  straight, 131
Manipulating factors
  effects upon tool wear, 28
Material-cutting tools, 1-87
Materials
  for gages, 336
  handling, 476
  tool, 19
  workpiece, 22
Maximum Material Condition (MMC), 359
Metal flow
  in draw dies, 246
  in rectangular shells, 248
Metal mold casting (*see* Permanent mold casting, Die casting)
Milling
  climb method, 76, 152
  conventional method, 76
  end mills, 77
  fixtures, evolution of, 152
  form-relieved cutters, 76
  horsepower for, 64
  number of cutter teeth, 77
  profile-sharpened cutters, 76
  straddle method, 156
Milling cutters (*see* Milling)
Mismatch, 301
Modulus of elasticity
  in bending, 447
  in torsion, 448
Multiple-point tools
  design, 60
  operation, 61
  types, 56

N

Nests, 104
Normalizing, 457
Nose radius, 4

O

OBI Press, 167

P

Parallelism, 328
Patterns
  coatings, 412
  colors, 411
  materials, 412
Peening, spin (*see* Spin peening)
Permanent molds
  gating systems for, 430
  materials, 431
  risers, 430
  vents, 430
Permanent mold casting, 427
Perpendicularity, 329
Pilots
  dimensioning of, 193
  direct-type, 217, 263
  indirect-type, 193, 263
  press-fit, 191
  retaining, 192
Plates, angle, 126
Pouring basins, 421
Pouring time, 422
Power for machining, 51
Presses
  coining, 291
  components of, 167
  forging, 287
  open-back-inclinable (OBI), 167
  selection of, 273
  swaging, 242
  types of, 168
Pressures
  bending, 223
  blankholder, 255
  center of, 170
  drawing, 255
  extrusion (Table), 278
  shearing, 176, 290
  stripping, 195, 213
  swaging, 243
Pressworking tools, 167-220
Progressive blanking dies
  evolution of, 213
Progressive dies
  elements, 269

490 Index

Progressive dies (*cont.*)
  evolution of, 269
  four-station, 273
  general design, 269
  selection, 260
  six-station, 274
  strip development, 260
Pump jigs, 121
Punches
  design, 213
  dimensioning, 182
  quilled, 184
  support methods, 182

Q

Quilled punches, 184

R

Radius
  bend, 221
  corner, 299
  draw, 257
  nose, 4
Ratio gating, 422
Reamers, 70
Reaming, 151
Resistance welding
  electrodes, 379
  types, 376
Reynolds' numbers, 419
Risers
  feeding distance, 424
  location, 424
  for permanent mold casting, 430
Riveting
  equipment for, 396
Rivets, 393
Rockwell hardness, 446
Roughness, surface, 327, 460
Route sheets, 441

S

Safety, 443
Sand casting
  patterns, 411
Scrap strip
  development, 206, 267
  disposition, 267
  layout, 206, 267
Shear action in die cutting, 169
Shearing forces, 176
Shearing process, 169
Shear strengths (Table), 177

Shell-molding, design factors, 426
Signature, tool, 4
Single-point tools
  angles for, 7, 8, 9
  types, 2
Soldering, tools for, 382
Speeds, cutting, 15
Spheroidizing, 457
Spin peening, 395
Springback, 223
Sprue base wells, 421
Squareness, 329
Staking, 402
Stapling, 400
Steel-rule dies, 202
Stitching, 400
Stock gathering, 317
Stock positioning, 263 (*see also* Pilots)
Stops, stock
  automatic, 186
  combination, 188
  escapement, 187
  latch-type, 185
  overhanging, 188
  pin-type, 186, 213
  punch-type, 188
  solid, 185
  starting-type, 185
  trigger-type, 185
  trim-type, 188
Straightness, 328
Strength
  compressive, 447
  impact, 448
  shear, 177
  tensile, 446
  yield, 447
Stress relieving, 457
Strip development, 206, 267
Strippers
  dimensioning of, 195
  fixed-type, 194, 213
  inside-type, 194
  spring-operated, 194
  springs, 197
Stripping forces, 195
Surface plates, 338
Swaging dies, 242

T

Tapping, 151
Taps, 78
Tempering, 459
Templates
  for drilling, 150
  for gaging, 339

Index

Tests, impact
  Charpy, 448
  Izod, 448
  torsion, 448
Thermal equilibrium, 428
Threaded fasteners, 391
Thrust, drilling, 62
Time and motion study, 476
Tolerances
  basic-hole, 465
  basic-shaft, 465
  bilateral, 333, 464
  conversion, 369
  definition, 325, 463
  gage, 332
  gagemaker's standard (Table), 333
  for MMC, 361
  positional, 327
  for sand castings, 413
  standard, 465
  systems, 333, 334, 463, 464
  unilateral, 334, 463
Tool materials
    (*see also* Die materials)
  aluminum, 455
  bismuth alloys, 455
  cast copper alloys, 455
  cast iron, 453
  effect of speeds, 21
  ferrous, 449
  lead-base alloys, 455
  magnesium, 455
  nonferrous, 455
  physical properties, 446
  selection criteria, 445
  zinc-base alloys, 455
Tool steels
  application, 449
  characteristics, 451
  classification, 450
  heat treatment, 452
Tooling economics, 469
Tools, angles for, 2
Tools, cost analysis, 469
Tools, cutting
  abrasive wear, 26
  built-up edge, 13
  causes of wear, 23
  chemical decomposition, 27
  diffusion, 27
  edge build-up, 13
  effect of cutting fluids, 21
  face wear, 23, 84
  failure, 22
  flank wear, 24, 84
  geometry, 4, 19
  life, 34-41
  machinability considerations, 29
  materials, 19
  multiple-diameter, 74
  multiple-point, 56
  nose wear, 24
  plastic deformation, 26
  positioning, 140
  power required, 51
  single-point, 2
  signature, 4
  wear control, 81
  wear mechanisms, 24
  workholder relations, 140
Tools, design analysis, 440
Tools, detailing, 442
Tools, effect of geometry, 19
Tools, general design considerations, 439-478
Tools, for joining processes, 373-405
Tools, layout, 441
Tools, life of, 34
Tools for mechanical joining, 390-404
Tools for physical joining, 373
Tools, pressworking (*see* Pressworking tools)
Tools, safe design, 443
Tools, signature, 4
Torque in drilling, 62
Twisting dies and operations, 239

U

Upset dies, 316 (*see also* Forging machine dies)

V

V bending, 223
V blocks, 124
Vises
  for assembly operations, 391
  jaws for, 90, 122, 124
  milling-machine, 153

W

Weld projections, 234
Welding of asperities, 27
Welding fixtures (*see* Fixtures, welding)
Workholders, 89-165
  actuation, 117
  cutting tool relations, 92, 94
  definition, 89
  design consideration, 92, 137

Workholders (cont.)
  elastic-type, 118
  elements of, 115
  evolution of, 144
  for flat workpieces, 112, 121
  function of, 92
  for irregular workpieces, 116, 121
  machine-tool mounting, 138
  nomenclature, 129
  rigid-type, 118
  for round surfaces, 116, 128
  selection considerations, 137
  self-actuating, 136
  standard, 140
  types of, 90, 115
  wedge-cam, 136
  wedge-roller, 136
Workholding devices (*see* Workholders)
Workpieces
  axial location, 133
  quality criteria, 327
Workplace conditions, 122